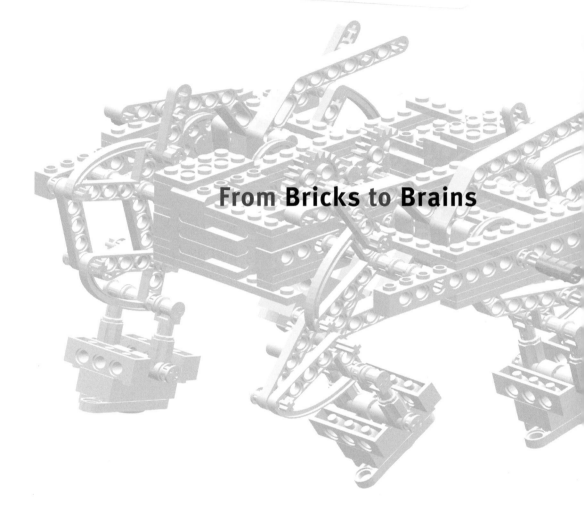

From Bricks to Brains

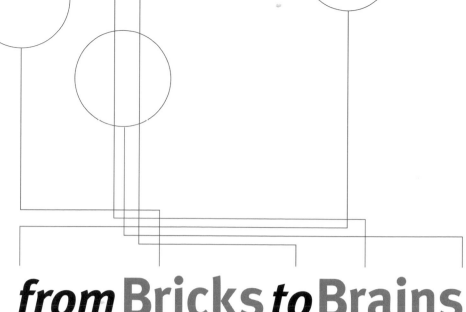

from Bricks *to* Brains
The Embodied Cognitive Science of LEGO Robots

Michael R.W. Dawson

Brian Dupuis

Michael Wilson

AU PRESS
Athabasca University

© 2010
MICHAEL DAWSON, BRIAN DUPUIS, AND MICHAEL WILSON

Published by AU Press, Athabasca University
1200, 10011 – 109 Street
Edmonton, AB   T5J 3S8

Library and Archives Canada Cataloguing in Publication

From bricks to brains : the embodied cognitive science of LEGO robots
/ editors, Michael Dawson, Brian Dupuis, Michael Wilson.

Includes bibliographical references and index.
Also available in electronic format (978-1-897425-79-4).
ISBN 978-1-897425-78-7

1. Robots—Programming. 2. Robots—Dynamics. 3. Cognitive science.
4. Robots—Design and construction. 5. Artificial intelligence. 6. LEGO
toys. I. Dawson, Michael Robert William, 1959- II. Dupuis, Brian, 1984-
III. Wilson, Michael, 1987-

TJ211.F76 2010   629.8'92   C2010-901786-2

Cover and book design by Natalie Olsen, kisscutdesign.com.
Printed and bound in Canada by Marquis Book Printing.

Acknowledgements

Contents

CHAPTER 2
Classical Music and the Classical Mind

CHAPTER 3
Situated Cognition and *Bricolage*

(57)

(93) **CHAPTER 4**
Braitenberg's Vehicle 2

CHAPTER 5
Thoughtless Walkers

CHAPTER 6
Machina Speculatrix

(199)

CHAPTER 7
The Subsumption Architecture

CHAPTER 8
Embodiment, Stigmergy, and Swarm Intelligence

CHAPTER 9
Totems, Toys — Or Tools?

Acknowledgements

This book manuscript was created with the support of an NSERC Discovery Grant, a SSHRC Standard Research Grant (particularly Chapter 2), and a 2007–08 McCalla Professorship from the Faculty of Arts at the University of Alberta, all awarded to MRWD.

Accompanying this book is additional web support that provides pdf files of traditional, "wordless" LEGO instructions for building robots, downloadable programs for controlling the robots that we describe, and videos that demonstrate robot behaviour. This web support is available at http://www.bcp.psych.ualberta.ca/~mike/BricksToBrains/.

The instructional images that are provided in this book were created by first building a CAD model of the robot using the LDRAW family of programs. This is a set of programs available as freeware from http://www.ldraw.org/. The CAD files were then converted into the instructional images using the LPUB4 program, available as freeware from www.kclague.net/LPub4.htm. Resources are available that provide detailed instructions on how to use such software tools (Clague, Agullo, & Hassing, 2002). The NXC code for the various robots described in this book was created, and downloaded to the robot, using the BricxCC utility, available as freeware at http://bricxcc.sourceforge.net/. This utility provides an excellent help file to describe the components and the syntax of the NXC programming language.

All of the photographs used in Chapter 1 were taken by Nancy Digdon, and are used with her permission. The beaver in Section 1.2 was in Astotin Lake at Elk Island National Park near Edmonton, Alberta. The beaver dam in the same section is located at the nature trail in Miramichi, New Brunswick. The two images of the bald-faced hornet nest were also taken at Miramichi. The wasp nest under construction in Section 1.10 was found at Elk Lake Park, near Victoria, British Columbia.

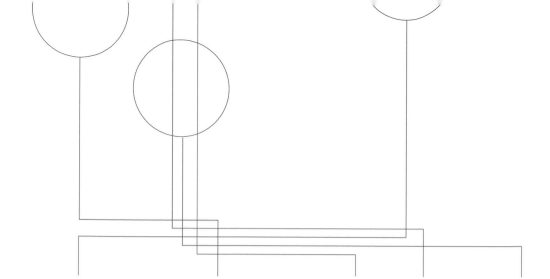

Chapter 1
Mind Control—Internal or External?

1.0 CHAPTER OVERVIEW

Classical cognitive science adopts the representational theory of mind (Pylyshyn, 1984). According to this theory, cognition is the rule-governed manipulation of internal symbols or representations. This view has evolved from Cartesian philosophy (Descartes, 1637/1960), and has adopted many of Descartes' tacit assumptions about the nature of the mind (Devlin, 1996). For example, it views the mind as a disembodied entity that can be studied independently of its relationship to the world. This view of the rational mind as distinct from the world has also been used to distinguish humans from other organisms. That is, rational humans are viewed as being controllers or creators of their environment, while irrational animals are completely under the environment's control (Bertalanffy, 1967; Bronowski, 1973; Cottingham, 1978).

The purpose of this chapter is to explore this view of the mind, in preparation for considering alternative accounts of cognition that will be developed in more detail as the book proceeds. We begin by considering the representational theory of mind, and how it is typically used to distinguish man from other organisms. We consider examples of animals, such as beavers and social insects, that appear to challenge this view because they create sophisticated structures, and could be viewed to some degree as controllers or builders of their own environment. A variety of theories of how they build these structures are briefly considered. Some of these theories essentially treat these animals as being rational or representational. However, more modern theories are consistent with the notion that the construction of elaborate nests or other structures is predominantly under the control of environmental stimuli; one prominent concept in such theories is *stigmergy*. The chapter ends, though, by pointing out that such control is easily found in prototypical architectures that have been used to model human cognition. It raises the possibility that higher-order human cognition might be far less Cartesian than classical cognitive science assumes, a theme that will be developed in more detail in Chapter 2. The notion of stigmergy that is introduced in Chapter 1 will recur in later chapters, and will be particularly important in Chapter 8's discussion of collective intelligence.

1.1 OUR SPECIAL INTELLIGENCE

We humans constantly attempt to identify our unique characteristics. For many our special status comes from possessing a soul or consciousness. For Descartes, the essence of the soul was "only to think," and the possession of the soul distinguished us from the animals (Descartes, 1637/1960). Because they lacked souls, animals could not be distinguished from machines: "If there were any machines which had the organs and appearance of a monkey or of some other unreasoning animal, we would have no way of telling that it was not of the same nature as these animals" (p. 41). This view resulted in Cartesian philosophy being condemned by modern animal rights activists (Cottingham, 1978).

More modern arguments hold that it is our intellect that separates us from animals and machines (Bronowski, 1973). "Man is distinguished from other animals by his imaginative gifts. He makes plans, inventions, new discoveries, by putting different talents together; and his discoveries become more subtle and penetrating, as he learns to combine his talents in more complex and intimate ways" (p. 20). Biologist Ludwig von Bertalanffy noted, "symbolism, if you will, is the divine spark distinguishing the poorest specimen of true man from the most perfectly adapted animal" (Bertalanffy, 1967, p. 36).

It has been argued that mind emerged from the natural selection of abilities to reason about the consequences of hypothetical actions (Popper, 1978). Rather than performing an action that would have fatal consequences, the action can be thought about, evaluated, and discarded before actually being performed.

Popper's position is central to much research in artificial intelligence and cognitive science. The fundamental hypothesis of such *classical* or symbolic research is that cognition is computation, that thinking is the rule-governed manipulation of symbols that represent the world. Thus the key role of cognition is planning: on the basis of perceptual information, the mind builds a model of the world, and uses this model to plan the next action to be taken. This has been called the *sense–think–act cycle* (Pfeifer & Scheier, 1999). Classical cognitive science has studied the thinking component of this cycle (What symbols are used to represent the world? What rules are used to manipulate these symbols? What methods are used to choose which rule to apply at a given time?), often at the expense of studying sensing and acting (Anderson et al., 2004; Newell, 1990).

A consequence of the sense–think–act cycle is diminished environmental control over humans. "Among the multitude of animals which scamper, fly, burrow and swim around us, man is the only one who is not locked into his environment. His imagination, his reason, his emotional subtlety and toughness, make it possible for him not to accept the environment, but to change it" (Bronowski, 1973, p. 19). In modern cognitivism, mind reigns over matter.

Ironically, cognitivism antagonizes the view that cognition makes humans special. If cognition is computation, then certain artifacts might be cognitive as well. The realization that digital computers are general purpose symbol manipulators implies the possibility of machine intelligence (Turing, 1950): "I believe that at the end of the century the use of words and general educated opinion will have altered so much that one will be able to speak of machines thinking without expecting to be contradicted" (p. 442).

However, classical cognitivism is also subject to competing points of view. A growing number of researchers are concerned that the emphasis on planning using representations of the world is ultimately flawed. They argue that the mind is not a planner, but is instead a controller that links perceptions with actions without requiring planning, reasoning, or central control. They would like to replace the "sense–think–act" cycle with a "sense–act" cycle in which the world serves as a model of itself. Interestingly, this approach assumes that human intelligence is largely controlled by the environment, perhaps making us less special than we desire. The purpose of this book is to explore this alternative view of cognition.

1.2 RODENTS THAT ENGINEER WETLANDS
1.2.1 *Castor canadensis*

Is our intelligence special? Perhaps the divide between ourselves and the animals is much smaller than we believe. Consider, for example, the North American beaver, *Castor canadensis*. A large rodent, a typical adult beaver usually lives in a small colony of between four and eight animals (Müller-Schwarze & Sun, 2003). Communication amongst animals in a colony is accomplished using scent marks, a variety of vocalizations, and the tail slap alarm signal (Figure 1-1).

1-1

In his classic study, Lewis Morgan noted that "in structural organization the beaver occupies a low position in the scale of mammalian forms" (Morgan, 1868/1986, p. 17). Nonetheless, the beaver is renowned for its artifacts. "Around him are the dam, the lodge, the burrow, the tree-cutting, and the artificial canal; each testifying to his handiwork, and affording us an opportunity to see the application as well as the results of his mental and physical powers" (p. 18). In short, beavers — *like humans* — construct their own environments.

1-2

To dam a stream (Figure 1-2), a colony of beavers will first prop sticks on both banks, pointing them roughly 30° upstream (Müller-Schwarze & Sun, 2003). Heavy stones are then moved to weigh these sticks down; grass is stuffed between these stones. Beavers complete the dam by ramming poles into the existing structure that sticks out from the bank. Poles are aligned with stream flow direction. The dam curves to resist stream flow; sharper-curved dams are used in faster streams. Beavers add mud to the upstream side of the dam to seal it. The dam is constantly maintained, reinforced and raised when water levels are low; it is made to leak more when water levels become too high (Frisch, 1974). Dams range in height from 20 cm to 3 m, and a large dam can be several hundred metres in length. A colony of beavers might construct more than a dozen dams to control water levels in their territory.

1.2.2 The Cognitive Beaver?

How does such a small, simple animal create these incredible structures? Morgan attributed intelligence, reasoning, and planning to the beaver. For instance, he argued that a beaver's felling of a tree involved a complicated sequence of thought processes, including identifying the tree as a food source and determining whether the tree was near enough to the pond or a canal to be transported. Such thought sequences "involve as well as prove a series of reasoning processes indistinguishable from similar processes of reasoning performed by the human mind" (Morgan, 1868/1986, pp. 262-263). If this were true, then the division between man and beast would be blurred. Later, though, we will consider the possibility that even though the beaver is manipulating its environment, it is still completely governed by it. However, we must also explore the prospect that, in a similar fashion, the environment plays an enormous role in controlling human cognition.

1.3 THE INSTINCTS OF INSECTS

To begin to reflect on how thought or behaviour might be under environmental control, let us consider insects, organisms that are far simpler than beavers.

1.3.1 The Instinctive Wasp

Insects are generally viewed as "blind creatures of impulse." For example, in his assessment of insect-like robots, Moravec (1999) notes that they, like insects, are intellectually damned: "The vast majority fail to complete their life cycles, often doomed, like moths trapped by a streetlight, by severe cognitive limitations." These limitations suggest that insects are primarily controlled by instinct (Hingston, 1933), where instinct is "a force that is innate in the animal, and one performed with but little understanding" (p. 132).

This view of insects can be traced to French entomologist J.H. Fabre. He described a number of experiments involving digger wasps, whose nests are burrows dug into the soil (Fabre, 1915). A digger wasp paralyzes its prey, and drags it back to a nest. The prey is left outside as the wasp ventures inside the burrow for a brief inspection, after which the wasp drags the prey inside, lays a single egg upon it, leaves the burrow, and seals the entrance.

While the wasp was inspecting the burrow, Fabre moved its paralyzed prey to a different position outside the nest (Fabre, 1919). This caused the wasp to unnecessarily re-inspect the burrow. If Fabre moved the prey once more during the wasp's second inspection, the wasp inspected the nest again!

In another investigation, Fabre (1915) completely removed the prey from the vicinity of the burrow. After conducting a vain search, the wasp turned and sealed the empty burrow as if they prey had already been deposited. "Instinct knows everything, in the undeviating paths marked out for it; it knows nothing, outside those paths" (Fabre, 1915, p. 211).

1.3.2 *Umwelt* and Control

The instincts uncovered by Fabre are not blind, because some adaptation to novel situations occurs. At different stages of the construction of a wasp's nest, researchers have damaged the nest and observed the ensuing repairs. Repaired nests can deviate dramatically in appearance from the characteristic nest of the species (Smith, 1978). Indeed, environmental constraints cause a great deal of variation of nest structure amongst wasps of the same species (Wenzel, 1991). This would be impossible if wasp behaviour were completely inflexible.

However, this flexibility is controlled by the environment. That is, observed variability is not the result of modifying instincts themselves, but rather the result of how instincts interact with a variable environment. Instincts are elicited by stimuli in the sensory world, which was called the *umwelt* by ethologist Jakob von Uexküll. The *umwelt* is an "island of the senses"; agents can only experience the world in particular ways because of limits, or specializations, in their sensory apparatus (Uexküll, 2001). Because of this, different organisms can live in the same environment, but at the same time exist in different *umwelten*, because they experience this world in different ways. The notion of *umwelt* is similar to the notion of affordance in ecological theories of perception (Gibson, 1966, 1979).

Some have argued that the symbolic nature of human thought and language makes us the only species capable of creating our own *umwelt* (Bertalanffy, 1967). "Any part of the world, from galaxies inaccessible to direct perception and biologically irrelevant, down to equally inaccessible and biologically irrelevant atoms, can become an object of 'interest' to man. He invents accessory sense organs to explore them, and learns behavior to cope with them" (p. 21). While the animal *umwelt* restricts them to a physical universe, "man lives in a symbolic world of language, thought, social entities, money, science, religion, art" (p. 22).

1.4 PAPER WASP COLONIES AND THEIR NESTS

Experiments have revealed the instincts of solitary wasps (Fabre, 1915, 1919) and other insects. However, social insects can produce artifacts that may not be so easily rooted in instinct. This is because these artifacts are examples of *collective intelligence* (Goldstone & Janssen, 2005; Kube & Zhang, 1994; Sulis, 1997). Collective intelligence requires coordinating the activities of many agents; its creations cannot be produced by one agent working in isolation. Might paper nests show how social insects create and control their own environment?

1.4.1 Colonies and Their Nests

For example, the North American bald-faced hornet (*Dolichovespula maculata*, which is not a hornet but instead a wasp) houses its colony in an inverted, pear-shaped "paper" nest. A mature nest can be as large as a basketball; an example nest is illustrated in Figure 1-3.

1-3
1-4

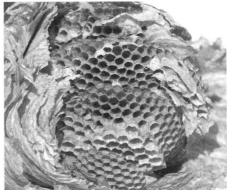

Inside the outer paper envelope is a highly structured interior (Figure 1-4). There are several horizontal layers, each consisting of a number of hexagonal combs. A layer of combs is attached to the one directly above it, so that the layers hang as a group from the top of the nest. Each comb layer is roughly circular in shape, and its diameter and shape match the outer contours of the nest. The walls of each comb are elongated, some being longer than others.

"In the complexity and regularity of their nests and the diversity of their construction techniques, wasps equal or surpass many of the ants and bees" (Jeanne, 1996, p. 473). There is tremendous variability in the size, shape, and location of social wasp nests (Downing & Jeanne, 1986). A nest may range from having a few dozen cells to having in the order of a million; some wasps build nests that are as high as one metre (Theraulaz, Bonabeau, & Deneubourg, 1998). As well, nest construction can involve the coordination of specialized labor. For example, *Polybia occidentalis* constructs nests using builders, wood-pulp foragers, and water foragers (Jeanne, 1996).

1.4.2 Scaling Up

The large and intricate nests constructed by colonies of social wasps might challenge simple, instinctive, explanations. Such nests are used and maintained by a small number of wasp generations for just a few months. Greater challenges to explaining nest construction emerge when we are confronted with other insect colonies, such as termites, whose mounds are vastly larger structures built by millions of insects extending over many years. Such nests "seem evidence of a master plan which controls the activities of the builders and is based on the requirements of the community. How this can come to pass within the enormous complex of millions of blind workers is something we do not know" (Frisch, 1974, p. 150). Let us now turn to considering termite nests, and ask whether these structures might offer evidence of cognitive processes that are qualitatively similar to our own.

1.5 THE TOWERS OF TERMITES
1.5.1 Termite Mounds

Termites are social insects that live in colonies that may contain as many as a million members. Though seemingly similar to bees and ants, they are actually more closely related to cockroaches. In arid savannahs termites are notable for housing the colony in distinctively shaped structures called mounds. One of the incredible properties of termite mounds is their size: they can tower over the landscape. While a typical termite mound is an impressive 2 metres in height, an exceptional one might be as high as 7 metres (Frisch, 1974)!

Termite mounds are remarkable for more than their size. One issue that is critical for the health of the colony is maintaining a consistent temperature in the elaborate network of tunnels and chambers within the mound. This is particularly true for some species that cultivate fungus within the mound as a source of food. Some termites regulate temperature by building a vertical ventilation system that enables hot air to rise and leave the structure via "chimneys" on the top of the mound.

Other termites adopt a different architectural solution to the problem of thermoregulation. The "compass" or "magnetic" termite *Amitermes laurensis* is found in Australia. Its mounds are wedge shaped, much longer than wide. Amazingly, the mound of this termite is oriented so that its long walls face north and south, and its narrow walls face east and west.

It has been suggested that the shape and orientation of the mound built by *Amitermes laurensis* helps protect the colony from the heat of the sun. When the sun rises in the east, only a single narrow wall is in direct sunlight. The west wall is actually shaded, and is insulated by the core of the mound (which is solid for this species of termite). In the morning, colony members will tend to congregate in the western side of the mound. Later in the day, when the sun is in the west, it is the eastern wall that is shaded, and the colony congregates on that side of the mound. The mound's wedge shape is such that at the hottest part of the day, when the sun is overhead, only its thin top edges are is exposed to direct heat. The wider northern and southern walls are never in direct sunlight, and have been shown to be in the order of 8° cooler than the others. As well, constituting the greatest areas of the outside of the mound, they provide a means for heat to dissipate outward. In short, *Amitermes laurensis* designs its mound for maximal coolness in severely hot conditions.

The shape of a "magnetic mound" also provides a solution to the problem of maintaining air quality for the colony. A large number of insects within a colony consume high volumes of oxygen, and produce high volumes of carbon dioxide. As a result, there is a pressing need to replenish the air within the mound. This must be accomplished via pores in the mound's outer wall. However, the effectiveness of these pores is reduced by moisture during the wet season. The shape of the magnetic mound results in a high ratio of wall surface area to mound volume. This increases the area over which air exchange is possible, helping the mound to "breathe," even during the wet season.

1.5.2 The Thinking Termite?

How do such tiny, simple animals as termites coordinate their activities to produce such amazing structures? Do termites exhibit intelligence that is similar in kind to our own? "One of the challenges of insect sociobiology is to explain how such colony-level behavior emerges from the individual decisions of members of the colony" (Jeanne, 1996, p. 473). There have been a wide variety of explanations proposed in the literature, ranging from rational insects, nest construction governed by blind instinct, colonies as intelligent superorganisms, and nest building controlled by the dynamic environment. In the following pages we will briefly consider a number of these different theories. We will see how environmental control may still be responsible for the construction of the elaborate nests of social insects. However, we must then consider whether a similar theory is applicable to human intelligence.

1.6 THE RATIONAL INSECT?

1.6.1 Computational Theory of Mind

The dominant perspective in cognitive science is the representational theory of mind (Fodor, 1975; Newell, 1980; Pylyshyn, 1984). According to this theory, external behaviour is guided or mediated by the contents of internal representations. Such representations are symbolic structures that have associated content, in the sense that they stand for states of affairs in the external world.

In the representational theory of mind, perceptual mechanisms are presumed to provide links between the external world and internal symbols. Thinking or cognition is the rule-governed manipulation of these internal representations in order to acquire new knowledge (e.g., by inference, by problem solving, by planning). The products of thinking are then responsible for producing behaviours, or actions upon the world. Thus, the computational theory of mind involves a continuous sense–think–act cycle (Pfeifer & Scheier, 1999). "Representation is an activity that individuals perform in extracting and deploying information that is used in their further actions" (Wilson, 2004, p. 183).

1.6.2 Are Insects Representational?

We have seen that social insects like termites and wasps are capable of monumental feats of engineering. What sort of intelligence guides the construction of such large, complex insect nests? "The problem before us is a very old one. Are the lower animals blind creatures of impulse or are they rational beings?" (Hingston, 1929). Can the computational theory of mind be applied to non-human agents? Can the nest of a colony of social insects be explained as the result of representational thought processes? Some accounts of insect behaviour, including nest construction, appeal to the notion of the rational insect.

Consider Major Richard Hingston, who was a doctor, a member of the 1924 expedition to Mount Everest, and an avid naturalist. He published accounts of his observations of insects, and of his experiments on their behaviour, including studies of nest building by solitary wasps (Hingston, 1929). While he was open to the notion that some insect behaviour was guided by instinct (Hingston, 1933), he also believed that insects were more rational or intelligent than many of us would expect: "So far as I can judge from the evidence given, we are not justified in making barriers between insect and human mentality. I mean we have no right to regard their minds as being totally different in kind" (p. 183).

Hingston's work was a direct reaction against studies demonstrating that insects were governed by blind instinct (Fabre, 1915, 1919). Four decades later, Hingston's naïve notion of the "rational insect" had evolved into one that was more sophisticated and representational.

For example, ethologist W.H. Thorpe reviewed studies of nest building in a variety of animals, including wasps, and proposed that nest construction behaviours were controlled by an *ideal releaser* (Thorpe, 1963). He did not describe the properties of ideal releasers in detail, but it is clear that to Thorpe they were representations of intended products. "The bird must have some 'conception' of what the completed nest should look like, and some sort of 'conception' that the addition of a piece of moss or lichen here and here will be a step towards the 'ideal' pattern, and that other pieces there and there would detract from it" (p. 22).

Thorpe's (1963) notion of the ideal releaser is consistent with heuristics used in models of problem solving that were being published around the same time (Newell & Simon, 1961). For instance, Newell and Simon's general problem solver (GPS) would maintain a representation of a goal state, compute differences between it and the current state of a problem, and then use these differences to solve the problem. Actions would reduce the differences between the current and goal states, and then differences would be recomputed until the problem was solved. The goal state in GPS served exactly the same role as the ideal releaser proposed by Thorpe.

1.7 INSECT AS SUPERORGANISM?
1.7.1 The Intelligent Whole

In spite of Hingston's evidence, the view that insects were rational was not endorsed by many researchers. They were instead interested in explaining how simple, non-rational beings were capable of impressive feats (such as nest building) that appeared to be intelligent. One approach was to attribute intelligence to a colony as a whole, not to its individual members. In the modern literature, this has become known as *swarm intelligence* (Bonabeau & Meyer, 2001; Sharkey, 2006; Tarasewich & McMullen, 2002).

1.7.2 Colonial Intelligence

The roots of swarm intelligence can be found in early-twentieth-century entomology (Wheeler, 1911). William Morton Wheeler argued that biology had to explain how organisms coped with complex and unstable environments. For Wheeler, "an organism is a complex, definitely coordinated and therefore individualized system of activities, which are primarily directed to obtaining and assimilating substances from an environment, to producing other similar systems, known as offspring, and to protecting the system itself and usually also its offspring from disturbances emanating from the environment" (p. 308).

Wheeler used this rather broad definition of "organism" because he proceeded to propose an unusual idea: that a colony of ants, considered as a whole, could be also classified as being an organism. "The animal colony is a true organism and not merely the analogue of the person" (Wheeler, 1911, p. 310). He then argued that insect colonies, considered as wholes, demonstrated each and every one of the properties listed in his definition of organism. These colonies became known as *superorganisms*.

Wheeler recognized that a superorganism's properties emerged from the actions of its parts (Wheeler, 1926). However, Wheeler also argued that higher-order properties could not be reduced to properties of the superorganism's components.

Wheeler's defended the notion that higher-order regularities could not be easily reduced to lower-order properties by applying ideas that were also in vogue in Gestalt psychology (Koffka, 1935; Köhler, 1947). Gestalt psychologists realized that many perceptual experiences could not be captured by appealing to the properties of their components. Instead, they proposed a number of perceptual laws that applied to the whole, and attempted to explain these higher-order principles by appealing to the notion of an organized perceptual field. Wheeler made arguments very similar to those made by Gestalt psychologists when arguing for a unique level of superorganismic properties: "The unique qualitative character of organic wholes is due to the peculiar non-additive relations or interactions among their parts. In other words, the whole is not merely a sum, or resultant, but also an emergent novelty, or creative synthesis." (Wheeler, 1926, p. 433).

Many modern theories in a number of different disciplines exploit the notion of emergence (Holland, 1998; Johnson, 2001; Sawyer, 2002). Holland argues that such modern theories, in order to be scientific, must exhibit a number of different properties. First and foremost, the higher-order patterns that emerge must be recognizable and recurring. These patterns are persistent at higher-order levels of analysis, in the sense that the higher-order pattern can remain even when the components underlying the phenomenon change. They are usually found in systems that are dynamic (i.e., that change over time) and adaptive (i.e., that change in response to demands). Most importantly, emergent patterns can be explained by appealing to laws or rules that explain how they are supported by the characteristics of system components. As noted by Wheeler in the quote in the preceding paragraph, the laws that explain emergence make explicit "the peculiar non-additive relations or interactions" between parts, and are often expressed in some formalism that can be related to dynamical systems theory (Port & Van Gelder, 1995).

1.8 THE ULTIMATE DEMOCRACY
1.8.1 Emerging Problems

Wheeler's notion of the superorganism, and the organizational principles of Gestalt psychology, are two examples of *holism* (Sawyer, 2002). Such theories recognize that the regularities governing a whole system cannot be easily reduced to a theory that appeals to the properties of the system's parts. For example, Gestalt psychology attacked psychological behaviourism because of its reductionist approach to explaining psychological phenomena. Unfortunately, holism has not had much success in being accepted as being scientific. "Holism is an idea that has haunted biology and philosophy for nearly a century, without coming into clear focus" (Wilson & Lumsden, 1991, p. 401). Gestalt psychology flourished in Germany from 1920 until just before World War II (Henle, 1977). By the end of the war, this school of thought had come to the end of its influence. Many of its students had been killed in the war, or had been displaced to a variety of different countries. Attempts to reignite Gestalt psychology in the United States failed because Gestalt ideas were in conflict with the then dominant school of behaviourism. One problem with Gestalt psychology was that it had difficulty being accepted as a form of *emergentism*. "Emergentism is a form of materialism which holds that some complex natural phenomena cannot be studied using reductionist methods" (Sawyer, 2002, p. 2). Gestalt psychologists had difficulty in providing materialist accounts of such concepts as perceptual fields, cortical currents, or isomorphic relations between objects in the world and objects in the mind (Henle, 1977). Ironically, Gestalt psychology was ahead of its time. Formal and algorithmic accounts that have appeared in some subdomains of modern cognitive science like connectionism

(Bechtel & Abrahamsen, 2002; Dawson, 2004; Rumelhart & McClelland, 1986) or dynamical systems theory (Port & Van Gelder, 1995) appear to offer approaches that could have converted the holism of Gestalt psychology into a more causally grounded emergentism (e.g., Sawyer, 2002).

Wheeler's notion of the superorganism has enjoyed an enduring popularity (Detrain & Deneubourg, 2006; Queller & Strassmann, 2002; Seeley, 1989; Wilson & Sober, 1989). However, in terms of biology, this idea suffered a fate similar to that of Gestalt psychology. The problem with the view that colonies are organisms is that it is very difficult to provide a scientific account of the laws that govern them. Where do the laws come from? How do laws governing the whole emerge from the actions of individual parts? Wheeler recognized that such questions posed "knotty problems," but was ultimately unable to provide adequate solutions to them (Evans & Evans, 1970). The result was that entomologists rejected the notion of the superorganism (Wilson & Lumsden, 1991).

1.8.2 From Whence Organization?

Rejecting the superorganism, however, does not remove the need for explaining how complex structures such as nests could be constructed by social insects. If the colony itself was not intelligent, then what was the source of amazing structures like termite mounds?

The alternative was to claim that colonial intelligence could be reduced to the actions of individual colony members. This view was championed by French biologist Etienne Rabaud, who was a contemporary of Wheeler. "His entire work on insect societies was an attempt to demonstrate that each individual insect in a society behaves as if it were alone" (Theraulaz & Bonabeau, 1999). Biologist E. O. Wilson has adopted a similar position. "It is tempting to postulate some very complex force distinct from individual repertories and operating at the level of the colony. But a closer look shows that the superorganismic order is actually a straightforward summation of often surprisingly simple individual responses" (Wilson & Lumsden, 1991, p. 402). In short, swarm intelligence wasn't real—it was just in the eye of the beholder. And the coordination of individuals might be accomplished via the environment, as is considered in the following pages.

1.9 PROGRAMS FOR NEST CONSTRUCTION
1.9.1 An Inherited Program

It has been proposed that wasps do not inherit an ideal releaser, but instead inherit a program for nest construction. One example of such a program is part of a general account of wasp behaviour (Evans, 1966;

Evans & West-Eberhard, 1970). In this model, a hierarchy of internal drives serves to release behaviours. For instance, high-level drives might include mating, feeding, and brood rearing. Such drives set in motion lower-level sequences of behaviour, which in turn might activate even lower-level behavioural sequences. For example, a brood-rearing drive might activate a drive for capturing prey, which in turn activates a set of behaviours that produces a hunting flight. So, for Evans, a program is a set of behaviours that are produced in a particular sequence, where the sequence is dictated by the control of a hierarchical arrangement of drives. However, these behaviours are also controlled by releasing stimuli that are *external* to the wasp. In particular, one behaviour in the sequence is presumed to produce an environmental signal that serves to initiate the next behaviour in the sequence. For instance, in Evans' (1966) model, the digging behaviour of a wasp produces loosened soil, which serves as a signal for the wasp to initiate scraping behaviour. This behaviour in turn causes the burrow to be clogged, which serves as a signal for clearing behaviour. Having a sequence of behaviours under the control of both internal drives and external releasers provides a balance between rigidity and flexibility: the internal drives serve to provide a general behavioural goal, while variations in external releasers can produce variations in behaviours (e.g., resulting in an atypical nest structure when nest damage elicits a varied behavioural sequence). "Each element in the 'reaction chain' is dependent upon that preceding it as well as upon certain factors in the environment (often gestalts), and each act is capable a certain latitude of execution" (Evans, 1966, p. 144).

1.9.2 Testing the Theory

Smith (1978) has provided compelling evidence of component of Evans' model, the external control of specific behaviours used by wasps to construct nests. Smith examined a particular mud wasp that digs a hole in the ground, lines the hole with mud, and then builds an elaborate funnel on top of the hole to keep parasites out. The funnel is a long straight tube, to which is added a marked curve, to which is attached a large bell-shaped opening. The existence of a mud-lined hole appears to be the stimulus that caused the wasp to build the straight tube. Smith demonstrated this by creating a hole in the curve that the wasp added to the straight tube. This caused the wasp to start creating a brand new tube out from the hole in the curve, resulting in a second funnel structure being built on top of the first. Importantly, and consistent with Evans' (1966) model, external stimuli are not the sole elicitors of behaviour (Baerends, 1959). Baerends studied digger wasps that provided for several nests at the same time. The nest is begun with a single egg and a single caterpillar as prey. Later, the wasp returns to the nest to inspect larval development. Depending upon the size of the larva, and upon the amount of food remaining in the nest, the wasp will hunt for more prey to be added to the nest. Baerends (1959) found that the state of the nest would affect behaviour only when the wasp made its first inspection. If he added or removed food after the inspection, the foraging behaviour of the wasp was not altered accordingly, even though the wasp was exposed to the new situation inside the nest when it returned to it with new prey. In other words, its foraging was not merely under the control of the nest-as-stimulus; foraging was also controlled by the internal state of the wasp during its first inspection. Nonetheless, in models like those of Evans (1966), the environment plays a key role in controlling the nest-building behaviour of insects.

1.10 THE ENVIRONMENT AS PROGRAM
1.10.1 A Complex Environment

Evans' (1966) theory of nest construction by solitary insects can easily be extended to insect societies. It has long been recognized that an insect colony provides a much more complex environment (i.e., a much richer set of stimuli) than would be available to asocial insects. The social insect "must respond not only to all the stimuli to which it reacted in its presocial stage but also to a great number of additional stimuli emanating from the other members of the society in which it is living" (Wheeler, 1923, p. 503). Clearly one sense in which this environment is more complex is with respect to the signals used by one colony member to communicate to others. Such signals include movements, such as the dance that one honeybee performs to communicate the location of a food source to others (Frisch, 1966, 1967, 1974), as well as with chemicals (Queller & Strassmann, 2002). "The members of an insect society undoubtedly communicate with one another by means of peculiar movements of the body and antennæ, by shrill sounds (stridulation) and by odors" (Wheeler, 1923, p. 506).

However, there is another sense in which an insect colony provides its individuals a complex and dynamic environment that affects their behaviour, even in the possible situation in which there is absolutely no direct communication between colony members using actions, sounds, or scents.

Consider wasps adding to a nest. Much of this process is parallel because more than one wasp works on the nest at the same time, as shown in Figure 1-5. Imagine an individual working on this nest, guided (as a working hypothesis) by a nest-building program (Evans, 1966). This wasp will perform some action governed by its internal state and by some triggering characteristic of the nest. At some point the wasp leaves to obtain new building materials. In its absence, the appearance of the nest will change because of the activities of other colony members. As a result, the behaviour performed by the returning wasp may be quite different than would have been the case had the nest been unaltered in its absence. In short, different colony members can communicate indirectly with one another by changing the nest, and as a result by changing the available releasing stimuli.

1.10.2 Stigmergy

French zoologist Pierre-Paul Grassé explained the mound-building behaviour of termites by appealing to the notion of indirect communication by changing the environment (Theraulaz & Bonabeau, 1999). Grassé demonstrated that the termites themselves do not coordinate or regulate their building behaviour, but that this is instead controlled by the mound structure itself. The term *stigmergy* was coined for this type of behavioural control (Grassé, 1959). The word *stigmergy* comes from the Greek *stigma*, meaning sting, and *ergon,* meaning work, capturing the notion that the environment is a stimulus that causes particular work (behaviour) to occur. Researchers describe quantitative stigmergy as involving stimuli that differ in intensity, but not quality, such as pheromone fields (Deneubourg & Goss, 1989). These stimuli modify the probability of individual responses. In contrast, qualitative stigmergy involves control of a variety of behaviours using a set of qualitatively different environmental stimuli (Theraulaz & Bonabeau, 1995).

1.11 STIGMERGY AND THE SYNTHETIC APPROACH
1.11.1 The Synthetic Approach

Stigmergy appeals to an environmental control structure that coordinates the performances of a group of agents. One of the appeals of stigmergy is that it explains how very simple agents create extremely complex products, particularly in the case where the final product (e.g., a termite mound) is extended in space and time far beyond the life expectancy of the organisms that create it. As well, it accounts for the building of large, sophisticated nests without the need for a complete blueprint and without the need for direct communication amongst colony members (Bonabeau et al., 1998; Downing & Jeanne, 1988; Grassé, 1959; Karsai, 1999; Karsai & Penzes, 1998; Karsai & Wenzel, 2000; Theraulaz & Bonabeau,

1995). One of the reasons that stigmergy can produce such complex products is because the behaviours of the agents, and the environmental stimuli that elicit these behaviours, are highly non-linear. As a result, it is very difficult to take a finished product, such as a completed wasp nest, and reverse engineer it to decipher the specific order of operations that produced it. However, it is also very difficult to look at a simple set of rules, such as a nest program, and to predict with any accuracy the final product that these rules could create for a colony of insects.

For this reason, stigmergy is often studied using a *synthetic methodology* (Braitenberg, 1984; Dawson, 2004; Pfeifer & Scheier, 1999). That is, researchers propose a small group of rules that are under stigmergic control, set these rules in motion in a computer simulation, and observe the products that the simulation creates.

1.11.2 Wasp Nest Examples

As an example, consider how the synthetic approach has been used to study nest construction by social paper wasps. A nest for such wasps consists of a lattice of cells, where each cell is essentially a comb created from a hexagonal arrangement of walls. When a large nest is under construction, where will new cells be added? This is a key issue, because the building activities of a large number of wasps must be coordinated in some manner to prevent the nest from growing predominately in one direction. Theraulaz and Bonabeau (e.g., 1999) used the synthetic approach to answer this question.

Theraulaz and Bonabeau (1999) proposed that an individual wasp's decision about where to build a new cell wall was driven by the number of already completed walls that were perceptible. If there is a location on the nest in which three walls of a cell already existed, then this was proposed as a stimulus to cause a wasp to add another wall here with high probability. If only two walls already existed, this was also a stimulus to add a wall, but this stimulus produced this action with a much lower probability.

The crucial characteristic of this approach is that it is stigmergic: when either of these rules results in a cell wall being added to the nest, then the nest structure changes. In turn, this causes changes in the appearance of the nest, which in turn causes changes in the locations where walls will be added next. Theraulaz and Bonabeau (1999) created a nest building simulation that only used these two rules, and demonstrated that it created simulated nests that were very similar in structure to real wasp nests. In addition to adding cells laterally to the nest, wasps must also lengthen walls that already exist. This is to accommodate the growth of a larva that lives inside the cell. Karsai (1999) proposed another stigmergic model of this aspect of nest building. His rule involved an inspection of the relative difference between the longest and the shortest wall of a cell. If the difference was below a threshold value, then the cell was untouched. However, if the difference exceeded a threshold, then this was a stimulus that caused a wasp to add material to the shortest wall. Karsai used a computer simulation to demonstrate that this simple stigmergic model provided an accurate account of the three-dimensional growth of a wasp nest over time.

1.12 STIGMERGY AND THE PARABLE OF THE ANT
1.12.1 Intelligence and Stigmergy

Stigmergy may explain how insects can be master architects, but still possess a lower intelligence than humans. It has certainly become an important concept in cognitive science and robotics (Goldstone & Janssen, 2005; Holland & Melhuish, 1999; Kube & Zhang, 1994; Sulis, 1997). However, researchers in cognitive science have been reluctant to apply stigmergy to explain the behaviours of higher organisms, including man (Susi & Ziemke, 2001). This is an important oversight. Consider beaver dams. Morgan (1868/1986) tacitly explained dam characteristics by appealing to the thoughts of beavers. He ignored the possibility, raised by stigmergy, that dams themselves play a large role in guiding their development and intricate nature.

The importance of the environment was a theme of early theoretical work in artificial intelligence (Simon, 1969). In Simon's famous *parable of the ant*, observers recorded the path traveled by an ant along a beach. How might we account for the complicated twists and turns of the ant's route? Cognitive scientists tend to explain complex behaviours by invoking complicated representational mechanisms (Braitenberg, 1984). In contrast, Simon noted the path might result from simple internal processes reacting to complex external forces — the various obstacles along the natural terrain of the beach. "Viewed as a geometric figure, the ant's path is irregular, complex, hard to describe. But its complexity is really a complexity in the surface of the beach, not a complexity in the ant" (Simon, 1969, p. 24).

A similar point can show how robot building can inform cognitive science. Braitenberg (1984) has argued that when we observe interesting behaviour in a system, we tend to ignore the environment, and explain all of the behaviour by appealing to internal structure. "When we analyze a mechanism, we tend to overestimate its complexity" (Braitenberg, 1984, p. 20). He suggested that an alternative approach, in which simple agents (such as robots) are built, and then observed in environments of varying complexity, can provide cognitive science with more powerful, and much simpler, theories. Such synthetic theories take advantage of the fact that not all of the intelligence must be placed inside an agent.

1.12.2 Are Mammals Stigmergic?

Might stigmergy account for the intelligence of "higher" organisms? Consider the beaver. Morgan (1868/1986) recounts a story of a colony that built a dam that threatened a railway line by the beaver pond. The track master had a hole cut through the middle of the dam to protect the track. "As this was no new experience to the beavers, who were accustomed to such rents, they immediately repaired the breach" (p. 102). The breaking and repairing of the dam went on repeatedly, 10 or 15 times, at this site. This story describes the tenacity of beavers, but perhaps is more revealing considered as a mammalian variant of Fabre's (1919) experiments revealing the blind instincts of digger wasps.

Beavers might respond to releasing stimuli in a fashion consistent with the preceding accounts of insect behaviour. The sound of running water brings them immediately to repair the dam. As a result, trappers would lure their prey by cutting holes into existing dams (Morgan, 1868/1986). Water levels around the lodge are stimuli to either raise the dam (to conserve water), or to make it leakier (to lower water levels (Frisch, 1974)). Researchers have had some success using environmental features to predict where dams will be constructed (Barnes & Mallik, 1997; Curtis & Jensen, 2004; Hartman & Tornlov, 2006). All of these observations suggest that it is plausible to hypothesize that stigmergy might have an important role in theories of the initiation, development, and final characteristics of beaver infrastructure.

If it is at least plausible that stigmergy guides some mammals, then is it possible that it might apply to theories of human cognition as well? A number of theories in modern cognitive science have opened the door to considering this idea.

1.13 EMBODIMENT AND POSTHUMANISM
1.13.1 Posthumanism

Our everyday experience of self-consciousness is inconsistent with the view of the mind held by modern cognitive scientists (Varela, Thompson, & Rosch, 1991). While we have a compelling sense of self, cognitive theories reject it. Many researchers believe that the mind is modular, incorporating a large number of independent machines that are isolated from consciousness (Fodor, 1983). We can be conscious of mental contents, but have no awareness of the mechanisms that represent them (Pylyshyn, 1981, 1984). Our sense of holistic consciousness is an illusion built from the activity of multiple, independent sources (Dennett, 1991, 2005). Entire theories of cognition begin with the foundational assumption that the mind is a society of simple, unconscious agents (Minsky, 1985, 2006).

These theoretical trends have resulted in a view that is known as *posthumanism* (Hayles, 1999). "The posthuman view configures human being so that it can be seamlessly articulated with intelligent machines. In the posthuman, there are no essential differences or absolute demarcations between bodily existence and computer simulation, cybernetic mechanism and biological organism, robot teleology and human goals" (p. 3). According to Hayles, posthumanism results when the content of information is more important than the physical medium in which it is represented, when consciousness is considered to be epiphenomenal, and when the human body is simply a prosthetic. Posthumanism is rooted in the pioneering work of cybernetics (Ashby, 1956, 1960; MacKay, 1969; Wiener, 1948), but it also flourishes in modern cognitivism.

The posthumanism that has developed from cybernetics and cognitive science denies our intelligence a special status. It proposes not only that that our thinking cannot be differentiated from that of animals, but also cannot be differentiated from that of machines.

1.13.2 Embodiment

Interestingly, a nascent theme in posthumanism (e.g., Hayles, 1999, Chapter 8) is not only blurring the distinction between different types of intelligence, but also blurring the distinction between mind, or body, and world. However, part of this blurring is motivated by the realization that the language of information can be applied equally easily to states of the world and to mental states. Hayles notes that one of the major implications of posthumanism is the resulting "systematic devaluation of materiality and embodiment" (p. 48). One of Hayles' goals is to resist the notion that "because we are essentially information, we can do away with the body" (p. 12).

One approach to achieving this goal is to explore the ideas in the new field of *embodied cognitive science* (Agre, 1997; Brooks, 1999; Clark, 1997, 1999; Gibbs, 2006; Pfeifer & Scheier, 1999). The theories of embodied cognitive science recognize that the individual can only be studied by considering his or her relationship to the environment, and that this relationship depends crucially upon embodiment (our physical structure) and situation (our sensing of the world). In other words, it places far more emphasis on the environment than has traditionally been found in the computational theories of modern cognitive science. It takes seriously the idea that Simon's (1969) parable of the ant might also be applicable to human cognition. "A man, viewed as a behaving system, is quite simple. The apparent complexity of his behavior over time is largely a reflection of the complexity of the environment in which he finds himself" (Simon, 1969, p. 25).

This raises two different kinds of questions. The first, explored in Chapter 2, is the degree to which higher-order cognitive phenomena in humans might be explained by such notions as embodiment, situation, and stigmergy. The second concerns the research methodologies required to study human cognition from the perspective of embodied cognitive science.

1.14 STIGMERGY AND CLASSICAL COGNITION

1.14.1 Classical Control

It is important to recognize that endorsing new ideas, such as the stigmergic control of cognition, does not require the complete abandonment of the representational theory of mind or of classical cognitive science. Indeed, stigmergy has a long history in prototypical models of higher-order human cognition.

Classical cognitive science views cognition as information processing (Dawson, 1998). An explanation in classical cognitive science thus requires that researchers propose a cognitive architecture (Pylyshyn, 1984; Van-Lehn, 1991). A cognitive architecture requires that the basic nature of an information processor's symbols or representations must be detailed. The set of rules that can manipulate these symbols must also be stipulated.

However, these two components are not enough to complete the architecture. In addition to specifying symbols and rules, a researcher must also specify the architecture's control structure (Hayes-Roth, 1985; Newell, 1973). A control structure is used to determine, at any given time, which particular rule should be used to manipulate the existing symbols in an information processor's memory. "The control problem is fundamental to all cognitive processes and intelligent systems. In solving the control problem, a system decides, either implicitly or explicitly, what problems it will attempt to solve, what knowledge it will bring to bear, and what problem-solving methods and strategies it will apply" (Hayes-Roth, 1985, p. 251).

One example of an information-processing proposal that includes an explicit control structure is the *production system* (Newell, 1973; Newell & Simon, 1961, 1972). Production systems have been used to simulate many higher-order cognitive phenomena (Anderson, 1983; Anderson et al., 2004; Meyer, Glass, Mueller, Seymour, & Kieras, 2001; Meyer & Kieras,

1997a, 1997b; Newell, 1990). The simplest production system consists of a working memory that holds symbolic expressions, and a set of productions that manipulate the expressions in memory. Each production is a condition-action pair: when it recognizes that its condition is true of working memory, it acts, by changing some symbols in memory.

Crucially, the control structure of a production system is stigmergic. That is, working memory can be viewed as being analogous to a nest for a colony of wasps, and each production can be seen as being analogous to an individual member of the colony. Each production is controlled by the symbolic expressions in working memory; changes in working memory made by one production can produce the conditions that call a different production into action. In short, stigmergic control is a key characteristic of an architecture that has made significant contributions to classical cognitive science for decades.

1.14.2 Externalizing Control

The key difference between the stigmergic control of nest construction and the stigmergic control of a production system is that the former is external (or environmental), while the latter is internal. Production systems strongly adhere to the sense–think–act cycle: even in modern production systems (Anderson et al., 2004; Meyer & Kieras, 1997a, 1997b), sensing is used to provide input to working memory, and working memory manipulations are used to plan actions to be carried out in the world. One of the main goals of embodied cognitive science is to replace the sense–think–act cycle with sense–act processing (Brooks, 1999; Clark, 1997, 2003; Pfeifer & Scheier, 1999). Embodied cognitive scientists recognize that the external world can be used to scaffold cognition (Hutchins, 1995), and that working memory — and other components of a classical architecture — have leaked into the mind. In Chapter 3 we will explore how these ideas can be incorporated into an architecture that is related to a production system. However, before doing so, we will first examine a high-order cognitive ability — the composition of classical music — to further motivate the need to externalize human cognition and its control.

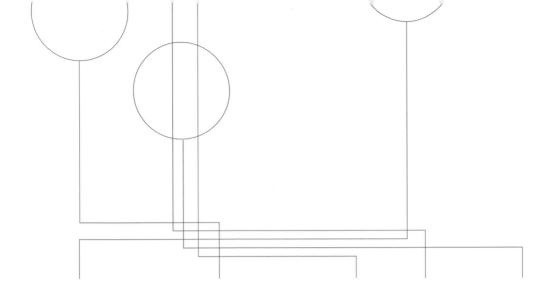

Chapter 2
Classical Music and the Classical Mind

2.0 CHAPTER OVERVIEW

The classical approach dominates modern cognitive science. The classical approach is based upon the assumption that cognition is the rule-governed manipulation of mental representations. It adopts the sense–think–act cycle (Pfeifer & Scheier, 1999), assuming that the primary purpose of cognition is to plan future actions by thinking about information provided by perceptual mechanisms. One purpose of the current chapter is to provide a brief introduction to some general characteristics of classical cognitive science: the prevalence of logicism, the manipulation of content-laden formal structures, the disembodiment of thought, and the emphasis on central control mechanisms. A second purpose of the current chapter is to draw a parallel between classical cognitive science and the traditional notion of classical music. This is done by showing that these central characteristics of classical cognitive science are also fundamental to the Austro-German tradition of classical music.

At the end of Chapter 1, we raised the question of whether higher-order human cognitive phenomena were less rational than might be expected. This question is one of the central motives for focusing on classical music in Chapter 2. As the first part of Chapter 2 explores the analogy between classical cognitive science and classical music, it seems evident that classical cognitive science is ideally suited to explain the act of musical composition. However, the second part of Chapter 2 pursues the analogy in such a way as to challenge "classical explanations." First, we explore a number of reactions against Austro-German classical music that are evident in modern classical music. We also show how these reactions are analogous to the reactions of embodied cognitive science against classical cognitive science. In the end, we see that many non-classical characteristics (decentralized control, embodiment, emergence, and stigmergy) are fundamental to modern music, and may also be evident in older musical traditions. In this situation, classical cognitive science may not be well suited to explain musical composition after all. Alternative traditions, such as embodied cognitive science, may be better suited. If this is true for explaining the composition of music, then is it not reasonable to expect that many less sophisticated human achievements might be better explained by alternative approaches to cognitive science? A main goal of Chapter 2 is to motivate this question; the chapters that follow then attempt to explore alternative approaches. For instance, Chapter 3 will introduce the notions of situated cognition and *bricolage,* which are central to the robots that we begin to describe in Chapter 4.

2.1 THE BOOLEAN DREAM
2.1.1 Cognitive Science

Cognitive science is an interdisciplinary approach to explaining mental phenomena (Bechtel, Graham, & Balota, 1998; Dawson, 1998; Goldman, 1993; Lepore & Pylyshyn, 1999; Pylyshyn, 1980; Simon, 1980; Thagard, 1996; Von Eckardt, 1993). Cognitive science is dominated by the representational theory of mind (Fodor, 1975; Newell, 1980; Pylyshyn, 1984), which is frequently called the *classical approach*, which assumes that external behaviour is mediated by the contents of internal representations. Such representations are intentional (Brentano, 1874/1995), in the sense that they stand for, or are about, states of affairs in the external world.

According to classical cognitive science, thinking is the manipulation of mental representations by rules. Rules are sensitive to the formal nature of mental symbols (Haugeland, 1985). That is, a symbol's form is used to identify it as being a token of a particular type. When identified in this way, only certain rules can be applied to it. While the rules are sensitive to the formal nature of symbols, they act in such a way to preserve the meaning of the information that the symbols represent. That is, if one manipulates the symbols according to the available formal rules, then one will not be able to create a meaningless representation. This property derives from classical cognitive science's logicism, which is discussed below.

2.1.2 Logicism

Aristotle claimed rationality separated men from beasts: humans have minds, and can reason; animals are mindless, and therefore cannot reason (Oaksford & Chater, 1998). This view is called *logicism*, and it is strongly linked to classical cognitive science.

Logicism is evident in George Boole's *An Investigation of the Laws of Thought* (Boole, 1854). Boole noted, "it is unnecessary to enter here into any argument to prove that the operations of the mind are in a certain real sense subject to laws" (p. 3). Boole's purpose was to "investigate the fundamental laws of those operations of the mind by which reasoning is performed; to give expression to them in the symbolic language of a Calculus, and upon this foundation to establish the science of Logic and construct its method" (p. 1).

Boole's (1854) logicism equated thinking with applying logical rules. "There is not only a close analogy between the operations of the mind in general reasoning and its operations in the particular science of Algebra, but there is to a considerable extent an exact agreement in the laws by which the two classes of operations are conducted" (p. 6). Modern versions of logicism endorse this view as well. For instance, a famous introductory logic book (Kalish & Montague, 1964) viewed logic as mirroring everyday reasoning. Formal logic is also a powerful and popular tool for knowledge representation and manipulation in the field of artificial intelligence (Genesereth & Nilsson, 1987; Levesque & Lakemeyer, 2000).

2.1.3 The Boolean Dream

Boole's (1854) formalism attempted to explain reasoning without appealing to underlying biological mechanisms. This has been called the Boolean Dream of classical cognitive science (Hofstadter, 1995). "The traditional holy grail of AI has always been to *describe thoughts at their own level* without having to resort to describing any biological (i.e. cellular) underpinnings of them" (p. 125). The Boolean Dream reflects the classical approach's emphasis on the roles or functions of various system components (Cummins, 1983; Fodor, 1968), and recognizes that many different physical devices can produce identical functions (the multiple realization argument). As a result, classical cognitive science is logicist in nature, because it attempts to explain cognition by appealing to formal rules or a "language of thought" (Fodor, 1975; Fodor & Pylyshyn, 1988; Pylyshyn, 1984). "It would not be unreasonable to describe Classical Cognitive Science as an extended attempt to apply the methods of proof theory to the modeling of thought" (Fodor & Pylyshyn, 1988, pp. 29–30).

2.2 CLASSICAL COGNITIVE SCIENCE
2.2.1 A Classical Device

There is a long history of performing logical operations with special purpose diagrams or machines (Gardner, 1982). The first special-purpose "logic machine" was Charles Stanhope's demonstrator, built sometime before 1816.

Classical cognitive science's logicism results because the modern representational theory of mind was inspired by a much more general symbol-manipulating device, the digital computer. Symbol-manipulating computers were proven to be capable of solving any computable problem (Turing, 1936). Philosopher Kenneth Craik linked the operations of such machines to laws of thought (Craik, 1943): "My hypothesis then is that thought models, or parallels, reality—that its essential feature is not 'the mind', 'the self', 'sense data' nor 'propositions', but is symbolism, and that this symbolism is largely of the same kind which is familiar to us in mechanical devices which aid thought and calculation" (p. 57). It did not take long for prominent researchers to become convinced of the inevitability of machine intelligence (Turing, 1950).

The prototypical "mechanical device" for manipulating symbols is the *Turing machine* (Hodges, 1983; Turing, 1936). It uses tickertape as a medium for storing data. The tape is divided into cells, each of which can store a single symbol from a finite alphabet. The Turing machine's *machine head* moves back and forth along the tape, reading symbols, and rewriting the tape according to rules stored in a *machine table*.

A Turing machine is a device for answering questions. A question is written on the machine's tape, and is then given to the machine head. When the machine head halts, it has written an answer to the question on the tape. The kind of question that a particular Turing machine can answer is dictated by its machine table. To answer different types of questions, different machine tables are required, requiring a *hardware* change in the machine head (Dawson, 1998).

However, one notable exception to this is the *universal Turing machine*. It can pretend to be any possible Turing machine using *software*. The machine table of the desired Turing machine is written as a program on the ticker tape, along with the to-be-answered question. The universal machine simulates the actions of the desired Turing machine by moving back and forth between the program and the question, answering the question as if the universal machine's table was identical to the one programmed on the tape. "It followed that one particular machine could simulate the work done by any machine ... It would be a machine to do everything, which was enough to give anyone pause for thought" (Hodges, 1983, p. 104).

One result of Turing's universal machine was the Church-Turing thesis: "Any process which could naturally be called an effective procedure can be realized by a Turing machine" (Minsky, 1972, p. 108). This in turn has led classical cognitive science to both the computer metaphor and the methodology of computer simulation. Early cognitive scientists predicted that psychological theories would be expressed as computer programs (Simon & Newell, 1958). This view has remained with the modern classical approach (Johnson-Laird, 1983): "In so far as there can be a science of the mind it will almost be certainly restricted to accounts that can be formulated as computer programs" (p. 8).

2.2.2 Three Key Characteristics

The link between logicism, classical cognitive science and the digital computer implies that classical theories will have three general characteristics. First, a classical theory will include a set of symbols for representing knowledge. Second, such a theory will include a set of rules for manipulating these symbols in a fashion analogous to a Turing machine. Third, a classical theory will stipulate some control procedure that chooses which rule to apply at any given moment in time.

2.3 CLASSICAL VIEWS OF MIND AND MUSIC
2.3.1 Mind and Music

Chapter 1 explored the notion that, in virtue of our rationality, humans create and control their environment, while (irrational) animals do not. Even when impressive architectural feats accomplished by social insects and other animals are considered, it can be argued that these accomplishments are products of mindless environmental control. This reassuringly supports the age-old idea that human mentality separates us from the beasts. However, it also raises an obvious question: to what extent might complex human behaviours also be the result of environmental control rather than rational thought? The current chapter explores this question by taking a prototypical example of human intellect—classical music—and examining the extent to which some of the ideas introduced in Chapter 1 can be applied to it. There are several reasons for using classical music to explore these ideas. First, classical music arguably embodies some of the highest accomplishments of human intellect. If such compositions could be explained by appealing to non-rational factors such as stigmergy, then this would provide a compelling reason to explore non-rational accounts of other cognitive phenomena. Second, classical music has been analyzed by a diverse range of thinkers, including composers, performers, critics, historians, philosophers, neuroscientists, and psychologists. As a result there is an extremely rich array of material that can be used to explore classical music as a cognitive product.

Third, there is an interesting analogy to draw between classical music and the classical notion of mind. This analogy is introduced below.

2.3.2 A Classical Analogy

There are many striking parallels between the classical mind and classical music, particularly the music composed in the Austro-German tradition of the eighteenth and nineteenth centuries. First, both rely heavily upon formal structures. That is, logicism can easily be found to apply to many aspects of classical music.

Second, both emphasize that their formal structures are content-laden. In classical cognitive science, mental representations have content, and we can predict the behaviours of others by ascribing mental content (Dennett, 1987). Classical music is also widely held to be meaningful (Meyer, 1956). Furthermore, its meaning is tightly linked to its formal structure: "One musical event (be it a tone, a phrase, or a whole section) has meaning because it points to and makes us expect another musical event" (Meyer, 1956, p. 35).

Third, both attribute great importance to abstract thought inside an agent (or composer), at the expense of contributions involving the agent's environment or embodiment. For instance, Mozart "carried his compositions around in his head for days before setting them down on paper" (Hildesheimer, 1983): in a letter that he wrote to his father in 1780, Mozart noted that "everything is composed, just not copied out yet." Fourth, both emphasize central control. In classical cognition, control is required to choose which rule to apply next. In classical music, a performance is also under strict control (Green & Malko, 1975): "The conductor acts as a guide, a solver of problems, a decision maker. His guidance chart is the composer's score; his job, to animate the score, to make it come alive, to bring it into audible being" (p. 7).

Fifth, the "classical" traditions of both mind and music have faced strong challenges, and many of the challenges in one domain can be related to parallel challenges in the other. Modern classical music is a reaction against the key attributes of Austro-German music (Griffiths, 1994, 1995), just as embodied cognitive science is a reaction against the central claims of classical cognitive science (Brooks, 1999, 2002; Clark, 1997; Varela et al., 1991).

2.4 MUSICAL LOGICISM
2.4.1 Musical Formalisms

Music's formal nature extends far beyond musical symbols on a sheet of staff paper. For example, some combinations of tones played in unison are pleasing to the ear while other combinations are not. This experience is easily related to the degree of separation between notes on a musical scale (Krumhansl, 1990). For instance, if one plays middle C on a piano, then a pleasing sound will be produced if the note F is played at the same time, because F is a perfect fifth (7 semitones) above C.

The consonance of two notes that are a perfect fifth apart can be explained by the physics of sound waves (Helmholtz & Ellis, 1954). Such physical relationships are ultimately mathematical. Indeed, there is an extensive literature on the mathematical nature of music (Assayag, Feichtinger, Rodrigues, & European Mathematical Society., 2002; Benson, 2007; Harkleroad, 2006). For instance, different approaches to tuning instruments reflect the extent to which tunings are deemed mathematically sensible (Isacoff, 2001).

2.4.2 Sonata-Allegro Form

Importantly, musical formalisms exist at levels beyond individual notes. A musical offering is expected to have a particular structure (Copland, 1939), "the planned design that binds an entire composition together" (p. 113). For Copland this structure is "one of the principal things to listen for," and much of his *What to Listen for in Music* describes structural variants. One important musical structure is *sonata-allegro form* (Copland, 1939). This is an example of three-part form in which there is an initial exposition of musical ideas, followed by their free development, and ending with their recapitulation as shown in Table 2-1.

Each of these parts has its own structure. The exposition introduces an opening theme in the tonic key (that is, the initial key signature of the piece), then follows with a second theme in the dominant key (a perfect fifth above the tonic), and finishes with a closing theme in the dominant key. The recapitulation uses the same three themes in the same order, but all are in the tonic key. The development section explores the exposition's themes, but does so using new material written in different keys.

The themes are expected to have certain characteristics as well. The opening theme is dramatic, the second theme is lyrical, and the third theme "may be of any nature that leads to a sense of conclusion ... This juxtaposition of one group of themes denoting power and aggressiveness with another group which is relaxed and more song like in quality is the essence of the exposition section and determines the character of the entire sonata-allegro form" (Copland, 1939, p. 185).

Sonata-allegro form was historically important because it foreshadowed the modern symphony, and it produced a market for purely instrumental music (Rosen, 1988). Importantly, it also provided a structure, shared by both composers and their audiences, which permitted instrumental music to be expressive. Rosen notes the sonata became popular because it "has an identifiable climax, a point of maximum tension to which the first part of the work leads and which is symmetrically resolved. It is a closed form, without the static frame of ternary form; it has a dynamic closure analogous to the denouement of eighteenth-century drama, in which everything is resolved, all loose ends are tied up, and the work rounded off" (p. 10). In short, its formal structure provided a logical structure that permitted the music to be meaningful.

Exposition (A)			Development (B)	Recapitulation (A)		
a	*b*	*c*	*abc*	*a*	*b*	*c*
First theme in tonic key	Second theme in dominant key	Closing theme in dominant key	Free combination of the three themes, and new material in foreign keys	First theme in tonic key	Second theme in tonic key	Closing theme in tonic key

Table 2-1 The hierarchial structure of sonata-allegro form.

2.5 A HARMONIOUS NARRATIVE
2.5.1 Representational Explanation

Classical cognitive science explains cognition by using mental representations. By noting that an agent has certain (semantically interpreted) goals, as well as beliefs about how these goals can be attained, it predicts the agent's future behaviour. "We can conclude that the representational, or semantic, level represents a distinct, autonomous level of description of certain kinds of systems" (Pylyshyn, 1984, p. 33).

Classical cognitive science's appeal to representations reveals its deep commitment to logicism. In order for the contents of representational states to predict behaviour, some general principles or laws must govern these states. One key principle is *rationality*, which is the notion that agents will use the content of their beliefs to achieve their goals. Rationality is at the foundation of Dennett's (1987) intentional stance: "This single assumption, in combination with home truths about our needs, capacities and typical circumstances, generates both an intentional interpretation of us as believers and desirers and actual predictions of behavior in great profusion" (p. 50). Similarly, Pylyshyn (1984, pp. 20–21) notes, "the principle of rationality … is indispensable for giving an account of human behavior."

Classical cognitive science recognizes that there are no causal principles relating semantics to behaviour. This is why the digital computer is so important, for it illustrates how a symbolic system preserves meanings while at the same time existing as a purely causal, physical machine. Symbols in a representational system have two lives: semantic and physical (Haugeland, 1985). Physical manipulations of the symbols are systematic, and preserve the meanings of symbolic expressions. Haugeland describes this with the formalist's motto: *take care of the syntax, and the semantics will take care of itself.*

2.5.2 Musical Expressions

One of the central questions in the philosophy of music is whether music can represent. As late as 1790, the dominant philosophical view of music was that it was incapable of conveying ideas, but by the time that E.T.A. Hoffman reviewed Beethoven's Fifth Symphony in 1810, this view was rejected (Bonds, 2006). Nowadays most philosophers of music agree that it is representational, and are concerned with *how* musical representations are possible (Kivy, 1991; Meyer, 1956; Robinson, 1994, 1997; Sparshoot, 1994; Walton, 1994).

Composers certainly believe that music can express ideas. Aaron Copland (1939, p. 12) notes that "my own belief is that all music has an expressive power, some more and some less, but that all music has a certain meaning behind the notes and that that meaning behind the notes constitutes, after all, what the piece is saying, what the piece is about." John Cage believed that compositions had intended meanings (Cage, 1961); "It seemed to me that composers knew what they were doing, and that the experiments that had been made had taken place prior to the finished works, just as sketches are made before paintings and rehearsals precede performances" (p. 7).

How do composers convey intended meanings with their music? One answer is by using the conventions of particular musical forms. Such forms provide a structure that generates expectations, expectations that are often presumed to be shared by the audience. Indeed, Copland's (1939) book on music listening—which places such a strong emphasis on musical form—is designed to educate the audience so that it can better understand his compositions, as well as those of others: "In helping others to hear music more intelligently, [the composer] is working toward the spread of a musical culture, which in the end will affect the understanding of his own creations" (p. vi). The extent to which the audience's expectations are toyed with, and ultimately fulfilled, can manipulate its emotion. These manipulations can be described completely in terms of the structure of musical elements (Meyer, 1956). In this sense, the formalist's motto also applies to classical music as traditionally conceived.

2.6 THE NATURE OF CLASSICAL COMPOSITION
2.6.1 The Disembodied Mind

Classical cognitive science attempts to explain cognitive phenomena by appealing to a sense–think–act cycle (Pfeifer & Scheier, 1999). In this cycle, sensing mechanisms provide information about the world, and acting mechanisms produce behaviours that might change it. Thinking—manipulating mental representations—is the interface between sensing and acting (Wilson, 2004). Interestingly, the sense–think–act cycle does not reflect the true nature of the classical approach. Classical cognitive science places an enormous amount of emphasis on the "thinking" part of the cycle, with an accompanying under-emphasis on sensing and acting (Clark, 1997). Sensors are merely providers of information that can be manipulated; actors are simply devices that are capable of carrying out a well-thought-out plan of action.

One can easily find evidence for the classical emphasis on representations. Autonomous robots developed following classical ideas devote most of their computational resources to using internal representations of the external world (Brooks, 2002; Moravec, 1999; Nilsson, 1984). Most survey books on cognitive psychology have multiple chapters on representational topics like memory and reasoning, and rarely mention embodiment, sensing, or acting (see Anderson, 1985; Best, 1995; Haberlandt, 1994; Robinson-Riegler & Robinson-Riegler, 2003; Solso, 1995; Sternberg, 1996). Classical cognitive science's sensitivity to the multiple realization argument (Fodor, 1968, 1975), with its accompanying focus on functional (not physical) accounts of cognition (Cummins, 1983), underlines its view of thinking as a disembodied process.

2.6.2 The Thoughtful Composer

Composing classical music is also viewed as abstract, disembodied, and rational. Does not a composer first think of a theme or a melody, and then translate this mental representation into a musical score?

One example of this is the story of Mozart carrying around completed compositions in his head prior to writing them out (Hildesheimer, 1983). Similar examples are easily found. Benson (2007, p. 25) notes that "Stravinsky speaks of a musical work as being 'the fruit of study, reasoning, and calculation that imply exactly the converse of improvisation'." In the liner notes of his Grammy award–winning *Symphony No. 1* (Sony Classical Music, 1999), Joe Jackson recalls that "I had a handful of very simple musical themes in my head and wanted to see if they could be developed and transformed throughout four whole movements."

There is a general prejudice against composers who rely on external aids (Rosen, 2002). Copland (1939, p. 22) observes that "a current idea exists that there is something shameful about writing a piece of music at the piano." Rosen traces this idea to Giovanni Maria Artusi's 1600 criticism of composers such as Monteverdi: "It is one thing to search with voices and instruments for something pertaining to the harmonic faculty, another to arrive at the exact truth by means of reasons seconded by the ear" (Rosen, 2002, p. 17). The expectation (then and now) is that composing a piece involves "mentally planning it by logic, rules, and traditional reason" (Ibid). This expectation is completely consistent with the disembodied, classical view of thinking.

To appreciate this, consider composing from the opposite perspective. From the early 1950s, John Cage's compositions were non-intentional (Griffiths, 1994); he increasingly used chance mechanisms to determine musical events. He worked toward removing the composer from the composition, perhaps succeeding most with his 1952 "silent piece" 4'33". Cage was reacting against the Austro-German tradition of composition (Nyman, 1999). He advocated "that music should no longer be conceived of as rational discourse" (Nyman, 1999, p. 32). He explicitly attacked the logicism of traditional music, declaring "any composing strategy which is wholly 'rational' is irrational in the extreme" (Ross, 2007, p. 371).

2.7 CENTRAL CONTROL OF A CLASSICAL PERFORMANCE
2.7.1 Central Control

Herbert Simon argued that "an adequate theory of human cognitive processes must include a description of the *control system* — the mechanism that determines the sequence in which operations will be performed" (Simon, 1979, p. 370). In classical cognitive science, such control is typically central. There is a centralized mechanism that controls, at any given time, which rule will manipulate the symbols in memory. This is consistent with the digital computer metaphor. In a modern digital computer, the central processing unit (CPU) is responsible for determining what operation will be used to modify a specific memory location at every tick of the CPU's clock cycle. Central control is also characteristic of classical music, as is discussed below.

2.7.2 Conductor as Central Controller

Within the Austro-German tradition, a musical composition is a formal structure intended to express ideas. A composer uses musical notation to signify the musical events that, when realized, accomplish this expressive goal. An orchestra's purpose is to bring the score to life, in order that the performance will deliver the intended message to the audience (Benson, 2003). "We tend to see both the score and the performance primarily as vehicles for preserving what the composer has created. We assume that musical scores provide a permanent record or embodiment in signs; in effect, a score serves to 'fix' or objectify a musical work" (Benson, 2003, p. 9). However, it is generally acknowledged that a musical score is vague; it cannot completely determine every minute detail of a performance (Benson, 2003; Copland, 1939). As a result, during a performance the score must be interpreted in such a way that the missing details can be filled in without distorting the composer's desired effect. In the Austro-German tradition of music an orchestra's conductor takes the role of interpreter, and controls the orchestra in order to deliver the composer's message: "The conductor acts as a guide, a solver of problems, a decision maker. His guidance chart is the composer's score; his job, to animate the score, to make it come alive, to bring it into audible being" (Green & Malko, 1975, p. 7).

The conductor provides another link between classical music and classical cognitive science, because the conductor is the orchestra's central control system. The individual players are expected to submit to the conductor's control. "Our conception of the role of a classical musician is far closer to that of self-effacing servant who faithfully serves the score of the composer. Admittedly, performers are given a certain degree of leeway; but the unwritten rules of the game are such that this leeway is relatively small and must be kept in careful check" (Benson, 2003, p. 5). It has been suggested — not necessarily validly — that professional, classically trained musicians are incapable of improvisation (Bailey, 1992)!

2.7.3 The Controlling Score

The conductor is not the only component of a performance's control structure. While it is unavoidably vague, the musical score of a composition also is designed to control the musical events generated by an orchestra. That is, if the score is a content-bearing formal expression, then it is reasonable to assume that it designates the musical events that the score is literally about.

Benson (2003, p. 5) describes this aspect of a score as follows: "The idea(l) of being 'treu' — which can be translated as true or faithful — implies faithfulness to someone or something. *Werktreue*, then, is directly a kind of faithfulness to the *Werk* (work) and, indirectly, a faithfulness to the composer. Given the centrality of musical notation in the discourse of classical music, a parallel notion is that of *Texttreue*: fidelity to the written score." Note Benson's emphasis on the formal notation of the score. It highlights the idea that the written score is analogous to a logical expression, and that converting it into the musical events that the score is about (in Brentano's sense) is not only desirable, but also rational. This logicism of classical music perfectly parallels the logicism found in classical cognitive science.

2.8 DISEMBODIMENT AND THE CLASSICAL AUDIENCE
2.8.1 Disembodiment

An infinite number of different physical substrates can deliver the same set of information-processing functions (Putnam, 1967). For example, Turing machines can be created from brass gears (Swade, 1993), toy train sets (Stewart, 1994), artificial neural networks (McCulloch & Pitts, 1943; Siegelmann, 1999), mixtures of chemicals (Hjelmfelt, Weinberger, & Ross, 1991), or LEGO (Agullo et al., 2003). This possibility of multiple realizations makes classical cognitive science adopt functionalism (Cummins, 1983). That is, it explains systems by describing the functional roles of system components, and not by appealing to their physical nature. This is why simulations can be used in cognitive science: the physical differences between computers and brains are irrelevant, as long as functional correspondences are maintained (Pylyshyn, 1984).

Classical cognitive science's functionalism is one example of its move toward disembodiment: representational theories can ignore the physical substrate that is instantiating symbols and the rules that manipulate them. However, there is another form of disembodiment that characterizes much of classical cognitive science.

Classical cognitive science explains psychological phenomena by appealing to the representational states of agents. Different behavioural predictions must be grounded in different representational states. A representational theory of mind must therefore be capable of individuating different representational states. This could be done in terms of their content (i.e., different states must refer to different entities in the world), but there are well-known philosophical problems with this approach (Pessin, Goldberg, & Putnam, 1996). An alternative approach that has strongly influenced classical cognitive science is *methodological solipsism* (Fodor, 1980). In methodological solipsism, representational states are individuated only in terms of their relations to other representational states. Relations of the states to the external world — the agent's environment — are not considered. "Methodological solipsism in psychology is the view that psychological states should be construed without reference to anything beyond the boundary of the individual who has those states" (Wilson, 2004, p. 77).

2.8.2 Audience and Composition

Methodological solipsism provides another link in the analogy between the classical mind and classical music. As was noted in Section 2.7, when a piece is performed it is brought to life with the intent of delivering a particular message to the audience. Ultimately, then, the audience is a fundamental component of a composition's environment. To what extent does this environment affect or determine the composition itself?

In traditional classical music, the audience has absolutely no effect on the composition. Composer Arnold Schoenberg believed that the audience was "merely an acoustic necessity — and annoying one at that" (Benson, 2003, p. 14). Composer Virgil Thompson defined the ideal listener as "a person who applauds vigorously" (Copland, 1939, p. 252). In short, the purpose of the audience is to passively receive the intended message. It too is under the control of the score: "The intelligent listener must be prepared to increase his awareness of the musical material and what happens to it. He must hear the melodies, the rhythms, the harmonies, the tone colors in a more conscious fashion. But above all he must, in order to follow the line of the composer's thought, know something of the principles of musical form" (Copland, 1939, p. 17).

To relate this to methodological solipsism, consider how compositions are to be identified or differentiated from one another. Traditionally, this is done by referring to a composition's score (Benson, 2003). That is, compositions are identified in terms of a particular set of symbols, a particular formal structure. The identification of a composition does not depend upon identifying which audience has heard it. A composition can exist, and be identified, in the absence of its audience-as-environment.

2.9 CLASSICAL REACTIONS
2.9.1 Reacting to Music

Another parallel between the classical mind and classical music is that there have been significant modern reactions against the Austro-German musical tradition (Griffiths, 1994, 1995). Interestingly, these reactions parallel many of the reactions of embodied cognitive science against the classical approach. In the pages that follow we will consider some of these reactions, and explore the idea that they make plausible the claim that "non-cognitive" processes are applicable to classical music.

2.9.2 Classical Competitors

The modern interdisciplinary study of mind begins with the science of cybernetics in the 1940s (Ashby, 1956; Conway & Siegelman, 2005; de Latil, 1956; Hayles, 1999; Wiener, 1948), and its famous Macy conferences through the 1950s. Cybernetics had waned by the end of the 1950s. It was replaced by cognitive science, whose origin occurred on September 11, 1956 (Gardner, 1984; Miller, 2003). Cognitive science has since flourished, dominated by the classical approach.

The classical view is not without its competitors. *Connectionists* believe that cognitive science is not best served by the digital computer metaphor (Bechtel & Abrahamsen, 2002; Medler, 1998; Quinlan, 1991; Schneider, 1987). They argue that the serial, rule-governed, centrally controlled processing performed by digital computers is too slow, too inflexible, and too divorced from biology to meaningfully account for human cognition. They insist that the information processing is the result of the parallel processing of multiple, simple, intercommunicating units. They model such information processing with artificial neural networks. Connectionists use such networks to support the claim that cognition does not require explicit rules and symbols, but instead emerges from neuronally inspired processes best described using statistical mechanics (Clark, 1989, 1993; McClelland & Rumelhart, 1986; McClelland, Rumelhart, & Hinton, 1986; Rumelhart & McClelland, 1986).

A second competitor to the classical approach, called *embodied cognitive science*, has concerns similar to connectionism, but develops them within a fundamental attack on methodological solipsism.

Classical and connectionist cognitive science are both fundamentally representational in nature (Dawson, 1998, 2004), emphasizing internal information processing at the expense of environmental influences. Embodied cognitive science views this as a serious mistake. Embodied cognitive scientists argue that a cognitive theory must include an agent's environment, as well as the agent's experience of that environment (Agre, 1997; Clancey, 1997; Clark, 1997; Pfeifer & Scheier, 1999; Varela et al., 1991). They recognize that this experience depends on how the environment is sensed (*situation*), that an agent's situation depends upon its physical nature (*embodiment*), and that an embodied agent can act upon and change its environment (Webb & Consi, 2001). The embodied approach replaces the notion that cognition is representation with the notion that cognition is the control of actions upon the environment. In embodied cognitive science, the environment contributes in such a way that it can be said that an agent's mind has leaked into the world (Clark, 1997; Wilson, 2004). For example, research in behaviour-based robotics eliminates resource-consuming representations of the world by letting the world serve as its own representation, one that can be accessed by a situated agent (Brooks, 1999). This robotics tradition has also shown that non-linear interactions between an agent and its environment can produce surprisingly complex behaviour, even when the internal components of an agent are exceedingly simple (Braitenberg, 1984; Grey Walter, 1950b; Webb & Consi, 2001). This observation reconnects cognitive science with cybernetics.

In short, to the notions of emergence and biological plausibility, embodied cognitive science adds the ideas of situation and embodiment. Interestingly, we will see that these ideas can also be found in reactions to classical music.

2.10 MODERN MUSIC
2.10.1 Out with the Old

By the end of the nineteenth century, classical music had reached its zenith. Composers had invented a market for instrumental music that was fueled by their discovery and refinement of particular musical forms (Rosen, 1988). For example, in the early seventeenth century, the symphony was merely a short overture played before the raising of the curtains at an opera (Lee, 1916). The more interesting of these compositions came to be performed to their own audiences outside the theater. The modern symphony, which typically consists of four movements that each has an expected form and tempo, begins to be seen in the eighteenth-century compositions of Carl Philip Emmanuel Bach. Experiments with this structure were conducted in the later eighteenth century by Haydn and Mozart.

When Beethoven wrote his symphonies in the early nineteenth century, symphonic form was established, and Beethoven proved its enormous expressive power. "No less a person than Richard Wagner affirmed that the right of composing symphonies was abolished by Beethoven's Ninth'" (Lee, 1916, p. 172). What are some general characteristics of prototypical classical music, such as a Beethoven symphony? Consider the properties of sonata-allegro form, which is always used to structure a symphony's first movement. First, this form is based upon particular musical themes or melodies. Second, these melodies are associated with a specific tonality: they are written in a particular musical key (such as the tonic or dominant mentioned in the discussion of Table 2-1). The tonality of the form dictates harmonic structure; that is, within a musical key certain combinations of notes (chords) will be concordant, while others will not be played because they will be discordant. The form itself indicates an expected order in which themes and musical keys will be explored, and an established rhythmic structure (related to a time signature) will be used throughout.

Perhaps the key feature from this list is tonality: the use of particular major and minor musical keys to establish an expected harmonic structure in a composition. "Harmony is Western music's uniquely distinguishing element" (Pleasants, 1955, p. 97). In the early twentieth century, strongly affected by both world wars, classical music found itself in a crisis of harmony (Pleasants, 1955). Composers abandoned most of the features listed above in an attempt to create a new music that better reflected modern times. "'Is it not our duty', he [Debussy] asked, 'to find a symphonic means to express our time, one that evokes the progress, the daring and the victories of modern days? The century of the aeroplane deserves its music'" (Griffiths, 1994, p. 98).

2.10.2 In with the New

Griffiths (1994) places the beginning of modern music with the flute solo that opens the *Prélude à 'L'après-midi d'un faune'* composed by Claude Debussy between 1892 and 1894. The *Prélude* begins to break away from the harmonic relationships defined by strict tonality. It fails to logically develop themes. It employs fluctuating tempos and irregular rhythms. It depends critically on instrumentation for expression. Debussy "had little time for the thorough, continuous, symphonic manner of the Austro-German tradition, the 'logical' development of ideas which gives music the effect of a narrative" (Griffiths, 1994, p. 9).

"Debussy had opened the paths of modern music — the abandonment of traditional tonality, the development of new rhythmic complexity, the recognition of color as an essential, the creation of a quite new form for each work, the exploration of deeper mental processes" (Griffiths, 1994, p. 12). In the twentieth century, composers experimented with new methods that further pursued these paths. We shall see, in the progression of these experiments, those reactions to traditional classical music parallel reactions to classical cognitive science, particularly in exploiting notions of emergence, embodiment, and stigmergy.

2.11 DODECAPHONY

2.11.1 Tonality and Atonality

The crisis of harmony was addressed by composing deliberately atonal music. The possibility of doing this is illustrated in Table 2-2. The top row of this table illustrates the keys on a piano from the note A to a second A that is an octave higher than the first. On the piano, some of these keys are white, and others are black; the colour of each key is depicted in the table. The interval between adjacent keys (e.g., from B to C, or from C to C#) is a *semitone*. If one were to play the top row in sequence, the result would be a *chromatic scale*, in which thirteen different notes were played (from A to A), and each note that is played is a semitone higher than the previous note.

A *major scale* is a sequence of notes that is associated with a particular key; for example, the A major scale is associated with the musical key (or tonal centre) of A. The A major scale is played by starting with the low A at the left of the table, and by playing notes in sequence until the high A at the right of the table is played. This scale has a distinctive sound because not all of the notes in the chromatic scale are used. The second row of the table illustrates the subset of notes that, when played, produces the A major scale. If there is a checkmark beneath the key, then it is played in the scale; if there is no checkmark then the key is not part of the scale. By choosing a different subset of notes a very different-sounding scale is produced. The third row of Table 2-2 shows the subset of piano notes that would be used to play a *minor scale* beginning with the note A.

A scale's tonality is the result of only including a subset of possible notes. That is, to compose a piece that had the tonal centre of A major, one would only include those notes that belonged to the A major scale.

In contrast, to produce music that is atonal, one would include all the notes from the chromatic scale. Because all notes are included, it is impossible to associate this set of notes with a tonal centre. One method of ensuring atonality is the "twelve-tone technique," or *dodecaphony*, invented by Arnold Schoenberg.

2.11.2 The Twelve-Tone Method

When dodecaphony is employed, a composer starts by listing all twelve notes in a chromatic scale in some desired order. This creates the *tone row*. The tone row is the basis for a melody: the composer begins to write the melody by using the first note in the tone row (for a desired duration, possibly with repetition). However, this note cannot be reused in the melody until the remaining notes have also been used in the order specified by the tone row. This ensures that the melody is atonal, because all the notes that make up a chromatic scale have been included.

Once all twelve notes have been used, the tone row is then shifted to create the next section of the melody. At this time, the tone row can be manipulated to produce musical variation (e.g., by reversing its order, or by inverting the musical intervals between adjacent tones).

The first example of a dodecaphonic composition was Schoenberg's 1923 *Suite for Piano, Op. 25*. Schoenberg and his students (Alban Berg, Anton Webern) composed extensively using the twelve-note technique. A later movement in music, *serialism*, used similar systems to determine other parameters of a score, such as note duration and dynamics. It was explored by Olivier Messiaen and his followers, notably Pierre Boulez and Karlheinz Stockhausen (Griffiths, 1995).

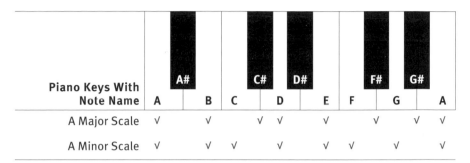

Piano Keys With Note Name		A	A#	B	C	C#	D	D#	E	F	F#	G	G#	A
A Major Scale		√		√		√	√		√		√		√	√
A Minor Scale		√		√	√		√		√	√		√		√

Table 2-2. The top row illustrates keys (white and black) on a piano from one A to the A an octave higher. The last two rows provide the subsets of notes that define A major and A minor scales; checkmarks are used to indicate which notes are included in a scale.

2.12 REACTIONS TO ATONAL STRUCTURE

2.12.1 From Structure to Structure

Schoenberg wrote his original atonal works without the aid of dodecaphony. The lack of a guiding structure made it difficult to create large, coherent, atonal works. His invention of dodecaphony solved this problem. Schoenberg was "troubled by the lack of system, the absence of harmonic bearings on which large forms might be directed. Serialism at last offered a new means of achieving order" (Griffiths, 1994, p. 81).

The twelve-tone technique provides an alternative to the traditional forms of classical music. However, this new form still followed the Austro-German tradition's need for structure. "The new rules must be applied to the construction of forms and textures in the old manner" (Griffiths, 1994, p. 85). Composer Philip Glass recognized this situation: "To me, it was music of the past, passing itself off as music of the present. After all, Arnold Schoenberg was about the same age as my grandfather!" (Glass, 1987, p. 13). Critics accused serialist compositions of being mathematical or mechanical (Griffiths, 1994). Indeed, serialism made computer composition possible: in 1964 Gottfried Koenig created Project 1, which was a computer program that composed serial music (Koenig, 1999).

Serialism also shared the traditional approach's disdain for the audience. American composer Steve Reich (Reich, 1974) notes that "in serial music, the series itself is seldom audible" (p. 10), which appears to be a serial composers' intent (Griffiths, 1994). This music's opacity, and its decidedly different or modern sound, frequently led to hostile receptions. One notable example is music critic Henry Pleasants' *The Agony Of Modern Music* (Pleasants, 1955): "The vein which for three hundred years offered a seemingly inexhaustible yield of beautiful music has run out. What we know as modern music is the noise made by deluded speculators picking through the slag pile" (p. 3).

That serial music was derived from a new kind of formalism also fuelled its critics. "Faced with complex and lengthy analyses, baffling terminology and a total rejection of common paradigms of musical expression, many critics — not all conservative — found ample ammunition to back up their claims that serial music was a mere intellectual exercise which could not seriously be regarded as music at all" (Grant, 2001).

2.12.2 Reducing Central Control

At issue was the fact that European composers were steeped in the centuries-old traditions of classical music, which made it difficult for them to break free of the old forms even when they recognized a need for new music (Griffiths, 1994). Schoenberg wrote, "I am at least as conservative as Edison and Ford have been. But I am, unfortunately, not quite as progressive as they were in their own fields" (Griffiths, 1995, p. 50).

American composers were certainly not drawn to the new atonal structures. Philip Glass describes his feelings about serialism: "A wasteland, dominated by these maniacs, these creeps, who were trying to make everyone write this crazy creepy music" (Schwarz, 1996). When Glass attended concerts, the only "breaths of fresh air" that he experienced was when works from modern American composers like John Cage were on the program (Glass, 1987). The new American music was more progressive than its European counterpart because American composers were far less shackled by musical traditions.

American composers were willing to relinquish the central control of the musical score, recognizing the improvisational elements of classical composition (Benson, 2003). Some were even willing to relinquish much of the composer's control over the piece (Cage, 1961). They recognized that many musical effects depended upon the audience's perceptual processes (Potter, 2000; Schwarz, 1996), and many examples of experimental music relied heavily upon the audience as an equal partner in bringing the composition to life (Nyman, 1999). It is these insights that provide links between the new music and the new cognitive science.

2.13 CONTROL AND EMERGENCE IN CAGE'S MUSIC

2.13.1 Silence

In 1937 composer John Cage declared, "The present methods of writing music, principally those which employ harmony and its reference to particular steps in the field of sound, will be inadequate for the composer, who will be faced with the entire field of sound" (Cage, 1961, p. 4). Cage was well versed in the twelve-tone technique (Ross, 2007), but did not see tonality (or atonality) as the defining characteristic of music. Instead, Cage emphasized rhythmic structuring, "since duration is the most fundamental musical characteristic, shared by both sound and silence" (Griffiths, 1994, p. 118). For Cage the entire field of sound included silence and sounds typically considered to be non-musical.

Cage's music was largely motivated by his desire to free it from the composer's will. He wrote, "when silence, generally speaking, is not in evidence, the will of the composer is. Inherent silence is equivalent to denial of the will" (Cage, 1961, p. 53). In Cage's compositions we see a composer who is willing to relinquish the central control so fundamental to traditional classical music.

Cage's most famous example of relinquishing control is in his composition 4'33", first performed by pianist David Tudor in 1952 (Nyman, 1999). It consists of three parts; the entire score for each part reads "TACET," which instructs the performer to remain silent. Tudor signalled the start of each part by closing the keyboard lid, and opened the lid when the part was over. When the composer relinquishes control in this way, what happens? 4'33" also illustrates Cage's desire to place more responsibility upon his audience. Nyman (1999, p. 24) quotes Cage on this subject: "Most people think that when they hear a piece of music, they're not doing anything but something is being done to them. Now this is not true, and we must arrange our music, we must arrange our art, we must arrange everything, I believe, so that people realize that they themselves are doing it."

2.13.2 Chance and Emergence

The intentional uses of silence, and the expectation of an actively involved audience, were not the only innovations that Cage pioneered as he decentralized control in his compositions. From the early 1950s onward Cage also made extended use of chance operations when he composed.

His 1951 piece *16 Dances* (BMG Music, 1994) paved the way for Cage's use of chance. *16 Dances* was composed using an 8 × 8 sound chart. Each entry on the chart was a particular musical event. Only one entry on the chart could be played at any given moment. Each movement of *16*

Dances involved playing one chart entry after another; Cage varied the contents of the chart for each movement of *16 Dances*. The tabular arrangement of this piece suggested the possibility of making arbitrary moves in the sound chart. Cage used dice rolls to determine the order of sounds in his 1951 piano piece *Music of Changes* (Ross, 2007).

The stochastic nature of Cage's compositional practices did not produce music that sounded random. This is because Cage put tremendous effort into choosing interesting sound elements. "In the *Music of Changes* the effect of the chance operations on the structure (making very apparent its anachronistic character) was balanced by a control of the materials" (Cage, 1961, p. 26). Cage relaxed his influence on control (that is, upon which element to perform next) with the expectation that this, coupled with his careful choice of elements, would produce surprising and interesting musical results. Cage intended novel results to *emerge* from his compositions.

The combination of well-considered building blocks to produce emergent behaviours that surprise and inform is characteristic of the robotics research (Braitenberg, 1984; Brooks, 1999; Pfeifer & Scheier, 1999; Webb & Consi, 2001) that has inspired embodied cognitive science. It is also found in the works of the minimalist composers inspired by Cage.

2.14 EMERGENCE IN MINIMALIST MUSIC
2.14.1 Tape as Medium

Cage's interest in expanding the field of sounds was fuelled by modern technology. Cage was enthused about using magnetic tape "to make a new music that was possible only because of it" (Cage, 1961, p. 9).

Tape compositions were prominent in early *minimalist* music. Composer La Monte Young, described minimalism as "that which is created with a minimum of means" (Schwarz, 1996, p. 9). Young created works that had nearly no musical notation, but were instead performance instructions that *might* lead to the production of musical sounds. Philip Glass was shocked at one of Young's performances: "He wasn't playing music; he was just drawing a line" (Schwarz, 1996, p. 111).

Minimalist pioneer Terry Riley began working with tape technology in 1960 (Potter, 2000). He recorded a variety of sounds and made tape loops from them. A tape loop permitted a sound segment to be repeated over and over. He then mixed these tapes using a device called an echoplex that permitted the sounds "to be repeated in an ever-accumulating counterpoint against itself" (Potter, 2000, p. 98). Further complexities of sound were produced by either gradually or suddenly changing the speed of the tape to distort the tape loop's frequency. Riley's tape loop experiments led him to explore the effects of repetition, which was to become a centrally important feature of minimalist music.

2.14.2 *It's Gonna Rain*

Riley's work strongly influenced another minimalist composer, Steve Reich. One of the most famous minimalist tape compositions is Reich's 1965 *It's Gonna Rain*. Reich recorded a sermon of a famous street preacher, Brother Walter, who made frequent Sunday appearances in San Francisco's Union Square. From this recording Reich made a tape loop of a segment of the sermon that contained the title phrase.

Reich played two copies of this tape loop simultaneously on different tape machines (Reich, 2002), and made a profound discovery: "In the process of trying to line up two identical tape loops in some particular relationship, I discovered that the most interesting music of all was made by simply lining the loops up in unison, and letting them slowly shift out of phase with each other" (p. 20). He recorded the result of phase-shifting the loops, and composed his piece by phase-shifting a loop of this recording. Composer Brian Eno describes Reich's *It's Gonna Rain*: "The piece is very, very interesting because it's tremendously simple. It's a piece of music that anybody could have made. But the results, sonically, are very complex. ... What you become aware of is that you are getting a huge amount of material and experience from a very, very simple starting point."

The complexities of *It's Gonna Rain* emerge from the dynamic combination of simple components, and thus are easily linked to the relinquishment of control that was begun by John Cage. However, they also depend to a large extent upon the perceptual processes of a listener when confronted with the continuous repetition of sound fragments.

"The mind is mesmerized by repetition, put into such a state that small motifs can leap out of the music with a distinctness quite unrelated to their acoustic dominance" (Griffiths, 1994, p. 167). From a perceptual point of view, it is impossible to maintain a constant perception of a repeated sound segment. During the course of listening, the perceptual system will habituate to some aspects of it, and as a result — as if by chance — new regularities will emerge. "The listening experience itself can become aleatory in music subject to 'aural illusions'" (Griffiths, 1994, p. 166). Minimalism took advantage of the active role of the listener, and exploited repetition to deliberately produce aural illusions. The ultimate effect of a minimalist composition is not a message created by the composer and delivered to a (passive) audience, but is instead a collaborative effort between musician and listener.

2.15 A MINIMALIST SCORE
2.15.1 *In C*

In the early days of minimalism, composers were able to discover the power of such techniques as repetition and phase shifting by working with electronic media. However, they felt a need to find a method that would transport these ideas into more traditional media (i.e., the creation of scores to be performed by musicians). The means of doing so was provided by Terry Riley.

Riley's 1964 composition *In C* is 53 bars of music written in the key of C major, indicating a return to tonal music. Each bar is extremely simple, and the entire score fits onto a single page. Performers were instructed to play each bar in sequence. However, they were to repeat a bar as many times as they liked before moving on to the next. When they reached the final bar, they were to repeat it until all of the other performers had reached it. At that time, the performance was to be concluded.

Riley's *In C* can be thought of as a tape loop experiment realized as a musical score. Each performer is analogous to one of the tape loops, and the effect of the music arises from their interactions with one another. The difference, of course, is that each "tape loop" is not identical to the others, because each performer controls the number of times that they repeat each bar. Performers listen and react to *In C* as they perform it. In his performance instructions, Riley notes "one of the joys of *In C* is the interaction of the players in polyrhythmic combinations that spontaneously arise between patterns. Some quite fantastic shapes will arise and disintegrate as the group moves through the piece."

2.15.2 Minimalism and Stigmergy

There are two compelling properties that underlie a performance of *In C*. First, each musician is an independent agent who is carrying out a simple act. At any given moment each musician is performing one of the bars of music. Second, what each musician does at the next moment is largely under the control of the musical environment that the ensemble of musicians is creating. A musician's decision to move from one bar to the next depends upon what they are hearing. In other words, the musical environment being created is literally responsible for controlling the activities of the agents who are performing *In C*. This is a musical example of *stigmergy*, a concept introduced in Chapter 1.

In stigmergy, the behaviours of agents are controlled by an environment in which they are situated, and which they also can affect. The performance of a piece like *In C* illustrates stigmergy in the sense that musicians decide what to play next on the basis of what they are hearing right now. Of course, what they decide to play will form part of the environment, and help guide the playing decisions of other performers.

The stigmergic nature of minimalist music contrasts with the ideal of a composer transcribing mental representations (see Section 2.6). One cannot predict what *In C* will sound like by simply examining the score. Only an actual performance will reveal what *In C*'s score represents. "Though I may have the pleasure of discovering musical processes and composing the musical material to run through them, once the process is set up and loaded it runs by itself" (Reich, 1974, p. 9).

Reich's idea of a musical process running by itself is reminiscent of the synthetic approach introduced in Section 1.11. In the synthetic approach, one includes a set of primitive processes in an agent. Typically there are non-linear interactions between these building blocks, and between the building blocks and the environment (in which the agent is embodied and situated). As a result, complex and interesting behaviours emerge — results that far exceed behavioural predictions based on knowing the agent's makeup (Braitenberg, 1984). Human intelligence is arguably the emergent product of simple, interacting mental agents (Minsky, 1985). The minimalists have tacitly (and presciently) adopted this view, and have created a mode of composition that reflects it.

2.16 MUSICAL STIGMERGY

The continual evolution of modern technology has had a tremendous impact on music. Some of this technology has created situations in which musical stigmergy is front and centre.

2.16.1 Musical Swarms

Consider a computer program called Swarm Music (Blackwell, 2003; Blackwell & Young, 2004a, 2004b). In Swarm Music there are one or more swarms of particles. Each particle is a musical event: it exists in a musical space where the coordinates of the space define musical parameters (e.g., pitch, duration, loudness); its position defines a particular combination of these parameters. The swarm of particles is dynamic, and is drawn to attractors that are placed in the space. The swarm can thus be converted into music. "The swarming behavior of these particles leads to melodies that are not structured according to familiar musical rules, but are nevertheless neither random nor unpleasant" (Blackwell & Young, 2004a, p. 124). Swarm Music is made dynamic by coupling it with human performers in an improvised — and stigmergic — performance. The sounds created by the human performers are used to revise the positions of the attractors for the swarms, causing the music generated by the computer system to change in response to the other performers. The human musicians then change their performance in response to the computer.

Performers who have improvised with Swarm Music provide accounts that highlight its stigmergic nature. Singer Kathleen Willison "was surprised to find in the first improvisation that Swarm Music seemed to be imitating her: '[the swarm] hit the same note at the same time — the harmonies worked'. However, there was some tension; 'at times I would have liked it to slow down ... it has a mind of its own ... give it some space'. Her solution to the 'forward motion' of the swarms was to 'wait and allow the music to catch up'" (Blackwell, 2003, p. 47).

2.16.2 The ReacTable

Another new technology in which musical stigmergy is evident is the *reacTable* (Jordà, Geiger, Alonso, & Kaltenbrunner, 2007; Kaltenbrunner, Jordà, Geiger, & Alonso, 2007). The reacTable is an electronic synthesizer that permits several different performers to play it at the same time. The reacTable gained widespread acclaim when Björk featured it in performances for the tour of her 2007 album *Volta*.

The reacTable is a circular, translucent table upon which objects can be placed. Some objects generate wave forms; some objects perform algorithmic transformations of their inputs; some objects control others that are nearby. Rotating an object, and using a fingertip to manipulate a visual interface that surrounds it, modulates a musical process (e.g., changes the frequency and amplitude of a sine wave). Visual signals displayed on the reacTable — and visible to all performers — indicate the properties of the musical event produced by each object, as well as the flow of signals from one object to another. At the time this section was written, a number of demonstrations of the reacTable were available on YouTube (e.g., ReacTable: Basic Demo #1, found at http://www.youtube.com/watch?v=0h-RhyopUmc).

The reacTable is an example of musical stigmergy because when multiple performers use it simultaneously, they are reacting to the existing musical events. These events are represented as physical locations on the reacTable itself (i.e., the positions of objects), the visual signals emanating from these objects, and the aural events that the reacTable as instrument is producing. By co-operatively moving, adding, or removing objects the musicians collectively improvise a musical performance. The reacTable is an interface intended to provide a "combination of intimate and sensitive control, with a more macro-structural and higher level control which is intermittently shared, transferred and recovered between the performer(s) and the machine" (Jordà et al, 2007, p. 145). That is, the reacTable—and the music that it produces—provides control analogous to that provided by the nest-in-progress of an insect colony as discussed in Chapter 1.

2.17 FROM HOT TO COOL
2.17.1 The Conduit Metaphor

Cybernetics began with the study of communication (Shannon, 1948; Wiener, 1948). Classical cognitive science developed when many cybernetic ideas were explored in a cognitivist context (Conrad, 1964; Leibovic, 1969; Lindsay & Norman, 1972; MacKay, 1969; Selfridge, 1956; Singh, 1966). As a result, the cybernetic notion of communication—transfer of information from one location to another—is easily found in the classical approach's literature.

The classical study of communication is dominated by the *conduit metaphor* (Reddy, 1979). According to the conduit metaphor, language provides containers (e.g., sentences, words) that are packed with meanings and delivered to receivers, who unpack them to receive the intended message. Reddy provides a large number of examples of the conduit metaphor, including: You still haven't *given me any idea* of what you mean; You have to *put each concept into words* very carefully; The *sentence was filled with emotion*.

The conduit metaphor also applies to the traditional view of classical music, which construes this music as a "hot medium" to which the listener contributes little (McLuhan, 1994). The composer places some intended meaning into a score, the orchestra brings the score to life exactly as instructed by the score, and the (passive) audience unpacks the delivered music to get the composer's message. If traditional music were a "cool medium," then much of the meaning would be contributed by an active audience. The conduit metaphor breaks down in modern music. If control is taken away from the score and the conductor; if the musicians become active contributors to the composition (Benson, 2003); if the audience is actively involved in completing the composition as well; if music is actually a cool medium, then what is the intended message of the piece?

2.17.2 Audible Processes

Minimalist composers adopt a McLuhanesque view of their compositions: the music doesn't deliver a message, but is itself the message. After being schooled in the techniques of serialism, which deliberately hid the underlying musical structures from the audience's perception, the minimalists desired to create a different kind of composition. In their compositions the audience would hear the musical processes upon which the pieces were built. Reich (2002, p. 34) is "interested in perceptible processes. I want to be able to hear the process happening throughout the sounding music."

Reich's made processes perceptible by making them gradual. But this didn't make his compositions any less musical. "Even when all the cards are on the table and everyone hears what is gradually happening in a musical process, there are still enough mysteries to satisfy all. These mysteries are the impersonal, unintended, psychoacoustic by-products of the intended process" (Reich, 2002, p. 35).

Reich's recognition that the listener contributes to the composition — that classical music is a cool medium, not a hot one — is fundamental to minimalist music (see also Section 2.14.2). Philip Glass was surprised to find that he had different experiences of different performances of Samuel Beckett's *Play* (for which Glass composed music) (Glass, 1987). He realized that "Beckett's *Play* doesn't exist separately from its relationship to the viewer, who is included as part of the play's content" (p. 36). Audiences of Glass's *Einstein on the Beach* had similar experiences. "The point about *Einstein* was clearly not what it 'meant' but that it was *meaningful* as generally experienced by the people who saw it" (p. 33).

In the cool medium of modern music, the composition has appeared to "leak" from the composer's mind, and requires contributions from both the performers and the audience. Imagine the goal of explaining the psychological processes that produced such a composition. Could classical cognitive science accomplish this goal, or would alternative theories and methods be required?

2.18 THE SHOCK OF THE NEW
2.18.1 Classical Value

While new theories seem necessary to explain how modern music is composed, a classical theory might be able to explain the composition of traditional classical music. Some consider modern music not to be music at all (Pleasants, 1955): perhaps there is no need for a non-classical cognitive science of music.

One reason for considering this view is that in the cool medium of modern music, where control of the composition is far more decentralized, a modern piece seems more like an improvisation than a traditional composition. "A performance is essentially an *interpretation* of something that already exists, whereas improvisation presents us with something that only comes into being in the moment of its presentation" (Benson,

2003, p. 25). Jazz guitarist Derek Bailey notes that the ability of an audience to affect a composition is expected in improvisation (Bailey, 1992). "Improvisation's responsiveness to its environment puts the performance in a position to be directly influenced by the audience" (p. 44). Such effects, and more generally improvisation itself, are presumed to be absent from the Austro-German musical tradition: "The larger part of classical composition is closed to improvisation and, as its antithesis, it is likely that it will always remain closed" (Bailey, 1992, p. 59).

Perhaps modern music will require alternative theories to explain composition because it is improvisational, while traditional classical music is not.

2.18.2 A Tradition of Improvisation

However, there is a problem with this dismissal. Modern music suggests that the composition, performance, and perception of music can involve processes that are not easily included in the theories of classical cognitive science. Are such possibilities true of traditional music as well? Perhaps one of the shocks delivered by modern music is that many of its characteristics also apply to traditional classical music.

For instance, Austro-German music has a long tradition of improvisation, particularly in church music (Bailey, 1992). A famous example of such improvisation occurred when Johann Sebastian Bach was summoned to the court of German Emperor Frederick the Great in 1747. The Emperor played a theme for Bach on the piano, and asked Bach to create a three-part fugue from it. The theme was a trap, probably composed by Bach's son Carl Philipp Emanuel (employed by the Emperor), and was designed to resist the counterpoint techniques required to create a fugue. "Still, Bach managed, with almost unimaginable ingenuity, to do it, even alluding to the king's taste by setting off his intricate counterpoint with a few *galant* flourishes" (Gaines, 2005, p. 9). This was pure improvisation, as Bach composed and performed the fugue on the spot.

Benson (2003) argues that much of traditional music is improvisational, though perhaps less evidently than in the example above. Austro-German music was composed within the context of particular musical and cultural traditions. This provided composers with a constraining set of elements to be incorporated into new pieces, while being transformed or extended at the same time. "Composers are dependent on the 'languages' available to them, and usually those languages are relatively well defined. What we call 'innovation' comes either from pushing the boundaries or from mixing elements of one language with another" (p. 43). Benson argues that improvisation provides a better account of how traditional music is composed than do alternatives like "creation" or "discovery," and then shows that improvisation also applies to the performance and the reception of pre-modern works.

The possibility that classical music shares many of the properties of modern music — improvisation, decentralized control, dependence upon the audience-as-environment (Benson, 2003) — raises a challenge to classical cognitive science's ability to explain musical composition. If much of classical music (modern or not) is cool, then what kinds of theories are required to explain how it is composed?

2.19 MUSICAL METHODS AND THE MIND
2.19.1 Characteristic Questions

Earlier in this chapter it was hypothesized that there was a strong analogy between classical cognitive science and classical music. As a result, the classical approach seemed ideally positioned to provide a cognitive science of music. However, the nature of modern music raises serious doubts about the validity of this analogy, and of what the analogy implies.

Classical music relied heavily upon formal structures (Copland, 1939). However, modern music rejects this reliance, and can develop in the absence of formal notation. "Stravinsky did not set out to produce a compendium of new rhythmic ideas; they came unbidden, and he found that he was inventing music which he did not at first know how to notate" (Griffiths, 1994, p. 41).

Classical music depended upon tonality, harmony, and conventional formats to communicate meanings to a passive audience (Meyer, 1956). Modern music abandons these ideas, content with communicating musical processes, but expecting aesthetic results to emerge from the interactions of a score, musicians, and an audience (Glass, 1987; Potter, 2000; Reich, 2002; Schwarz, 1996). "Minimalism was marked by a spirit of discovery: the discovery of models in extra-European music [...], and the discovery of how extended musical structures could be created out of rudimentary ideas" (Griffiths, 1994, p. 188).

Classical music adhered to the ideal of the disembodied composer, capable of creating themes in his or her mind alone, to be later committed to a score (Hildesheimer, 1983). To modern music this is not an ideal, but a myth. "It is also enlightening that Mozart refused to compose without a keyboard at him, for the traditional view is that he was able to compose everything in his head'" (Benson, 2003, p. 59). Modern music recognizes that the responsibility of a composition has "leaked out" of the composer's mind (Clark, 1997; Wilson, 2004) into an environment that includes the conductor, the performers, and even the audience.

Classical music placed total control of a performance in the hands of a conductor whose mission was to deliver the message contained in a score. "Presence of mind is among his [the conductor's] central attributes; law-breakers must be curbed instantly. The code of laws, in the form of the score, is in his hands" (Canetti, 1962). Modern music recognizes that there is no central control, and uses this recognition to explore the full range of musical possibilities. "One way of thinking about a musical work is that it provides a world in which music making can take place. Performers, listeners, and even composers in effect dwell within the world it creates. And their way of dwelling is best characterized as 'improvisation'" (Benson, 2003, p. 32). If ideas like decentralized control, emergence from musical agents, and stigmergy can plausibly be applied to a complex psychological phenomenon like the composition of classical music, then is it not also plausible that these ideas can be applied to more mundane aspects of cognition? If this is the case, then what new kinds of theories are needed in cognitive science? And what new kinds of methods are required to permit these theories to flourish?

2.19.2 The Synthetic Approach

Classical cognitive scientists prefer to locate the source of complicated behaviour within the organism, and not within its environment (Braitenberg, 1984). This has been called the *frame of reference problem* (Pfeifer & Scheier, 1999), whose implications were long ago highlighted in the parable of the ant (Section 1.12). A consequence of the frame-of-reference problem is that relatively simple systems can surprise us, and generate far more complicated behaviour than we might expect. To take advantage of this, Braitenberg has called for the adoption of the *synthetic approach*. In the synthetic approach, one takes an interesting set of building blocks, creates a working system from them, and then sees what the system can or cannot do (Dawson, 2004). The next chapter explores the relationship between the synthetic approach and the new alternatives to classical cognitive science.

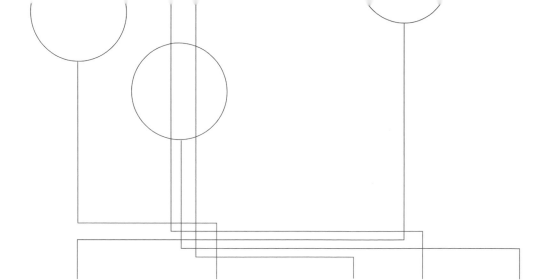

Chapter 3
Situated Cognition and *Bricolage*

3.0 CHAPTER OVERVIEW

One example of a prototypical architecture for classical cognitive science is the production system (Newell & Simon, 1972). Early production systems were used to explore the manipulation of mental representations. Thus, logicism is one central characteristic of production system models. As production systems evolved, researchers included elements that modelled perception and action (Anderson et al., 2004; Meyer & Kieras, 1997a, 1997b). However, there were no direct interactions between sensing and acting. As a result, modern production systems preserve logicism and maintain a strict adherence to the sense–think–act cycle. However, some theories that endorse logicism, such as Piaget's theory of cognitive development, are open to the idea that logicism is founded upon actions in the world. Indeed, computational arguments, as well as numerous results from cognitive neuroscience, support the claim that human cognition involves both sense–think–act and sense–act processing. As a result, new theories are arising in cognitive science in which the external world plays a more important role in cognition. Cognitive processing is seen to be aided or scaffolded by worldly support, and for this reason some researchers claim that the mind has leaked into the world. As well, rather than identifying a unifying theory of mind (a principal goal of production system architectures), some researchers argue that cognition is mediated by a diverse collection of cognitive agents (Minsky, 1985, 2006). Thinking thus is not viewed as the rational application of logical rules, but is instead viewed as a form of *bricolage* in which thinkers choose subsets of processes at hand, some sense–think–act and other sense–act, and interact with the world to solve problems. How does one study a mind that is both leaky and composed of a collection of non-linear agents? One approach is to adopt a synthetic methodology in which a set of interesting primitive processes is selected, organized into a system, and then observed in action as the embodied and situated system interacts with its environment. It has been claimed that this synthetic approach can lead to simpler theories than would be the case if more traditional analytic methodologies were employed (Braitenberg, 1984). But how, then, do we train students to consider the mind in this different way, and how do we prepare them to study the mind using synthetic methodologies? At the end of this chapter it is proposed that we train them to be *bricoleurs* to construct and observe simple robots in action. Specific examples of such training are then introduced in this book, beginning with Chapter 4.

3.1 THREE TOPICS TO CONSIDER
3.1.1 Review to This Point

Chapter 1 began by exploring the age-old view that man is different from, and superior to, beast. Many have suggested that the source of this difference is human rationality. It is by virtue of rationality that man controls his environment, while animals are controlled by their environment. We considered potential exceptions to this view, such as the large and complex nests constructed by the social insects. However, we saw that many researchers believe that these intricate structures emerge from the stigmergic control of insects by their environment. But, if the giant mounds of termites can be explained by appealing to straightforward environmental control mechanisms, then why would it be unreasonable to offer similar explanations for human creations?

Chapter 2 explored this question by considering a prototypical example of human intelligence: composing classical music. We saw that the traditional view of composition treated music as the ultimate product of a rational, representational mind. However, these ideas were challenged when the characteristics of modern classical music were examined. We saw that modern music removed itself from the traditions of classical music by abandoning structure, by eliminating the goal of communicating particular messages, by exploring the complex products that could emerge from the interaction of simple processes, and by removing central control from the score in the conductor and distributing control to the performers and the audience. Furthermore, we encountered the argument that these characteristics of modern classical music could also be found in more traditional music if one took the care to search for them. If it is possible for the composition of classical music to involve processes that are not completely rational, then is it not also possible that this is true of more mundane cognitive activities?

3.1.2 New Headings

Importantly, questions about human rationality do not hinge upon the structure of insect nests or the nature of modern music. We have seen that classical cognitive science views rationality as a fundamental principle of cognition; this is consistent with its reliance upon logicism. To its credit, classical cognitive science has developed influential and powerful theories of reasoning and problem solving that have rationality and logicism as their foundation. Nonetheless, a large number of experimental results concerning judgment, decision making, or problem solving have shown human cognition to depart from rational norms (Hastie, 2001; Mellers, Schwartz, & Cooke, 1998; Oaksford & Chater, 1998; Piattelli-Palmarini, 1994; Tversky & Kahneman, 1974). Many attempts have been made to revise these norms, or to propose alternative representational processes that incorporate additional constraints and move beyond these norms, to accommodate these irrational results within a rational cognitive science.

Of course, these results have also motivated alternative views of the mind. For instance, embodied cognitive science might attempt to explain apparently irrational results by noting these results make perfect sense in the context of an agent and its environment. One purpose of the current chapter is to explore an alternative view of mind that acknowledges the contributions of the environment to cognition.

The emergence of new views of mind has also led to the development of new approaches to its study. In particular, new approaches to modelling have appeared that involve novel relationships between the model and the system being modelled. A second purpose of the current chapter is to discuss the general characteristics of such models. New views of the mind, and new approaches to modelling cognitive phenomena, also raise questions about how these ideas should be introduced to students of cognitive science. A third purpose of the current chapter is to propose a general approach to instruction that is adopted and illustrated in more detail in subsequent chapters.

3.2 PRODUCTION SYSTEMS AS CLASSICAL ARCHITECTURES
3.2.1 The Production System

Let us start by describing the general characteristics of a classical information processing architecture, the *production system*. Production systems are powerful architectures that have been used to model many psychological phenomena (Anderson, 1983; Anderson et al., 2004; Anderson & Matessa, 1997; Meyer et al., 2001; Meyer & Kieras, 1997a, 1997b; Newell, 1990; Newell & Simon, 1972). One aim of this chapter is to dispute some of the traditional assumptions of this architecture. The intent is not to attack production systems in particular, or to attack classical cognitive science in general. Rather, we will explore the idea that the foundational assumptions of production systems can easily be modified to show how this presumed classical system can incorporate the general characteristics of alternatives to classical cognitive science.

A production system is a general purpose symbol manipulator (Anderson, 1983; Newell, 1973; Newell & Simon, 1972). Like the Turing machine, and like most digital computers, production systems are defined by a sharp distinction between data and process. In the Turing machine, data are the symbols stored on the tickertape, while process is the set of rules that reside in the machine head. In a modern digital computer, data are the representations stored in random access memory (RAM), while process consists of the basic operations built into the central processing unit (CPU). In a production system, data are symbolic expressions that are stored in a working memory (and in some versions also stored in a long-term memory), while process consists of a set of rules (productions) that are capable of manipulating these expressions to achieve a desired information-processing end. Each production in a production system is a condition–action pair that scans the expressions in working memory for a pattern that matches its condition. If a match is found, then the production's action is carried out. An action usually involves manipulating symbols in working memory (e.g., adding new symbols).

In general, all the productions scan working memory in parallel. When one production finds its condition, it takes control for a moment, disabling the other productions while the controlling production changes memory. Then control is released, and the parallel scan of working memory is reinitiated. Additional control mechanisms can be added to deal with the situation in which more than one production finds its condition at the same time, or when one production finds its condition at different places in memory at the same time. As well, some productions might write goals into the working memory, and these goals can be included as conditions for some of the productions. This permits a hierarchical set of goals and subgoals to control the order in which productions are activated (Anderson, 1983, pp. 7–10).

3.2.2 Classical Characteristics

In addition to contributing to cognitive science by modelling many higher-order cognitive phenomena, production systems have all of the prototypical characteristics of classical information processing. First, they distinguish symbols from processes. Second, they are serial information processors in the sense that only one production can alter working memory at any given time. Third, the entire purpose of a production system is to manipulate symbolic expressions. Fourth, production systems have the requisite computational power. It has been proven that a production system is capable of solving the same problems that can be solved by a universal Turing machine (Newell, 1980).

The purpose of presenting production systems at this point in the chapter is to provide a concrete example to which we can later return. In the pages that follow we will consider some positions that challenge the classical assumptions that production systems embody. We will then see how the traditional notion of the production system can be modified in response to these challenges.

3.3 SENSE–THINK–ACT WITH PRODUCTIONS
3.3.1 An Early Production System

Production systems are prototypical classical architectures in the sense that they adhere to a strict sense–think–act cycle (Pfeifer & Scheier, 1999). That is, the goal of perceptual mechanisms is to generate symbolic expressions about the world to be stored in working memory. Internal mechanisms manipulate these expressions, producing other expressions that can represent plans for action that might affect the outside world.

The role of working memory (and operations upon it) as a mediator between sensing and acting is illustrated in Figure 3-1A below. This figure illustrates the main properties of early production systems used to model human cognition (Newell & Simon, 1972). The "thinking" component of the model is the large grey box that contains both a working memory and a procedural memory (i.e., a set of productions). The double-headed arrow indicates that control can flow from working memory to procedural memory, and vice versa.

The single-headed arrows illustrate that sensing adds content to the working memory, and that the contents of working memory later cause actions upon the world. Early production system models did not elaborate theories about sensing or acting, in spite of the fact that their developers recognized a need to do so. "One problem with psychology's attempt at cognitive theory has been our persistence in thinking about cognition without bringing in perceptual and motor processes" (Newell, 1990, p. 15).

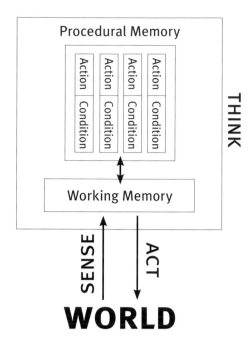

3-1a

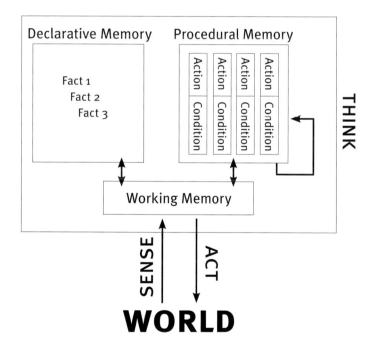

3-1b

3.3.2 The Next ACT

Figure 3-1B illustrates the general properties of the next stage of production system models, Anderson's *adaptive control of thought* (ACT) architecture (Anderson, 1983). Two of its major innovations are represented in the figure: the introduction of a declarative memory to serve as a store of knowledge that was independent of productions, and the introduction of learning mechanisms (indicated by the one-headed arrow from the procedural memory to itself) that permitted new productions to be added. Of course, ACT included other innovations, such as new formats for the elements that were represented in the "thinking" part of the system. However, the early ACT architectures remained true to their antecedents by acknowledging the existence of sensing and acting, but also by failing to elaborate the nature of these components. The ACT architecture "historically was focused on higher level cognition and not perception or action" (Anderson et al., 2004, p. 1038).

3.4 LOGIC FROM ACTION
3.4.1 Productions and Logicism

Researchers who employ production systems are searching for a unified theory of cognition (Anderson, 1983; Anderson et al., 2004; Newell, 1990). "The unitary approach holds that all higher-level cognitive functions can be explained by one set of principles" (Anderson, 1983, p. 2). Of course, in a production system some behaviours (e.g., errors, latencies to perform various tasks) are the result of memory limitations, or the timing of certain operations (Meyer & Kieras, 1997b).

Nevertheless, the basic claim of a production system theory "is that underlying human cognition is a set of condition-action pairs" (Anderson, 1983, p. 5). The crucial behaviours to be explained by a production system are not found in (for instance) memory limitations, but are instead grounded in how the productions themselves rationally manipulate representational content. "All the behavioral flexibility of universal machines comes from their ability to create expressions for their own behavior and then produce that behavior. Interpretation is the necessary basic mechanism to make this possible" (Newell, 1980, p. 158). Production systems realize a commitment to logicism, a logicism instantiated as a particular representational theory of mind.

3.4.2 Logic as Internalized Action

However, logicism can be rooted in non-logical action. Consider Piaget's theory of cognitive development (Inhelder & Piaget, 1958, 1964; Piaget,

1970a, 1970b, 1972; Piaget & Inhelder, 1969). According to Piaget, children achieve adult-level cognitive abilities at the stage of *formal operations* sometime in their early teens. This stage is formal in the sense that the child operates on symbolic representations. Furthermore, these operations are logical in nature. Formal operations involve completely abstract thinking, where relationships between propositions that represent the full range of possibilities are considered. "Considering possibilities" involves representing potential states of affairs in an organized combinatorial matrix. Different locations in this matrix encode different combinations of values of whatever variables are critical to the problem at hand. The INRC group is a set of four logical operations (identity, negation, reciprocal, and correlation) that permit the child to manipulate the combinatorial matrix, moving from one location to another, organizing it into logically significant groups. Clearly, the stage of formal operations is representational, and is an expression of logicism.

However, the route to formal operations begins with direct interactions with objects in the world (the *sensorimotor stage*). These objects are later internalized as symbols in the *preoperational stage*. In the next stage (*concrete operations*) these symbols can be manipulated, but these manipulations are not abstract: they bear "on manipulable objects (effective or immediately imaginable manipulations), in contrast to operations bearing on propositions or simple verbal statements (logic of propositions)" (Piaget, 1972, p. 56). According to Piaget, the roots of logic are the child's physical manipulation of his or her world. "The starting-point for the understanding, even of verbal concepts, is still the actions and operations of the subject" (Inhelder & Piaget, 1964, p. 284).

For example, through their actions, children naturally group (classify) and order (seriate) objects. Classification and seriation are operations that can be defined in precise, formal terms. Piagetian theory attempts to explain how, as a child develops, their classifications and seriations conform to formal specifications provided by logic or mathematics. Piaget concludes that such formal abilities are "closely linked with certain actions which are quite elementary: putting things in piles, separating piles into lots, making alignments, and so on" (Inhelder & Piaget, 1964, p. 291).

Production systems, like most classical theories, emphasize thinking at the expense of sensing and action. Theories like those of Piaget provide alternatives in which humans are agents who act upon their world. This doesn't mean that all sense–think–act cycles should be abandoned. Rather, it raises the possibility that they should be supplemented with sense–act processes.

3.5 AN EPIC EVOLUTION
3.5.1 Productions, Sensing, and Action

Production system researchers should, rightly, object to the criticism that they ignore sensing and acting. The omission of sensing and acting from such models was merely historical: central processes were simply modelled first (Anderson et al., 2004; Newell, 1990). Modern production system architectures do include sensing and acting.

3.5.2 The EPIC Architecture

Consider the EPIC (for *executive-process interactive control*) architecture (Meyer & Kieras, 1997a, 1997b). EPIC is designed to model temporal regularities observed when humans perform single or multiple tasks. One main focus of EPIC has been modelling the *psychological refractory period* (PRP). Imagine a subject who is simultaneously performing two tasks (e.g., making one response to stimulus 1, and a second response to stimulus 2). The *stimulus onset asynchrony* (SOA) for this subject is the duration from the onset of stimulus 1 to the onset of stimulus 2. When the SOA is long (1 second or more), the time taken to make the response for either task is similar. However, when the SOA is short (0.5 seconds or less) the latency for the second response is longer than that for the first. This increase in response latency when SOA is short is the PRP.

EPIC is very similar to the production system that was illustrated in Figure 3-1B. EPIC consists of declarative, procedural, and working memories. The major innovation of EPIC as a production system is that it permits productions to act in parallel. That is, at any time cycle in EPIC processing, *all* productions that have matched their conditions in working memory will act to alter working memory. This is important; when multiple tasks are modeled there will be two different sets of productions in action, one for each task.

EPIC also extends earlier production systems by including sensory processors, such as virtual eyes and ears. These sensory processors use table lookups to classify some physical aspect of the world, and then add symbols to working memory to represent this physical quality. For example, at first the virtual ear will indicate the presence of a signal by writing the string "AUDITORY DETECTION ONSET" to working memory. Later, the identity of the signal will be represented (e.g., by writing "AUDITORY TONE 800 ON" to memory).

Another EPIC extension is the inclusion of motor processors. This recognizes that action can provide constraints on performing cognitive tasks. For example, in EPIC a single motor processor controls both "virtual hands." This permits interference between two tasks that involve making responses with different hands.

Motor processors take symbols from working memory. For instance, the string "LEFT INDEX" in working memory is an instruction from the "cognitive processor" (i.e., the three memories working in unison) to perform some action with the left index finger. The motor processors convert working memory symbols into symbols that can be used to control a motor system, and place these into a memory buffer devoted to the motor system. This permits sequences of motor actions to be planned and stored, and also permits actions to be repeated by using the motor buffer directly to run a motor program that has already been created. Motor processors also send information back to working memory for central processes to monitor the progress of requested actions.

The preparation and activation of motor commands, as well as the transfer of information from sensory processors, are all operations that take specific amounts of time. When the EPIC architecture is used to simulate human performance on multiple tasks (producing what Meyer and Kieras (1997a) call the strategic response-deferment model), additional assumptions are used to coordinate the different tasks, and these assumptions also have temporal implications. As a result, this architecture can successfully simulate the reaction time regularities observed in a number of experimental studies of the PRP.

3.6 PRODUCTIONS AND FORMAL OPERATIONS
3.6.1 Sense — Think — Act

While the inclusion of sensing and acting components in models like EPIC (Meyer & Kieras, 1997a, 1997b) is commendable, it is accomplished by shoehorning them into the classical conception of mind. Figure 3-2 provides an illustration of an EPIC-like production system to help make this point. To simplify matters, it only illustrates a single sensor and a single motor processor.

It is evident in Figure 3-2 that both sensing and acting are mediated by central cognitive processing. Sensing transduces properties of the external world into symbols to be added to working memory. Working memory provides symbolic expressions to be interpreted by motor processors. Thus working memory centralizes the "thinking" that maps sensations onto actions. There are no direct connections between sensing and acting that bypass working memory.

When sensing and acting are placed in a sense–think–act cycle, embodied cognitive science (sometimes called situated action [Vera & Simon, 1993]) becomes rooted in symbol manipulations. "It follows that there is no need, contrary to what followers of SA [situated action] seem sometimes to claim, for cognitive psychology to adopt a whole new language and research agenda, breaking completely from traditional (symbolic) cognitive theories. SA is not a new approach to cognition, much less a new school of cognitive psychology" (Vera & Simon, 1993, p. 46).

3.6.2 Formal Operations

Making situated action "cognitive" by forcing sensing and acting to be mediated by symbol manipulations is analogous to the formal operations proposed by Piaget that were discussed in Section 3.4. In models such as EPIC, the bulk of sensation and action have been internalized as symbol manipulations. EPIC uses completely virtual sensors and actors. The assumption is that if these virtual components were replaced with real sensors and actors, then the main results of EPIC would still hold.

However, this is not necessarily the case. First, in order for the sense–think–act cycle to work, a detailed internal model of the external world must be constantly maintained. As the complexity of this model increases (e.g., by increasing the number or complexity of sensors), a number of logical and practical problems emerge (Brooks, 1999, 2002; Clark, 1997; Pylyshyn, 1987). In general, more and more resources are required for modeling, and as a result action becomes slower and slower.

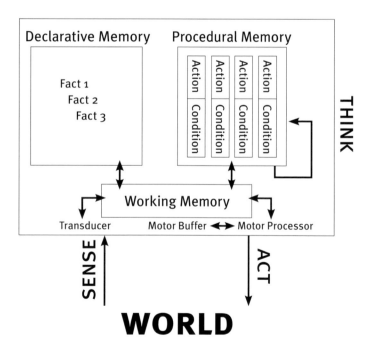

Second, the sense–think–act cycle usually requires the use of a common language (e.g., the symbols in working memory). However, simple and fluid interactions between sense and action may result when one considers specialized coordinate systems (Hutchins, 1995). "Intelligence and understanding are rooted not in the presence and manipulation of explicit, language-like data structures, but in something more earthy: the tuning of basic responses to a real world that enables an embodied organism to sense, act, and survive" (Clark, 1997, p. 4). We see next that neuroscientists have uncovered evidence for such earthy processing in the human brain.

3.7 EVIDENCE FOR SENSING AND ACTING WITHOUT THINKING

3.7.1 Classical Modularity

An influential idea in cognitive science is that of *modularity* (Fodor, 1983). A module is a domain-specific system that solves a very particular problem, is incapable of solving other information-processing problems, and is usually associated with fixed neural architecture.

The modularity proposal is usually incorporated into the "sense–think–act" cycle of classical cognitive science (Dawson, 1998). Specifically, most modules solve particular perceptual problems (sense). The output of these modules is then passed on to visual cognition or higher-order cognition for inferential or semantic processing (think). The results of this higher-order processing are then used to generate actions (Wright & Dawson, 1994). However, this is not the only way in which modularity has been incorporated into cognitive science.

3.7.2 Visuomotor Modules

Research on the vision of the frog has established the existence of processors that do not appear to feed into higher-order thinking mechanisms, but instead serve directly as "sense–act" or visuomotor modules (Ingle, 1973). Ingle surgically removed one hemisphere of the optic tectum of a frog; the optic tectum is the part of the frog brain most responsible for visual processing. Ingle's lesion produced a particular form of blindness in which the frog pursued prey presented to the eye that was connected to the remaining tectum, but did not respond to prey presented to the eye that was originally connected to the ablated tectum.

Ingle found that, over time, the nerve fibres from the tectumless eye grew to be connected to the remaining optic tectum on the "wrong" side of the animal's head. As a result, when a target was presented to this eye, the frog was no longer blind to it. However, the animal's motor responses were aberrant! The frog always moved toward a location that was mirror-symmetrical to the actual location of the target, and this incorrect response was shown to be due to the topography of the regenerated nerve fibres. This result demonstrated that the frog optic tectum converts a visual sensation directly into a motor response. "The visual system of most animals, rather than being a general-purpose network dedicated to reconstructing the rather limited world in which they live, consists instead of a set of relatively independent input-output lines, or visuomotor 'modules', each of which is responsible for the visual control of a particular class of motor outputs" (Goodale & Humphrey, 1998, p. 183).

Such results, and conclusions, are not limited to the frog. Parallel results have been found in the gerbil (Mlinar & Goodale, 1984). As well, studies of brain-injured patients have demonstrated that the human visual system may also be organized into visuomotor modules (Goodale, 1988, 1995; Goodale & Humphrey, 1998; Goodale, Milner, Jakobson, & Carey, 1991).

For instance, Goodale and his colleagues have studied one patient, DF, who suffered irreversible brain damage that dramatically impaired the ability to recognize visual shapes or patterns. However, DF's visuomotor abilities were not impaired at all. Another patient, VK, had the exact opposite pattern of dysfunction after a series of strokes. VK had normal form perception, but her visuomotor control—in particular, her ability to form her hand to grasp objects of different shapes—was severely impaired.

Goodale and Humphrey (1998) have argued for the existence of two complementary visual systems, one responsible for controlling object-directed action, the other responsible for creating an internal model of the external world. "Although there is clearly a division of labor between the perception and action systems, this division reflects the complementary role the two systems play in the production of adaptive behavior" (p. 203). This view, called the *duplex theory*, supports roles for both sense–act and sense–think–act processing. Such a view is not evident in modern production system models such as EPIC, which require that all perception and action be mediated by central cognitive processing.

3.8 ACTION WITHOUT REPRESENTATION?
3.8.1 Multiple Visual Pathways

There are two parallel physiological pathways in the human visual system (Livingstone & Hubel, 1988; Maunsell & Newsome, 1987; Ungerleider & Mishkin, 1982). One, the ventral stream, seems to process visual form (i.e., specifying *what* an object is), while the other, the dorsal stream, seems to process visual motion (i.e., specifying *where* an object is).

These pathways are distinct: brain damage can produce deficits in motion perception, but not affect form perception, or vice versa (Botez, 1975; Hess, Baker, & Zihl, 1989; Zihl, von Cramon, & Mai, 1983). Similarly, cell recordings have revealed neurons that are sensitive to stimulus movement, but not to form (Albright, 1984; Dubner & Zeki, 1971; Maunsell & van Essen, 1983; Rodman & Albright, 1987; Zeki, 1974).

Historically, these two streams were considered to be representational: the ventral stream represented information about form, while the dorsal stream represented information about motion or location. Furthermore, the two streams were only sensitive to the information that they could represent. Goodale's contribution to this literature is to reconceptualize the pathways.

In the duplex approach to vision (Goodale & Humphrey, 1998), the two pathways are sensitive to similar information, but use different transformations to perform distinct, though complementary, functions. The ventral stream creates perceptual representations, while the dorsal stream mediates the visual control of action. "The functional distinction is not between 'what' and 'where', but between the way in which the visual information about a broad range of object parameters are transformed either for perceptual purposes or for the control of goal-directed actions" (Goodale & Humphrey, 1998, p. 187).

3.8.2 Blindsight

Additional evidence supports Goodale and Humphrey's (1998) position. Consider the phenomenon called *blindsight* (Weiskrantz, 1986, 1997; Weiskrantz, Warrington, Sanders, & Marshall, 1974). A patient who exhibits blindsight claims to be unable to see presented stimuli. That is, visual experiences have been ablated as the result of brain injury. However, these patients still demonstrate some ability to point to or detect visual stimuli. Blindsight occurs in human patients who have had damage to their primary visual cortex, and can be created experimentally in primates by surgically removing their visual cortex (Stoerig & Cowey, 1997).

Blindsight must be mediated by neural pathways not affected by the damage to primary visual cortex. Such damage has severe effects upon the ventral stream of processing. However, a variety of results suggest that much of the functionality of the dorsal stream remains intact (Danckert & Rossetti, 2005). "Action-blindsight depends on the integrity of residual pathways that terminate in the dorsal 'action' stream" (Danckert & Rossetti, 2005, p. 1041). These results, and the phenomena associated with blindsight in general, are completely consistent with

Goodale and Humphrey's (1998) claim for a non-representational stream responsible for visually guided action.

Healthy subjects can also provide support for the duplex theory. In one study (Pelisson, Prablanc, Goodale, & Jeannerod, 1986), subjects reached toward an object while detailed measurements of their movement were recorded. However, the experimental method was such that as subjects reached, the object's position was changed — but only when a subject's eyes performed a saccadic eye movement. This manipulation resulted in subjects not being consciously aware that the object had actually moved. Nevertheless, their reach was adjusted to compensate for the object's new position. "No perceptual change occurred, while the hand pointing response was shifted systematically, showing that different mechanisms were involved in visual perception and in the control of the motor response" (Pelisson et al., 1986, p. 309).

3.9 A NEED FOR ACTION
3.9.1 Incorporating Action

The earliest production systems were prototypical examples of classical cognitive science (Newell, 1973; Newell & Simon, 1972). They have evolved into architectures that explicitly include sensing and acting (Anderson et al., 2004; Meyer & Kieras, 1997a, 1997b). However, this is accomplished via an explicit sense–think–act cycle, and ignores the possibility of sense–act processing.An alternative to the classical approach is called embodied cognitive science (Agre, 1997; Brooks, 1999, 2002; Clancey, 1997; Clark, 1997, 2003, 2008; Pfeifer & Scheier, 1999; Robbins & Aydede, 2009; Varela et al., 1991). Embodied cognitive science recognizes the importance of sensing and acting, but reacts against central cognitive control. Its more radical proponents strive to completely replace the sense–think–act cycle with sense–act mechanisms. "In particular I have advocated situatedness, embodiment, and highly reactive architectures with no reasoning systems, no manipulable representations, no symbols, and totally decentralized computation" (Brooks, 1999, p. 170).

That some behaviour results from sense–act processing is supported by research on multiple perceptual streams in the brain (Livingstone & Hubel, 1988; Maunsell & Newsome, 1987; Ungerleider & Mishkin, 1982), the hardwiring of sense–act reflexes (Ingle, 1973; Mlinar & Goodale, 1984), and the cognitive neuroscience of visually guided action (Goodale, 1988, 1990, 1995; Goodale & Humphrey, 1998; Goodale et al., 1991; Pelisson et al., 1986). However, embodied cognitive scientists are not primarily motivated by such evidence. Instead, they see good computational reasons for removing central cognitive control. "The realization was that the so-called central systems of intelligence—or core AI as it has been referred to more recently—was perhaps an unnecessary illusion, and that all the power of intelligence arose from the coupling of perception and actuation systems" (Brooks, 1999, p. viii).

3.9.2 Advantages of Action

Survival depends upon swift action. Attempts to employ the sense–think–act cycle in early mobile robots (Moravec, 1999; Nilsson, 1984) produced systems that took too long to think (see the discussion of Shakey in Section 7.2.2), and therefore could not act appropriately in real time. "The disparity between programs that calculate, programs that reason, and programs that interact with the physical world holds to this day. All three have improved over the decades, buoyed by a more than million fold increase in computer power in the fifty years since the war, but robots are still put to shame by the behavioral competence of infants or small animals" (Moravec, 1999, p. 21).

Some argue that this is due to the computational bottleneck caused by the cost of maintaining an internal model of the world. This bottleneck might be removed by using the world as its own model. This eliminates the costly need to internalize it (e.g., Brooks, 1999), making actions faster and more adaptive.

Slow action might also characterize unifying theories of mind (Anderson et al., 2004; Newell, 1990). This is because they need to mediate all sensing and acting via a common symbolic framework. In addition to the cost of maintaining this framework, its generality may ignore faster solutions that are possible with more specialized processing. An alternative is to use specialized external devices to reduce cognitive demands, and to speed information processing up (Dourish, 2001; Hutchins, 1995). "By failing to understand the source of the computational power in our interactions with simple 'unintelligent' physical devices, we position ourselves well to squander opportunities with so-called intelligent computers" (Hutchins, 1995, p. 171).

This raises another crucial advantage of action: extending the capacity of central cognition by using external objects to support computation. This is called *cognitive scaffolding* (Clark, 1997), and we will now turn to exploring its implications.

3.10 THE EXTERNAL WORLD AND COMPUTATION
3.10.1 Worldly Support for Cognition

Nobel laureate physicist Richard Feynman once took an advanced biology course (Feynman, 1985). He presented a seminar about a paper on nerve impulses in cat muscles. To understand the paper, he had gone to the library to consult "a map of the cat." Feynman began his presentation naming various muscles in a drawing of an outline cat. He was interrupted by his classmates' claiming they already knew this. "'Oh,' I say, 'you *do*? Then no *wonder* I can catch up with you so fast after you've had four years of biology.' They had wasted all their time memorizing stuff like that, when it could be looked up in fifteen minutes" (Feynman, 1985, p. 59).

Feynman's tale illustrates that there are different approaches to solving information-processing problems. One could memorize all of the information that might be required. Or, one could reduce cognitive strain by using the external world. One doesn't have to (internally) remember all of the details if one knows where they can be found in the environment.

The world is more than just a memory. Hutchins (1995) describes a task in which a navigator must compute a ship's speed using the measure of how far the ship has travelled over a recent interval of time. One solution to this task involves calculating speed based on internalized knowledge of algebra, arithmetic, and conversions between yards and nautical miles.

A second approach is to draw a line on a three-scale representation called a *nomogram*. The top scale of this tool indicates duration, the middle scale indicates distance, and the bottom scale indicates speed. The user marks the measured time and distance on the first two scales, joins them with a straight line, and reads the speed from the intersection of this line with the bottom scale. In this case "it seems that much of the computation was done by the tool, or by its designer. The person somehow could succeed by doing less because the tool did more" (Hutchins, 1995, p. 151).

3.10.2 Scaffolding

The use of external structures like the nomogram to support cognition is called *scaffolding* (Clark, 1997, 2003). The exploitation of the external world to support thinking has a long history in developmental psychology. For example, in Piagetian theory sensorimotor stage processing predominantly involves scaffolding, while concrete operations represent a stage in which previously scaffolded thought is internalized.

Such a view is more explicit in other theories of cognitive development (Vygotsky, 1986). Vygotsky, for example, emphasized the role of assistance in cognitive development. He defined the difference between a child's ability to solve problems without aid and their level of ability when assisted as the *zone of proximal development*. Vygotsky argued that the zone of proximal development was crucial, and noted that children with larger zones of proximal development did better in school. He criticized methods of instruction that required children to solve problems without help. "The true direction of the development of thinking is not from the individual to the social, but from the social to the individual" (p. 36).

Vygotsky is also important for broadening the notion of what resources were available in the external world for scaffolding. For example, he viewed language as a tool for supporting cognition: "Real concepts are impossible without words, and thinking in concepts does not exist beyond verbal thinking. That is why the central moment in concept formation, and its generative cause, is a specific use of words as functional 'tools'" (Vygotsky, 1986, p. 107). Clark (1997, p. 180) argues that intelligence depends upon scaffolding in this broad sense: "Advanced cognition depends crucially on our abilities to dissipate reasoning: to diffuse knowledge and practical wisdom through complex social structures, and to reduce the loads on individual brains by locating those brains in complex webs of linguistic, social, political, and institutional constraints."

3.11 SOME IMPLICATIONS OF SCAFFOLDING
3.11.1 The Leaky Mind

The scaffolding of cognition causes cognitive scientists to face a number of important theoretical issues. First, where is the mind located (Wilson, 2004)? The traditional view — typified by the production system models that we have briefly considered — is that thinking is inside the individual, and that sensing and acting involve the world outside. However, if cognition is scaffolded, then some thinking has moved from inside the head to outside in the world. "It is the human brain *plus* these chunks of external scaffolding that finally constitutes the smart, rational inference engine we call mind" (Clark, 1997, p. 180). From this perspective, Clark describes the mind as a leaky organ, because it has spread from inside our head to include whatever is used as external scaffolding.

The leaky mind has a profound impact on classical cognitive science. For example, in the classical approach it is standard to assume that mental states are realized as brain states (Wilson, 2004). Leaky minds mean that this commonplace view of realization has to be revisited.

3.11.2 Group Cognition

The scaffolding of cognition also raises the possibility of public cognition, in which more than one cognitive agent manipulates the world that is being used to support information processing. Hutchins (1995) provides an excellent example of this in describing how a team of individuals is responsible for navigating a ship. He argues that "organized groups may have cognitive properties that differ from those of the individuals who constitute the group" (p. 228). For instance, in many cases it is very difficult to translate the heuristics used by a solo navigator into a procedure that can be implemented by a navigation team.

The possibility that group abilities are qualitatively different from those of a group's component individuals is an example of a central theme in embodied cognitive science, *emergence* (Holland, 1998; Johnson, 2001; Sawyer, 2002). We have already been introduced to emergence in both the behavioural products of social insects (Detrain & Deneubourg, 2006) and in the musical processes of minimalist music (Reich, 2002).

The computational power of groups over individuals is growing in importance, and can be found in discussions on collective computation and swarm intelligence (Deneubourg & Goss, 1989; Goldstone & Janssen, 2005; Holland & Melhuish, 1999; Kube & Zhang, 1994; Sulis, 1997). For cognitive science, it raises the issue of whether there might exist a "group mind" that cannot be associated with an individual (Wilson, 2004). It also raises the possibility of collective human cognition that is a product of stigmergy.

3.11.3 Specialized Cognition

Public cognition can proceed in a variety of ways. The most obvious is when two or more individuals collaborate on a task using a shared environment (Hutchins, 1995). Less obvious is the contribution, over time, of specialized environmental tools used in scaffolding. Hutchins stresses the cultural and historical nature of these tools. For instance, he notes that navigation is impacted by the mathematics of chart projections that was worked out centuries ago, as well as by number systems that were developed millennia ago.

Hutchins (1995) suggests extending the parable of the ant (Simon, 1969) that was introduced in Section 1.12. Instead of watching a single ant for a brief period, Hutchins argues that we should instead arrive at a beach after a storm, and watch many generations of ants working on this *tabula rasa*. As the ant colony matures, the ants appear smarter, because their behaviours are more efficient. But this is because "the environment is not the same. Generations of ants have left their marks on the beach, and now a dumb ant has been made to appear smart through its simple interaction with the residua of the history of its ancestor's actions" (Hutchins, 1995, p. 169).

3.12 STIGMERGY OF THOUGHT
3.12.1 Environmental Import

Scaffolded cognition disrupts the classical approach's reliance on the sense–think–act cycle. The purpose of this cycle is to use central cognitive processes to control sensing and acting. However, with scaffolding, such central (and internal) control is lost. Thinking becomes the result of action on the world, not the control of it. Scaffolding raises the possibility that the sense–think–act cycle can be replaced with sense–act processes that interact directly with the world, and not with an internal representation of it (Brooks, 1999). This emphasizes a completely different notion of control: the world elicits actions upon itself. This is another example of stigmergy (Theraulaz & Bonabeau, 1999) that was introduced in Chapter 1. Indeed, traditional production system control is internally stigmergic, because the contents of working memory determine which production (or productions, as in the case of EPIC [Meyer & Kieras, 1997a]) will act at any given time. When working memory leaks into the world via scaffolding, cognitive control becomes as stigmergic as a wasp nest's control of its own creation.

We can now reformulate our earlier production system illustrations. Figure 3-3 is a recasting of Figure 3-1B; its working memory has leaked into a scaffolding world. This is shown by extending the working memory "box" so that it includes at least a subset of the external world.

The second alteration evident in Figure 3-3 is that procedural memory has been generically described as a set of primitives that sense and act. Their behaviour is identical to that of the productions in classical theories (Anderson, 1983, 1985; Meyer & Kieras, 1997a, 1999; Newell, 1990; Newell & Simon, 1972): when a triggering condition is sensed, then some action is carried out. However, there is no need to commit to the claim that these are productions in the traditional sense.

Rather, the key claim to make is that there are four general types of these primitives, and these types are defined in terms of whether they interact with internal or external memory. One primitive is like the traditional production: it senses information in working memory, and also acts on this working memory. The other three are less traditional. One type senses information in working memory, but acts on the world. One type senses information in the world, but acts on working memory. The final type senses and acts on the world.

Figure 3-3 is extremely simple, and is not intended to illustrate a complete architecture. Its purpose is to highlight the coexistence of two different types of processing, involving two different notions of control. One is the sense–think–act cycle, represented by primitives that sense internal information. The other is sense–act processing, represented by primitives that sense external information. It is this second type of processing that brings the environment (and stigmergy) to the forefront via scaffolding. It is also this second type of processing that is excluded from the production system architectures that have been discussed in preceding pages.

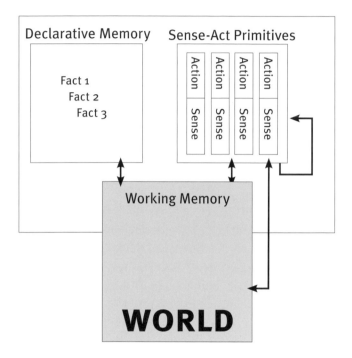

3.13 *BRICOLAGE*
3.13.1 Resource Allocation

Figure 3-3 indicates that there are two different styles of processing available: sense–think–act or sense–act. Scaffolding raises the possibility of sense–act processing, but also raises the issue of the degree to which such processing might be combined with the more traditional sense–think–act cycle. Even when sensing and acting are included in production systems (Anderson et al., 2004; Meyer & Kieras, 1997a) they are under cognitive control. Unified theories of mind (Anderson, 1983; Newell, 1990) rely on sense–think–act processing; sense–act processing is absent from such models. Behaviour-based robotics architectures (Brooks, 1989, 1999, 2002), which reacted against classical theories, rely completely upon sense–act processing, and deliberately exclude the sense–think–act cycle. In short, radical classical approaches, and radical reactions to them, deny the simultaneous existence of both processes.

More moderate views, such as the duplex theory that we have introduced (Goodale & Humphrey, 1998), and Figure 3-3, acknowledge the complementary existence of both types of processes, and must therefore go on to consider their allocation. "Minds may be essentially embodied and embedded and *still* depend crucially on brains which compute and represent" (Clark, 1997, p. 143).

One example of both types of processing being active at the same time is *horizontal décalage* from Piagetian developmental theory (Flavell, 1963). A horizontal *décalage* occurs when a child can use a more advanced level of operations to solve one problem, but cannot do so for a related problem. For example, children conserve quantity or mass earlier than they conserve weight. Given that it can be argued that Piaget's earlier stages of development involve more sense–act processing than do later stages, the existence of horizontal *décalages* suggest that cognitive development can exhibit periods during which sense–act and sense–think–act cycles coexist.

3.13.2 Thought as *Bricolage*

Cognitive scientists are not the only ones faced with allocating resources between sense–act and sense–think–act cycles. If both are available to a cognitive agent, then the agent itself has to flexibly allocate these resources as well. An agent might solve a problem with sense–think–act processing at one time, yet solve it with sense–act processing at another, and therefore be able to choose processing types. Even more plausibly, both types of processes might be in play simultaneously, but applied in different amounts when the same problem is encountered at different times and under different task demands (Hutchins, 1995). Resource allocation might depend upon something like the *007 principle* (Clark, 1989): "Creatures will neither store nor process information in costly ways when they can use the structure of the environment and their operations upon it as a convenient stand-in for the information-processing operations concerned. That is, know only as much as you need to know to get the job done" (p. 64).

The sense–act operators in Figure 3-3 comprise a finite set of "tools" available for information processing. At a given point of time, a subset of these tools is employed. Depending upon the subset that is selected, a problem could be solved with sense–think–act cycles, with sense–act processes, or with some combination of the two.

Such information processing — the selection of a subset of available operators — is akin to the notion of *bricolage* (Lévi-Strauss, 1966). A *bricoleur* is an "odd job man" in France. "The *'bricoleur'* is adept at performing a large number of diverse tasks, but unlike the engineer, he does not subordinate each of them to the availability of raw materials and tools conceived and procured for the purpose of the project. His universe of instruments is closed and the rules of his game are always to make do with 'whatever is at hand'" (Lévi-Strauss, 1966, p. 17).

3.14 THE POWER OF *BRICOLAGE*
3.14.1 *The Savage Mind*

The notion of *bricolage* was introduced in *The Savage Mind* (Lévi-Strauss, 1966). Lévi-Strauss was interested in explaining the practice of totemism, in which individuals or groups in a society are given names of animals or plants. He found that totemism was based upon sophisticated classification systems. "Native classifications are not only methodical and based on carefully built up theoretical knowledge. They are also at times comparable from a formal point of view to those still in use in zoology and botany" (p. 43). Furthermore, the logic of totemism involved mapping relationships between classified items in the world to analogous relationships between groups.

For example, imagine that one clan was assigned an eagle totem, while another was assigned a bear totem. These totems capture second-order properties, or *differences* between pairs of categories: a characteristic used to distinguish the two clans is mapped onto an observed difference between eagles and bears in a detailed classification scheme (Lévi-Strauss, 1966, Chapter 4).

Lévi-Strauss (1966) argued that the regularities governing totemism established that such thought was not primitive. Nevertheless, when he used the analogy of the *bricoleur* to illustrate "primitive" thinking as being different from scientific thought, Lévi-Strauss still cast it in a negative light. "The 'bricoleur' is still someone who works with his hands and uses devious means compared to those of a craftsman" (pp. 16−17). The problem was that the *bricoleur* is limited to a fixed set of materials at hand. These components or tools can be rearranged, but cannot be extended. "The engineer is always trying to make his way out of and go beyond the constraints imposed by a particular state of civilization while the 'bricoleur' by inclination or necessity always remains within them" (p. 19).

3.14.2 Power from Non-linearity

The view that the *bricoleur* is constrained by finite materials fails to recognize that, in particular circumstances, finite resources provide surprising power. Consider a set of sense−act operators in Figure 3-3 as a finite resource. One characteristic of these particular operators is their *non-linearity*: they follow an "all or none law," and only carry out their action when particular triggering information has been sensed (Dawson, 2004). Such non-linearity is a source of incredible computational power; the collective power of simple non-linear operators is huge.

For example, artificial neural networks are comprised of very simple, non-linear components. As a result they can solve the same problems as a universal Turing machine (Cybenko, 1989; Hornik, Stinchcombe, & White, 1989; Lippmann, 1987; McCulloch & Pitts, 1943). Even critics of connectionism (Fodor & Pylyshyn, 1988) have noted that "the study of Connectionist machines has led to a number of striking and unanticipated findings; it's surprising how much computing can be done with a uniform network of simple interconnected elements" (p. 6). In general, interactions between non-linear components produce complex emergent phenomena such that the behaviour of the whole goes beyond, or cannot be predicted from, the behaviour of the component parts (Holland, 1998; Luce, 1999).

Bricolage is receiving renewed respect (Papert, 1980; Turkle, 1995). Papert notes that "if *bricolage* is a model for how scientifically legitimate theories are built, then we can begin to develop a greater respect for ourselves as *bricoleurs*" (p. 173). Turkle describes *bricolage* as a sort of intuitive tinkering, a dialogue mediated by a virtual interface. "As the computer culture's center of gravity has shifted from programming to dealing with screen simulations, the intellectual values of *bricolage* have become far more important. ... Playing with simulation encourages people to develop the skills of the more informal soft mastery because it is so easy to run 'What if?' scenarios and tinker with the outcome" (p. 52). We will shortly argue for the use of *bricolage* to further our understanding of embodied cognitive agents.

3.15 THE SOCIETY OF MIND
3.15.1 Agents and Agencies

Unified theories of mind aim to provide an account of cognition that explains a diversity of phenomena by appealing to a single set of rules. "All the higher cognitive processes, such as memory, language, problem solving, imagery, deduction and induction, are different manifestations of the same underlying system" (Anderson, 1983, p. 1). However, theories that view the mind as a collection of non-linear operators do not necessarily share this goal. Consider, for example, Minsky's *society of mind* (Minsky, 1985, 2006).

The society of mind is a theory that grows from a basic assumption:

"Any brain, machine, or other thing that has a mind must be composed of smaller things that cannot think at all" (Minsky, 1985, p. 322). Minsky then proceeds to generate hypotheses about what these smaller things might be, how they might interact, and what these interactions can produce.

Minsky (1985, 2006) proposes that the basic building blocks of the mind are *agents*. An agent is given a very vague definition: "Any part or process of the mind that by itself is simple enough to understand" (Minsky, 1985, p. 326). However, in practice an agent is analogous to a production, a sense–act operator from Figure 3-3, or a unit in an artificial neural network. That is, an agent is a simple device that receives input, makes a decision on the basis of this input (i.e., it is non-linear), and then sends an output signal.

Minsky (1985, 2006) proposes a large number of different types of agents, each associated with performing different kinds of tasks. His collection of agents includes censors, demons, direction-nemes, memorizers, micronemes, nemes, nomes, paranomes, polynemes, pronomes, recognizers, sensors, and suppressors. Luckily, Minsky (1985) also provides a glossary to help manage the diverse nature of his theory!

The power of a society of mind comes from organizing a number of agents into groups called *agencies*. Again, Minsky (1985, 2006) proposes a diversity of agencies, involving different organizational principles, and designed to accomplish different higher-order tasks: A-brains, B-brains, cross-exclusions, cross-realm correspondences, frames, frame arrays, interaction-squares, k-lines, picture-frames, transframes, and uniframes.

An agency is an explicit example of a whole transcending the computational power of its parts. Early on, Minsky (1985) describes an example agency for building with toy blocks called *Builder*. "If you were to watch *Builder* work, from the outside, with no idea of how it works inside, you'd have the impression that it knows how to build towers. But if you could see *Builder* from the inside, you'd surely find no knowledge there. You would see nothing more than a few switches, arranged in various ways to turn each other on and off" (p. 23).

Interestingly, though the society of mind is untraditional in its construal of thinking, it is still presented as a traditional sense–think–act model. However, sense–think–act processing is not a necessary characteristic of this theory. If agents could interact with the environment as well as with each other, then even greater computational surprises would emerge from a society of mind.

3.15.2 Explaining Mental Societies

The diversity at the heart of the society of mind poses problems for explaining cognition. Minsky (1985, p. 322) argues against unified theories modelled after physics: "The operations of our minds do not depend on similarly few and simple laws, because our brains have accumulated many different mechanisms over aeons of evolution. This means the psychology can never be as simple as physics, and any simple theory of mind would be bound to miss most of the 'big picture'. The science of psychology will be handicapped until we develop an overview with room for a great many smaller theories." However, Minsky leaves an unanswered question: How do we make such room in our theories?

3.16 ENGINEERING A SOCIETY OF MIND

What is required to explain a society of mind? Minsky (1985, p. 25) sketches a general strategy: "First, we must know how each separate part works. Second, we must know how each part interacts with those to which it is connected. And third, we have to understand how all these local interactions combine to accomplish what that system *does* — as seen from the outside." What tactics might we employ to carry out Minsky's strategy?

3.16.1 Reverse Engineering

One popular approach is called *reverse engineering*. Reverse engineering takes Minsky's strategy in the opposite order. It begins with observations of what the system does from the outside, and then uses these observations (usually collected with clever experimental methodologies) to infer interactions between parts. Ultimately, it attempts to ground this analysis in a set of primitives — that is, the basic parts of the system.

Classical cognitive science makes extensive use of reverse engineering. Most cognitive theories are the product of a general approach called functional analysis (Cummins, 1975, 1983). Functional analysis proceeds as follows: First, a general function of interest is defined. Second, this general function is decomposed into an organized system of subfunctions capable of carrying out the general function. Subfunctions themselves might be further decomposed into organized systems of sub-subfunctions. This analysis proceeds until the final stage of subsumption. When a subfunction is subsumed, it is explained by appealing to physical laws, and cannot be decomposed into any smaller functions.

The earliest production system models of cognition were achieved by functional analysis. For instance, human subjects solved cryptarithmetic problems, thinking aloud as they worked. Their verbalizations were transcribed and analyzed to identify a likely set of productions being used to solve the problem. In one famous example, a set of only 14 different productions produced a remarkable fit to how a single subject solved a single cryptarithmetic problem (Newell & Simon, 1972).

The problem with reverse engineering is that it often runs into the frame of reference problem (Pfeifer & Scheier, 1999). The frame of reference problem occurs when the parable of the ant (Simon, 1969) is ignored, as was briefly discussed in Chapter 1. When functional analysis is performed, it is typical to place all of the complexity inside the cognitive system, and ignore potential accounts of this complexity that might be provided by including environmental factors. As a result, it has been argued that theories that are produced via analysis are more complicated than necessary (Braitenberg, 1984).

3.16.2 Forward Engineering

An alternative approach that is deliberately designed to avoid the frame of reference problem is to perform *forward engineering*. In forward engineering, one follows Minsky's strategy in his stated order. A set of building blocks is created, and a system is built from them. The system is then observed to determine whether it generates surprising or complicated behaviour. This has also been called the synthetic approach (Braitenberg, 1984). It is not as widely practised as reverse engineering. "Only about 1 in 20 'gets it' — that is, the idea of thinking about psychological problems by inventing mechanisms for them and then trying to see what they can and cannot do" (Minsky, 1995, personal communication).

Forward engineering addresses the frame of reference problem because when complex or surprising behaviours emerge, pre-existing knowledge of the components — which were constructed by the researcher — can be used to generate simpler explanations of the behaviour. "Analysis is more difficult than invention in the sense in which, generally, induction takes more time to perform than deduction: in induction one has to search for the way, whereas in deduction one follows a straightforward path" (Braitenberg, 1984, p. 20).

3.17 SYNTHESIS IN ACTION
3.17.1 Cricket Phonotaxis

In later chapters we will explore historical examples of forward engineering, including Braitenberg's Vehicle 2 (Braitenberg, 1984) and Grey Walter's cybernetic animal *Machina speculatrix* (Grey Walter, 1950a, 1950b, 1963). For the time being, though, let us briefly consider how the synthetic approach is used to study an interesting insect behaviour. Female crickets track down a mate by listening to, and following, a male cricket's song. Crickets have an ear located on each foreleg, and can use differences in stimulation of each ear to compute directional information. This ability is called *phonotaxis* (Webb, 1996). Many researchers are interested in determining the mechanisms that mediate cricket phonotaxis.

Phonotaxis depends crucially upon the structure of the male cricket's song. The cricket's familiar "chirps" are pure tone signals that have a frequency of 4–5 kHz and are delivered in bursts or syllables that are 10–30 ms in duration (Webb & Scutt, 2000). If the frequency of the song, or the interval between repetitions of syllables, is disrupted then so too is phonotaxis. Webb notes that researchers typically propose that phonotaxis is mediated by mechanisms used to localize a call, as well as additional mechanisms that operate in parallel to analyze the signal. Signal analysis is assumed, for instance, to explain how a female cricket chooses to follow one song when several male crickets are attempting to attract her at the same time.

3.17.2 Robot Phonotaxis

Barbara Webb and her colleagues have used forward engineering to study cricket phonotaxis by building call-following robots, beginning with wheeled devices (Webb, 1996; Webb & Scutt, 2000) and later using six-legged machines (Horchler, Reeve, Webb, & Quinn, 2004; Reeve, Webb, Horchler, Indiveri, & Quinn, 2005). This research is motivated by a simple guiding hypothesis: the female cricket moves toward a song by sensing whether it is coming from the left or the right, and by turning in the sensed direction. Consistent with the synthetic approach, these robots begin with very simple circuits: "a more powerful way to explore the actual functional roles of the neurons is to look at what behavior it is possible to obtain with gradual elaborations of simpler circuits" (Webb & Scutt, 2000, p. 250). Surprisingly, a four-neuron circuit can model cricket phonotaxis without requiring separate song analysis.

The model uses two auditory neurons, each receiving signals from one side of the cricket. Two motor neurons cause the robot to turn in a particular direction. An auditory neuron excites the motor neuron on the same side, and inhibits the other motor neuron. All four components are highly non-linear, generating action potentials at frequencies that are governed by external stimulation, and by internal signals, which vary over time. To cause a turn, one auditory neuron must repeatedly "spike" before the other in order to produce a motor neuron spike.

The robot successfully demonstrates phonotaxis (Webb & Scutt, 2000). However, it also behaves as if songs are being analyzed. When song syllable interval is modified, the robot's phonotaxis is impaired. As well, the robot meanders to non-directional songs presented from above, which is usually taken as evidence of song analysis in real crickets. The robot can also choose, moving in the direction of a single song when other similar songs are being played from other speakers at the same time.

How is this simple circuit capable of performing song analysis? Responses of the robot depend upon the temporal properties of the model, which are affected by the separation of ears on the robot, and the latencies of the model's neurons. These temporal properties result in sensitivity to very particular temporal properties of songs. That is, the circuit analyzes song structure "for free" because of its dynamic, temporal properties. "Thus it is clear from our results that much of the evidence for the standard 'recognize and localize' model of phonotaxis in crickets is insufficient to rule out an alternative, simpler model" (Webb & Scutt, 2000, pp. 265–66).

3.18 *VERUM-FACTUM*
3.18.1 Synthetic Psychology

The use of robots to study cricket phonotaxis (Webb & Scutt, 2000) illustrates synthetic psychology (Braitenberg, 1984). In synthetic psychology, a system is first constructed from a set of interesting components. The behaviour of the system is then observed and explored, usually by embedding the system in an interesting or complicated environment. For instance, Webb and Scutt manipulated the location, number, and nature of calls being presented to their cricket robot. Non-linear interactions between components, or between the system and its environment, can produce behaviour that is more complicated than expected. For example, Webb and Scutt's system was designed to localize sounds, but also behaved as if it analyzed sound properties. Finally, the fact that the system is both simple and constructed by the researchers means that simpler theories can be proposed to account for complex or surprising behaviour.

3.18.2 Vico's Philosophy

The synthetic approach has been described as "understanding by building" (Pfeifer & Scheier, 1999). The idea that the route to understanding a system comes from our ability to construct it is not new. It is rooted in the philosophy of Giambattista Vico, who was an early-eighteenth-century philosopher. Vico reacted against the seventeenth-century philosophy of René Descartes, which inspired the logicism of classical cognitive science (Devlin, 1996). For example, Vico believed that if

Cartesian methods were taught too early, they would stifle a student's imagination and memory (Vico, 1990). Indeed, Vico developed a metaphysics and theory of mind that attempted to replace Cartesian views (Vico, 1988), and attempted to explain societal creations, such as law (Vico, 1984).

Vico's philosophy is based on the central assumption that the Latin term for truth, *verum*, was identical to the Latin term *factum*. As a result, "it is reasonable to assume that the ancient sages of Italy entertained the following beliefs about the true: 'the true is precisely what is made'" (Vico, 1988, p. 46). This assumption leads to an epistemology that resonates nicely with forward engineering. "To know (*scire*) is to put together the elements of things" (p. 46). Vico believed that humans could only understand the things that they made, which is why he turned his philosophical studies to societal inventions, such as the law. A famous passage (Vico, 1984, p. 96) highlights Vico's philosophical position: "The world of civil society has certainly been made by men, and its principles are therefore to be found within the modifications of our own human mind. Whoever reflects on this cannot but marvel that the philosophers should have bent all their energies to the study of the world of nature, which, since God made it, he alone knows."

This view also resulted in an embodied view of mind that stood in stark contrast to the disembodied view espoused by Descartes. For instance, Vico recognized that the Latins "thought every work of the mind was sense; that is, whatever the mind does or undergoes derives from contact with bodies" (Vico, 1988, p. 95). Indeed, Vico's *verum-factum* principle is based upon embodied mentality. Because the mind is "immersed and buried in the body, it naturally inclines to take notice of bodily things" (Vico, 1984, p. 97).

Classical cognitive science, with emphasis on the rule-governed manipulation of symbols, has evolved from Descartes' view of the rational, disembodied mind (Descartes, 1637/1960). Reactions against classical cognitive science are in essence reactions against the Cartesian mind. "The lofty goals of artificial intelligence, cognitive science, and mathematical linguistics that were prevalent in the 1950s and 1960s (and even as late as the 1970s) have now given way to the realization that the 'soft' world of people and societies is almost certainly not amenable to a precise, predictive, mathematical analysis to anything like the same degree as is the 'hard' world of the physical universe" (Devlin, 1996, p. 344). Perhaps it is fitting that the synthetic approach is rooted in a philosophy that reacted against Descartes, and which attempted to explain regularities in this "softer" domain.

3.19 MIND AND METHOD
3.19.1 Mind

Consider the possibility that intelligence must be explained using a society of mind (Minsky, 1985, 2006). In such a theory, the mind is composed of a diversity of agents that are non-linear in nature, that interact with one another, and through these interactions produce surprising, complex, emergent results.

That the mind might be of this nature is not a radical idea. Production systems, old and new (Anderson, 1983; Anderson et al., 2004; Meyer & Kieras, 1997a, 1997b; Newell, 1973, 1990; Newell & Simon, 1961, 1972), can be considered to be particular instances of societies of mind. Individual productions serve the role of agents; they are non-linear in the sense that they only perform their actions when their conditions are precisely matched. Productions interact with one another via their manipulation of working memory. Production systems are powerful in the sense that small numbers of productions are capable of producing sophisticated behaviour. Production systems are surprising because working memory's stigmergic control of productions make it impossible to predict what a complete system will do without actually running a simulation.

Consider now that human intelligence might result from a society of mind that is not limited to the sense–think–act cycle of classical cognitive science (Pfeifer & Scheier, 1999). At least some of cognition is likely scaffolded (Clark, 1997; Hutchins, 1995; Wilson, 2004) and involves sense–act processes. In this view, some of the mind has leaked into the world, and the world can be directly accessed so that computational resources are not used to build internal representations of it (Brooks, 1999).

In embodied cognitive science, the environment is part of intelligence (Varela et al., 1991). Problem solving occurs by seeking solutions in actions on the world. This is exploited when, for example, children learn about geometry by programming a LOGO Turtle (Papert, 1980, 1993). "To solve the problem look for something like it that you already understand. The advice is abstract; Turtle geometry turns it into a concrete, procedural principle: *Play Turtle. Do it yourself.* In Turtle work an almost inexhaustible source of 'similar situations' is available because we draw on our own behavior, our own bodies" (Papert, 1980, p. 64). Again, this idea is not new. For instance, we have seen that the developmental theories of Piaget and Vygotsky recognize that thinking develops from action on the world.

The importance of the embodied view is its increased emphasis on the environment, which seems missing from the classical sense–think–act cycle. Embodied cognitive science recognizes that complex behaviour can result when a cognitive system interacts with its environment, as typified by the parable of the ant (Simon, 1969).

This view that intelligence is the product of an embodied society of mind suggests that thinking is *bricolage*. Available are a set of primitive operations, some sense–think–act and others sense–act, which can be drawn upon to solve information-processing problems as they arise. Learning to think becomes learning to choose what operations to use at any given time. This in turn may depend upon internally represented goals, or upon externally present stimuli or aids. "The process reminds one of tinkering; learning consists of building up a set of materials and tools that one can handle and manipulate. Perhaps most central of all, it is a process of working with what you've got" (Papert, 1980, p. 173).

3.19.2 Method

If intelligence is the product of an embodied society of mind, if cognizing systems are *bricoleurs*, then how should cognitive science proceed? One promising approach is for researchers to think like the systems that they study — to become *bricoleurs* themselves. The synthetic approach, which assembles available elements into embodied agents whose surprising behaviour exceeds what might be expected of their simple components, is an example of a cognitive science that depends upon 'tinkering'.

3.20 SYNTHESIS AS PROCESS, NOT AS DESIGN
3.20.1 Synthesis Is Not Design

Most models in classical cognitive science are derived from the analysis of existing behavioural measurements (Dawson, 2004). In contrast, models created using the synthetic approach involve making some assumptions about primitive capacities, building these capacities into working systems, and observing the behaviours that result. In synthetic psychology, model construction *precedes* behavioural analysis. With the synthetic approach, "the focus of interest shifts from reproducing the results of an experiment" (Pfeiffer & Scheier, 1999, p. 22).

What does the focus of interest shift to when the synthetic methodology is employed? Pfeiffer and Scheier (1999) argue that a key element of synthetic psychology is *design*. "What we are asking is how we would design a system that behaves in a particular way that we find interesting" (p. 30). However, this design perspective is in conflict with the spirit of synthetic psychology. This is because if one designs a system with particular goal behaviours in mind, then one might be blind to interesting *unintended* behaviours that the system produces.

A former student's project in a robot-building course illustrates the problems with synthesis as design. She built two Braitenberg Vehicle 2s (Braitenberg, 1984) from LEGO components, as will be described in detail in the next chapter. These robots can produce behaviour in which they move toward light. Her design goal was to mount a light on the back of one of these robots in order to cause the other robot to follow it closely. That is, her desire was to create a "robot convoy."

However, this goal proved difficult to achieve. The behaviour that she desired (i.e., following) was never produced. As a result, she crafted new lighting systems, powered by battery packs that she created herself, to create more potent stimuli for one of the robots to follow. This engineering did not produce the desired result either, and this student became very frustrated.

However, when others watched her machines in action at this point, they saw complicated interactions between robots that were clearly affected by the mounted lights, but which were quite different from the (intended) following behaviour. It was only when an outside observer — not committed to her design perspective — pointed out these interesting emergent behaviours that the student was able to break free from the constraints of her design, and document the robot behaviour that she definitely did not intend.

The synthetic approach, when dominated by design, is analogous to the serialist reaction to Austro-German music (see Chapter 2). In this musical example, one rigid set of rules was replaced with a different set that was no less rigid. Conducting synthetic psychology with design in mind is no different than using analytic models to reproduce experimental results. That is, both modelling approaches evaluate the quality of the model in terms of its ability to meet predefined criteria (e.g., fit extant data, or accomplish a design's objective).

3.20.2 Synthesis as Process

Synthetic psychology might better be conducted in a fashion analogous to the minimalist reaction to both Austro-German and serialist music. Recall from Chapter 2 composer Steve Reich's notion of choosing musical processes, and then setting them in motion. In this musical approach, Reich was content to let the resulting composition run itself. As a result, he was able to experience auditory effects that emerged from the processes from which the composition was constructed. This approach was successful because of the care that Reich and other minimalists took to choose the musical processes in the first place. Rather than begin with an overarching design, synthetic psychologists should begin with a carefully selected set of processes (e.g., agents in a society of mind), set them in motion, and be open to the surprising behaviours that are most certain to emerge.

3.21 BUILDING *BRICOLEURS*

3.21.1 Cartesian Alternatives

There is a deep Cartesian bias underlying most of modern cognitive science (Devlin, 1996). Descartes' dualism now exists as the distinction between the internal self and the external world, a distinction that agrees with our everyday experience. However, some have argued that our notion of a holistic internal self is illusory (Clark, 2003; Dennett, 1991, 2005; Minsky, 1985, 2006; Varela et al., 1991). "We are, in short, in the grip of a seductive but quite untenable illusion: the illusion that the mechanisms of mind and self can ultimately unfold only on some privileged stage marked out by the good old-fashioned skin-bag" (Clark, 2003, p. 27).

For researchers, the frame of reference problem is one consequence of not challenging this illusion (Pfeifer & Scheier, 1999). Our Cartesian bias, coupled with traditional analytic methodologies, causes us to ignore the parable of the ant (Simon, 1969), to assign too little credit to the world, and to assign too much credit to internal processes.

Classical cognitive science views thought as the rational, goal-driven manipulation of symbols by a mind that can be studied as an abstract, disembodied entity. New approaches in cognitive science react against this Cartesian-rooted position. Embodied cognitive scientists comfortably view thought as *bricolage* involving a collection of non-linear processes that have leaked outside of the "skin-bag" into the world.

This alternative view has led to methodologies that replace analysis with synthesis. Systems are first constructed from collections of non-linear components, and are then situated in an interesting world (Braitenberg, 1984). Do unexpected behaviours arise when simple agents are in a world that they can sense and manipulate?

3.21.2 Students as *Bricoleurs*

If seasoned researchers frequently face the frame of reference problem, then imagine the challenge facing students who are beginning to learn about embodied cognitive science. Everyday experience provides convincing support for our self-concept, and our brains are so proficient at exploiting the world that we are often unaware of scaffolding. Clark (2003, p. 48) asks "how can we alter and control that of which we are completely unaware?" Further to this, how can we teach students about a view of mind that they may not naturally experience?

One approach would be to merely tell students about this view. Clark (2003, p. 33), however, is of the opinion that this will not suffice: "We cannot understand what is special and distinctively powerful about human thought and reason by simply paying lip service to the importance of the web of surrounding structure. Instead, we need to understand in detail how brains like ours dovetail their problem-solving activities to these additional resources, and how the larger systems thus created operate, change, and evolve." Our view in the current book is that to "understand in detail" is to experience.

One can provide hands-on experience of embodied cognitive science. Tools such as LEGO Mindstorms enable students to build simple robotic agents, situate them in manipulable environments, and observe the surprising results. Such agents provide concrete examples of scaffolding and the perils of the frame of reference problem. Robot design and exploration also provides first-hand experience of *bricolage*: students use the materials at hand to build robots and their environments.

The next few chapters provide some examples of robot projects that have been used to allow students to experience embodied cognitive science. They provide theoretical and historical contexts, as well as detailed instructions for construction. For those readers not able or not inclined to build these machines, the following chapters also provide example videos of robot behaviour that illustrate key themes. However, consistent with Vico's *verum-factum* principle, building and exploring the projects that follow is much more rewarding than merely reading about them.

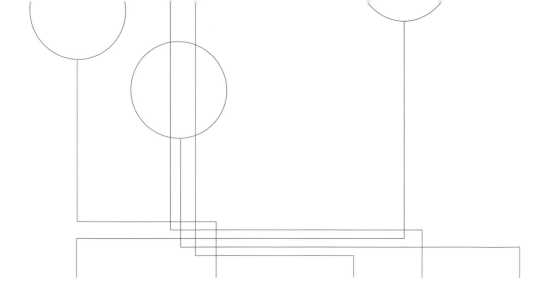

Chapter 4
Braitenberg's Vehicle 2

4.0 CHAPTER OVERVIEW

Three general themes have been developed in the preceding chapters. The first has been the general nature of the classical approach in cognitive science. Classical cognitive science views thinking as the rule-governed manipulation of symbols, inspired by the workings of the digital computer. As a result, its general characteristics include logicism, the manipulation of content-laden formal structures, the disembodiment of thought, and the emphasis on central control mechanisms. A second theme has been that alternative approaches are arising; these new views are reactions against classical cognitive science. In embodied cognitive science, an agent's world is seen as an important contributor to its intelligence. The mind is said to have leaked into a world that scaffolds intelligence or thinking. Thought is not disembodied rationalism, but is instead *bricolage* in which different processes — some scaffolded, others not — are selected to solve problems at hand. A third theme has been that theories in embodied cognitive science might be best developed by using a synthetic approach. In this approach, systems are first constructed from interesting components, and then observed in action to see what kinds of interesting behaviours they can produce, and how these behaviours are affected by changing the environment in which the system is embedded.

At the end of Chapter 3, it was suggested that not only were *bricolage* and the synthetic approach important to the theories of embodied cognitive science, but these notions were also central to teaching students about the embodied approach. The purpose of Chapter 4 is to provide a concrete example of this. It provides detailed instructions about building, programming, and observing a simple robot constructed from LEGO Mindstorms components. This robot is used to provide some hands-on experience with the themes that have been introduced in the first three chapters. This robot also sets the stage for slightly more advanced machines that are discussed in later chapters. Of particular note is the antiSLAM robot that is presented in Chapter 9, which demonstrates the navigational capabilities of a Vehicle 2 that has "evolved" to have additional sensory mechanisms.

4.1 A ROBOT'S PARABLE
4.1.1 Path of a Robot

In the parable of the ant (Simon, 1969), a researcher's task was to explain the complex path taken by an ant along a beach. Figure 4-0 illustrates another path; one traced by a pen attached to a robot that wandered along a sheet of paper. What mechanisms are responsible for the shape of the robot's path? This chapter explains these mechanisms.

4.1.2 Analysis and Synthesis

Braitenberg has argued that "when we analyze a mechanism, we tend to overestimate its complexity" (Braitenberg, 1984, p. 20). There is an overwhelming tendency to explain complex behaviour by attributing complex mechanisms to a behaving agent. Contributions of the agent's environment are ignored. Noting this, Braitenberg proposed the *law of uphill analysis and downhill synthesis*. According to this law, theories produced by analyzing agent behaviours will be more complicated than theories created by building a situated system and observing what surprising and complex behaviours it can produce. The goal of building a robot to produce the path in Figure 4-0 can be used to illustrate Braitenberg's point.

For example, one could analyze Figure 4-0 with the goal of writing a LOGO program that would cause the LOGO turtle (Papert, 1980) to reproduce it. The program would tell the turtle when to move forward, when to turn, when to put the pen down, and so on. Comparisons would be made between the drawing made by the turtle and Figure 4-0. Any discrepancies between the two would result in the program being modified until the LOGO turtle produced a satisfactory rendering. At this point, this program would likely be long, complex, and would make the LOGO turtle completely responsible for the drawing. The turtle's environment would play no role.

An alternative synthetic approach would be to ignore Figure 4-0 altogether and instead build a simple system that was attracted or repelled by stimuli in the environment. In exploring the behaviour of this system, it might be discovered that a complex environment would cause the simple robot to follow the path illustrated in the figure below. In the following pages, we will show that a robot that uses two light sensors to control the speeds of two motors will follow a moving light around, and can produce the path below — in a complex environment. The path of Figure 4-0 is the result of a simple robot following the more complex path of a moving light.

4.2 BRAITENBERG'S THOUGHT EXPERIMENTS
4.2.1 A Thought Experiment

In his classic book *Vehicles*, neuroscientist Valentino Braitenberg explores synthetic psychology by describing, as thought experiments, a number of different robots (Braitenberg, 1984). One, called Vehicle 2, propelled itself underwater. It had two separate engines, one on each side, and two separate sensors (e.g., for measuring temperature), again on each side of the agent. The output of one sensor was used to control one motor, and the output of the other sensor was used to control the other motor, as follows: motor speed was directly proportional to the value detected by its sensor, so that when this value increased, the motor sped up, and when this value decreased, the motor slowed down.

This chapter provides instructions for building Vehicle 2 out of LEGO NXT Mindstorms components. This robot is a land-based agent that has been inspired by Braitenberg's (1984) thought experiment.

Braitenberg (1984) argued that Vehicle 2 would generate complicated behaviour if it were embodied and situated in an interesting world. Furthermore, the kind of behaviour generated would depend upon whether a sensor was attached to the motor on the same side of the robot or to the motor on the other side.

Our incarnation is a "tractor-like" robot (Figure 4-1) that is intended to move around fairly flat surfaces.

4.2.2 Goals

Vehicle 2 is a simple robot that is fairly easy to build, to program, and to observe. It provides an ideal platform to introduce some of the concepts and skills that are central to this book.

With respect to skills, this robot provides hands-on experience with construction, including basic principles of sensors and motors. It requires the builder to also learn some simple programming skills in order to bring Vehicle 2 to life. Finally, it introduces the builder to the process of observing the agent's behaviour, and as well as the manipulation of this behaviour by varying both the robot's environment and some basic aspects of its design.

With respect to concepts, this robot begins to reveal the complexities of behaviour that can emerge when a simple, embodied agent is situated in an interesting, dynamic environment. Might human intelligence be derived from similar principles?

4-1

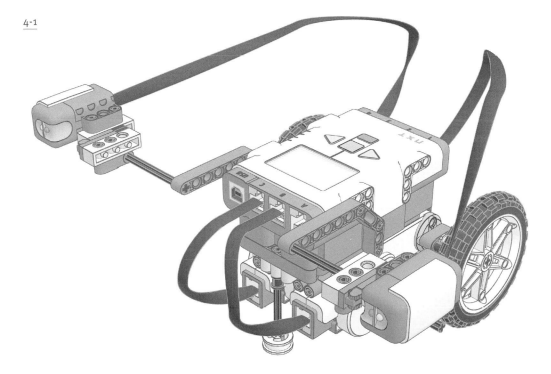

4.3 FORAGING FOR PARTS

4.3.1 Parts and Foraging

Figure 4-2 depicts the parts required to construct the version of Vehicle 2 that was illustrated in Figure 4-1. For our students, one approach to creating the robot might be to gather all of these parts prior to construction, foraging amongst the bins of available LEGO pieces.

4.3.2 Robot *Bricolage*

However, it is important to remember that some of these parts are not as plentiful as others, and that other robot builders are foraging for them as well. In some instances a desired part might be unavailable. In that case, the robot builder's — the *bricoleur's* — ingenuity must take over, and other (less desirable?) parts must be used instead. Slight deviations from the instructions might be required. As well, consistent with Braitenberg's (1984) recognition of natural selection as a robot design principle, these deviations might result in the construction of a better robot than the one that was originally used to create this chapter of instructions.

The pages that follow provide instructions for constructing the LEGO Vehicle 2. If the reader would prefer to use wordless, LEGO-style instructions, they are available as a pdf file from the website that supports this book (http://www.bcp.psych.ualberta.ca/~mike/BricksToBrains/).

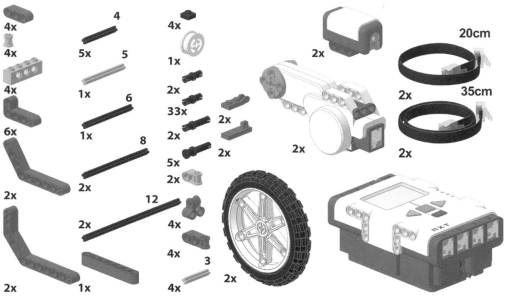

4-2

4.4 CHASSIS DESIGN (STEPS 1 THROUGH 4)

4.4.1 General Design

Our NXT version of Braitenberg's Vehicle 2 will employ two light sensors, which in turn will control two motors, which in turn will rotate two rear wheels. The chassis of this robot is a rigid, central "spine" to which all of the other robot parts will be attached. The chassis is essentially constructed from a set of different liftarms that are held together by pins.

4.4.2 Initial Chassis Construction

4-3

The first four steps for building the chassis are illustrated in Figure 4-3, and are labelled on the left of the figure. In Step 1, a black pin (with friction) is inserted into a bent 9-hole liftarm.

In Step 2, a second bent 9-hole liftarm is attached to the pin inserted in Step 1. A length-4 axle is inserted through the axle hole on each end of the joined liftarms so that an equal amount of axle protrudes from each side. Then two pins are attached as shown.

For Step 3, slide a 2 × 4 L-shaped liftarm on the axle on the long end of the bent liftarm. At this point it will hang loosely and easily fall off, but will be attached more firmly soon. Insert a length-4 axle into the axle hole of the L-shaped liftarm as shown.

Step 4, shown at the bottom of Figure 4-3, involves sliding two more L-shaped liftarms onto the axle that was added in Step 3, and inserting two pins into each as shown in the image.

4-4

4.5 CONSTRUCTING THE CHASSIS (STEPS 5 THROUGH 7)
4.5.1 General Design

Figure 4-4 illustrates the next three steps involved in creating the chassis. In Step 5, attach a 2 × 4 L-shaped liftarm as illustrated, and insert two pins into it. Now the pieces cannot slip off the axle but will still swing freely. Complete this step by attaching two more pins in the same position in the liftarm on the other side.

To begin Step 6, slide two perpendicular axle joiners onto a length-4 axle side by side and centred on the axle. Slide the axle into the axle hole of a 2 × 4 L-shaped liftarm so that the axle joiners protrude in the opposite direction of the L and attach the L-shaped liftarm to the centerpiece as shown in Figure 4-4. Secure it with a second L-shaped liftarm and then insert a pin into each of the axle joiners.

For Step 7, connect two perpendicular axle joiners with 2-holes with a length-4 axle and attach them to the pins added in Step 6.

4.6 THE NXT INTERACTIVE SERVO MOTOR
4.6.1 The Evolution of LEGO Motors

Vehicle 2 will move on its own by activating two NXT servo motors that will rotate wheeled axles. These are the latest generation of motors provided for LEGO robots, and they have distinct advantages over their ancestors, the 9V Technic mini-motor.

One problem with this older LEGO motor is that it rotates very quickly (about 340 rpm when there is no load), but it supplies very little rotational force. As a result, even a moderate load on an axle will stall the motor. In order for the motor to supply sufficient torque to move Vehicle 2, this problem must be overcome, usually by using a gang of gears to increase the torque that is required to use the wheels to drive the robot forward.

A second problem with this older motor is that it is designed to be attached to other LEGO pieces using studs. However, half of the bottom of the motor is smooth, stud-free, and extends below the pips on the other part of the motor's bottom. As a result, it is impossible to securely attach the mini-motor to a chassis on its own. Several additional parts are required to reinforce the motor; because of this, the motor itself does not contribute to the structural integrity of the robot.

4.6.2 The NXT Servo Motor

The NXT servo motor (Figure 4-5) that we will use in this chapter has been explicitly designed to solve both of these problems (Astolfo, Ferrari, & Ferrari, 2007). In terms of its internal structure, this motor has a built-in gear train of eight gears that produces a substantial gear reduction. As a result, while the maximum rotational speed of this motor is half of the older mini-motor (170 rpm versus 340–360 rpm, as reported on http://www.philohome.com/motors/motorcomp.htm), the NXT servo motor delivers 3–4 times the mechanical power. This reduces the need for additional gears to be added to the robot.

In terms of its external structure, an NXT servo motor was designed not only to be incorporated into a studless design (i.e., a design that uses studless liftarms or beams instead of studded bricks), but also to provide structural support when used (Astolfo et al., 2007). As can be seen in

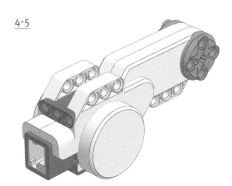

4-5

Figure 4-5, the motor has one built-in 3-hole beam near its narrow end, and two such built-in beams at the opposite end of the motor. As well, there are three holes perpendicular to the pair of built-in beams that accept pins.

In addition to the built-in gear reduction and studless connectivity, the NXT servo motor has an internal rotation sensor. It is an optical encoder that counts the rotations of the motor shaft, and is accurate to 1° of rotation (Astolfo et al., 2007). The motor is interactive in the sense that while the NXT brick can send commands to turn the motor on, it can also receive signals from this rotation sensor, and use these signals to offer precise motor control. For example, one could use this sensor to determine when the motor is in a particular rotational position, or to impose relational properties, such as synchronization, on two or more motors. Our Braitenberg Vehicle 2 will not require exploiting this internal rotation sensor, but we will take advantage of its existence for more complex robots that we will discuss later in this book.

4.7 ADDING MOTORS TO THE CHASSIS (STEPS 8 AND 9)

Construction Step 8, shown in Figure 4-6 below, completes the chassis of our NXT Braitenberg Vehicle 2. It involves attaching a perpendicular axle joiner with double holes on to each end of the axle added in Step 7. Then insert pins into the holes as shown in the figure. Later, these pins will be used to attach the NXT brick.

The chassis is now ready to have two NXT servo motors attached to it, as shown in Step 9 below. Push the two neighbouring pins on one side of the chassis into the holes in the motor near the orange motor output. Then insert the pin and adjacent axle into the holes on the rear. With the motor attached the centrepiece will no longer swing freely. The second motor is then added to the other side of the chassis, mirroring the first. Note how the structural integrity of the chassis is due to the rigidity of the motors — the motors are part of the chassis too!

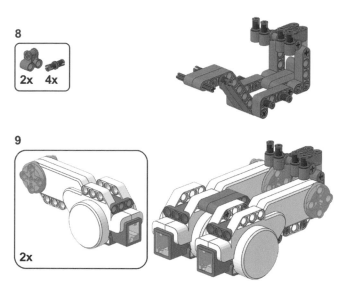

4.8 ADDING A FRONT SLIDER (STEP 10)

4.8.1 Passive Front Support

The front slider attaches to the rear of the motors, which will be the front of the robot. It further secures the two motors, and will also be used to help attach the NXT brick. The front slider supports the robot as it moves. Importantly, using a slider (instead of a front wheel) means that all of the robot's turning will be due to differences in the speeds of the two motors.

4.8.2 Constructing the Front Slider

Step 10, the construction of the front slider, is illustrated in Steps 1 and 2 of the subassembly shown in Figure 4-7. To begin, slide three perpendicular axle joiners onto a length-5 axle, as shown. Then attach a double-hole perpendicular axle joiner on to each end of the axle so that the double holes are oriented perpendicular to the holes of the three single-hole axle joiners (see image). In the centre single-hole axle joiner insert a pin with bush so that the bush is on the same side as the double-hole axle joiners' holes. Insert a long pin into the other two axle joiners and two long pins into each of the double-hole axle joiners as in the image.

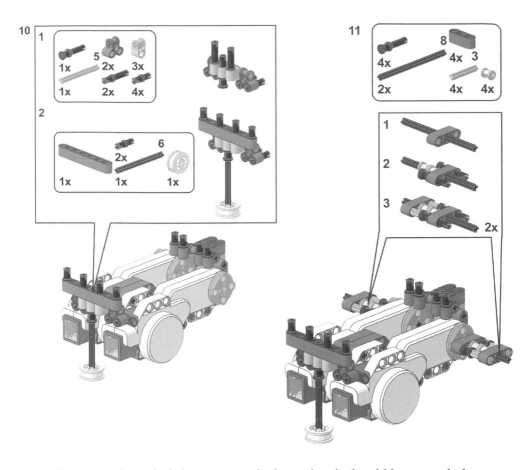

Next, attach a 7-hole beam on to the long pins; it should have two holes protruding on either side. Insert a pin into each of the holes on the end. The pins protruding from the beam will serve as the second attachment point for the NXT brick. The last step in the centrepiece is to insert a length-6 axle into the stop bush, to add a large wheel centre to the end of the axle, and to attach the entire slider to the chassis as illustrated.

4.9 CONSTRUCTING REAR AXLES (STEP 11)

4.9.1 Wheel Axle Design

Step 11 of our robot construction involves building the wheel axles. The orange motor output on the NXT motors has an axle hole in the centre and four pin holes around it. We can take advantage of the pin holes to add some extra strength to our wheel axles.

4.9.2 Constructing the Wheel Axles

There are three steps to constructing wheel axles, as shown in the sub-assembly illustrated in Figure 4-8. In Step 1 slide a 3-hole beam on to a length-8 axle. In Step 2, slide two bushes onto the axle and insert two long pins with stop bushes into the empty holes of the beam as shown in the Figure. For Step 3, insert a length-3 axle into each of the stop bushes and add a second beam of the same length. **Remember to use these steps to build two axle assemblies, one for each motor.** Once constructed, they can be attached to the robot as shown in Figure 4-9.

4.10 ATTACHING THE NXT BRICK (STEP 12)
4.10.1 The NXT Brick

Our Braitenberg Vehicle 2 requires that light sensor inputs be converted into motor speed outputs. In our robot, such sense–act connections are mediated by the NXT brick. This brick is a small computer that has four different input ports and three different output ports; it also has a USB port to connect it to a computer in order to download programs onto it. The heart of this device is a 32-bit ARM7 microprocessor. The NXT brick can be powered by six AA batteries, or by a rechargeable lithium battery pack. It is possible to connect three different NXT bricks together, and to have communication between them, in order to develop a more complicated robot. However, our Vehicle 2 is simple enough that only one of these bricks is required.

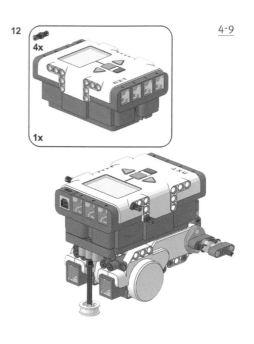

12

4-9

4.10.2 Attaching the Brick

In Step 12, shown in Figure 4-9, the NXT brick is connecting to the chassis by attaching the holes on its back to the pins on top of the chassis. Make sure that the screen is on the same side as the front slider! Add two pins to each side of the NXT brick as illustrated. They will serve as attachment points for the sensor mounts.

4.11 ATTACHING LIGHT SENSOR SUPPORTS (STEP 13)
4.11.1 Sensor Mount Design

For Vehicle 2 to be situated in its environment it requires sensors. Braitenberg's (1984) description of Vehicle 2 is general enough to permit a variety of different sensors to be employed. For our robot, we will use light sensors because they are very responsive to changes in light and give a simple output that is easily used to control motor output. The light sensors are mounted on axles and can be slid back and forth and angled inward. They are easily changeable to allow for experimentation on how the embodiment of the robot affects its behaviour. Step 13, shown in Figure 4-10, illustrates how to construct the basic structure to which light sensors will be attached. A 1 × 11.5 double bent liftarm is attached to each side of the NXT brick using the pins that were inserted in Step 12. Then a length-12 axle is inserted into the end of each liftarm as shown in the figure.

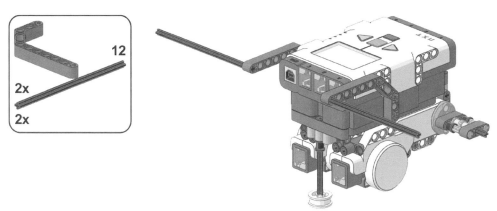

4.12 ADDING LIGHT SENSORS (STEP 14)
4.12.1 Mounting Light Sensors

In Step 14, shown in Figure 4-11 below, a mount for each light sensor is constructed and then is mounted to the axles added in Step 13.

There are four steps involved in building a light sensor mount; these steps are illustrated in the subassembly part of Figure 4-11. Start by connecting a stack of two 1 × 1 plates to the end stud of a 1 × 4 brick with holes (Step 1). Duplicate this piece for the second mount.

Proceeding to Step 2, attach a 1 × 3 locking joint to the each brick and add a second 1 × 4 brick on top. In Step 3, insert 2 pins into each of the bricks as shown in Figure 4-11. Note that the position of the pins in the right sensor mount mirror the position of the pins in the left sensor mount.

Finally, attach a light sensor and a perpendicular axle joiner with two holes as shown in the image. The light sensors have two diodes protruding from the front. The pale pink one emits red light when the sensors are in reflected light mode but does not when in ambient light mode. For this robot the sensors will be set to measure the ambient light.

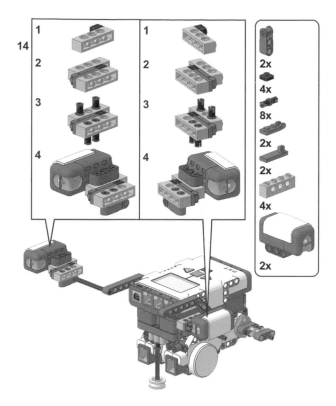

4.13 WHEELS AND CABLE CONSIDERATIONS

4.13.1 Completing the Robot

Step 15 (shown in Figure 4-12) is the final stage of constructing Vehicle 2, in which we add wheels onto the axles and connect the cables. The right motor should be connected to port C and the left motor should be connected to port B.

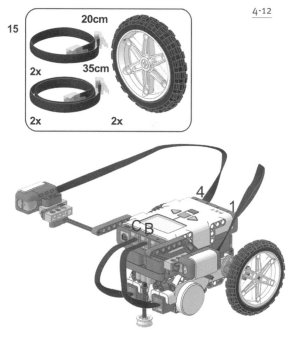

The options for connecting the light sensor cables are more interesting. One can make the relationship between sensors and motors *contralateral*—the sensor on the robot's left controls the speed of the motor on the robot's right, and the right sensor controls the left motor's speed. This is accomplished by connecting the left sensor to input port 4 and the right sensor to input port 1 (Figure 4-13). Note the crossing of the cables in Figure 4-13.

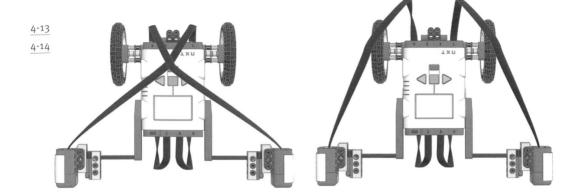

A second approach to connectivity would be to enforce an *ipsilateral* relationship between sensors and motors, where the left sensor controls the left motor, and the right sensor controls the right motor. To accomplish this, connect the left sensor to input port 1 and the right sensor to input port 4 (Figure 4-14 or Figure 4-12). Note that in these two figures the sensor cables do not cross.

We will see that whether the sensor-motor relation is contralateral or ipsilateral has a huge impact on the behaviour of our Vehicle 2. However, in order to see this in our assembled robot, we must first create a simple program that mediates the relationship between sensed light and motor speed.

4.14 SENSING, ACTING, AND THE NXT BRICK
4.14.1 The NXT Brick

What is the point of building Vehicle 2? It is to start to explore the kinds of surprising or complex behaviours that might emerge from a simple embodied agent that is situated in an interesting environment. The

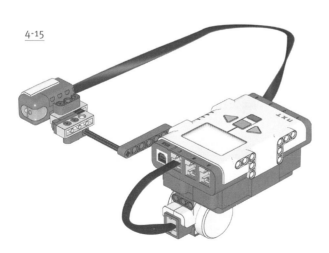

notion of agent at this point is intended to mean a constructed robot that is capable of sensing some properties of its environment, and to directly convert these properties into certain actions. That is, the agent is defined by its sense–act cycles (Pfeifer & Scheier, 1999) — essentially, by its reflexes, which directly link sensing to acting and do not require any intermediary thinking.

The implementation of these reflexes in Vehicle 2 is the job of the NXT brick that is mounted on top of it. That is, the NXT serves as the interface between LEGO sensors and motors (Figure 4-15). While, as described below, the NXT is a small digital computer, we will avoid considering it as a "thinking" component of a sense–think–act cycle. This is because the simple code that we develop to link sensing and acting could be implemented by replacing the NXT with hardware components that perform the same function as the NXT and our code.

At the heart of the NXT is a small computer, which employs a 32-bit central processing unit (CPU) — Atmel's ARM7 microprocessor. This system has access to four input ports and three output ports, is capable of analog-to-digital conversion using an ARV coprocessor. This computer is small — it has available only 64 kb of random-access memory (RAM). The NXT's operating system (its firmware) is held in a 256 kb FLASH memory.

In Vehicle 2, analog-to-digital converters built into the NXT convert the sensor readings into a usable form. That is, the sensor readings will be converted into a form that the program we write for the robot can translate into an output signal to be delivered from the NXT's output ports to the motors mounted on the rear of the robot.

The NXT uses pulse-width modulation (PWM) to control motor speed. When a signal is sent from the output port, it is a pulse of constant amplitude. Motor speed is varied by altering the duration of this pulse. For example, to run the motor at 35% speed, a pulse is sent that is on for 35% of 128μs cycle that is standard for the brick, and then the output is switched off for the remaining 65% of the cycle (Gasperi, Hurbain, & Hurbain, 2007). In order to increase the speed of the motor, there is a decrease in the proportion of the cycle during which a pulse is sent. Because the speed of the motor is determined by the average voltage that it receives, there is a linear relationship between speed and applied voltage.

Pulse-width modulation works quite nicely to achieve turning behaviour in our Vehicle 2 when the two motors are running at different speeds because of differences between the two light sensors. As noted earlier, the servo characteristics of the motor provide more sophisticated means of control that will be exploited in later chapters, but which are not required for this particular robot.

4.15 NXT LIGHT SENSOR PROPERTIES
4.15.1 The LEGO Light Sensor

Programming the desired sense–act cycles into Vehicle 2 is helped by having a reasonable understanding of the workings of the device that situates the robot, the NXT light sensor shown in Figure 4-16. This device is an analog sensor that measures the intensity of light that it receives. Its sensitivity covers a fairly broad range, from 0.5 lux (where a lux is the measure of lumens per square meter) to 500 lux (Prochnow, 2007).

4-16

While this device is treated as a light sensor when used in Vehicle 2, it actually is more sophisticated. It can be used an active sensor. As an active sensor, it consists of two functional components. One is a phototransistor that measures incoming light. The other is a light-emitting diode that generates light. In active mode, the light sensor activates the LED for a short period of time, and then it measures returning light with the phototransistor. It is continually oscillating between sending light and receiving light.

The light sensor can also be configured as a passive sensor, which is its typical usage in Vehicle 2. As a passive sensor, the LED is never turned on, and the light sensor uses its phototransistor to measure the intensity of the ambient light in the environment.

The LEGO light sensor has several properties that can make it tricky to work with. It has peak sensitivity to light in the infrared range of the spectrum, and is less sensitive to shorter wavelengths of light. This means, for example, that it will see incandescent light bulbs as being brighter than would be experienced by a human observer. As well, as battery power changes, the behaviour of the light sensor is affected. As a result, light sensors are affected by changes in ambient light, by the reflective properties of environmental objects, and by decreasing battery power. Some of these sensing nuances might be sources of an interesting set of robot behaviours. However, it might be desirable to control some of these properties to some extent to reduce light sensor fluctuations.

One approach to controlling the light sensor involves deciding how it will be read by the NXT brick. The analog-to-digital conversion that the NXT brick performs on its input ports can be processed in a number of different ways.

For instance, one mode for taking light sensor readings is RAW. When this is done, the digital representation of sensor output is used directly. When the light sensor is detecting very bright light, readings will be values in the order of 300, while very dark conditions will produce readings around 1023.

Alternatively, one can set the NXT brick to deliver readings in PERCENT mode. When this mode is used, the brightest light produces a value of 100, while darkness produces a value of 0. Another interesting mode for the light sensor is BOOLEAN; in this mode the light sensor only provides a signal when a change in light is detected!

While the default mode for light sensors is PERCENT, other modes might provide greater control over the variability of light sensor behaviour, and their effects on robot behaviour are worthy of exploration.

Of course, in addition to exploring the modes in which a sensor is read, Vehicle 2's behaviour can be affected by whether the light sensors are active or passive. For instance, in a highly reflective environment, active sensors will provide additional light sources, and will produce different behaviour than will passive sensors!

4.16 PROGRAMMING THE NXT BRICK
4.16.1 Programming Steps

With the robot constructed, and armed with some understanding of the brick and the light sensor, we are now in a position to create a program that will mediate light sensor readings and motor speeds. This requires choosing a programming language, and creating some code in that language. This merely involves writing a text file on a desktop computer, where the contents of the file are the lines of the program. This text file is then processed by a compiler, which converts the text into a form that can be executed by the brick. Finally, the executable code is downloaded from the computer to the NXT brick using a cable that LEGO provides.

4.16.2 Programming Environment

Because of the popularity of LEGO Mindstorms robots, there are a number of different programming environments that integrate code creation, downloading, and other activities involving the brick. LEGO provides one environment with a Mindstorms kit, and many others are available on the internet as freeware.

We have elected to program the NXT brick using a language called Not eXactly C (NXC), and to do so in a programming environment called BricxCC, which is available from http://bricxcc.sourceforge.net/. This environment permits NXC code to be typed and saved, provides aids for debugging code, permits the code to be compiled and downloaded, and provides a number of other useful tools for programming and examining the brick. It also comes with a comprehensive set of help files that provide instruction in using the various menu items for BricxCC, as well as a complete manual describing the NXC language. Figure 4-17 illustrates BricxCC loaded with the Vehicle 2 program to be described in the following pages.

4.17 A SIMPLE MAIN TASK
4.17.1 The Main Task

Any NXC program requires that one or more *tasks* are defined. A task defines a standalone operation, in the sense that it is assumed that more than one task can be running at the same time.

One of the tasks in the program must be called "main." For Vehicle 2, the main task initializes the light sensors, controls what information is displayed on the LCD screen of the NXT brick, and starts two other tasks.

The listing at the bottom of this page provides the main task written for our robot, and begins with the declaration "task main() {". The first two lines of the task tell the NXT that two of its input ports are going to be connected to light sensors, and that these sensors are passive. The next two lines request the NXT brick to process the signals from these two sensors in PERCENT mode. The next two lines start two additional tasks, called "DriveLeft" and "DriveRight," which will be discussed shortly. The only point to note here is that they are initiated by the main task.

4.17.2 Defining Variable Names

The main task that is listed below uses variable names like "LeftEye" and "RightEye" that make the program easier to read. These "plain English" terms are established using a set of #define statements that are also part of the program. Our Vehicle 2 program uses four of these statements, which are given below. For instance, the first one lets us use the term

LeftEye in place of S1, which is how NXC typically names input port 1. The third line lets us use the term LeftMotor instead of OUT_B, which is what NXC usually uses to represent output port B.

```
//Plain English definitions.
#define LeftEye  S1
#define RightEye  S4
#define LeftMotor  OUT_B
#define RightMotor  OUT_C
```

Note that these variable definitions will occur outside of any task, and are usually the first bit of code in any program.

4.17.3 Miscellaneous Syntax

The listing below provides many examples of NXC syntax; the semicolons and the use of parentheses are critical, and more information about syntax can be found in the NXC documentation. BricxCC also provides utilities to help keep proper NXC syntax. The indenting and commenting of the code is helpful for understanding it, and is recommended practice, but is not required for the code to function properly.

```
//Main task. Turn on the eyes (but not their LEDs) and start the tasks.
task main(){
  SetSensorType(LeftEye, SENSOR_TYPE_LIGHT_INACTIVE);
  SetSensorType(RightEye, SENSOR_TYPE_LIGHT_INACTIVE);
  SetSensorMode(LeftEye, SENSOR_MODE_PERCENT);
  SetSensorMode(RightEye, SENSOR_MODE_PERCENT);
  start DriveLeft;
  start DriveRight;
} // end task
```

4.18 LINKING LIGHT SENSORS TO MOTORS
4.18.1 Two More Tasks

The main task described in Section 4.17.1 started two additional tasks, two variables DriveLeft and DriveRight. These two tasks are used to convert a light sensor reading into a motor speed; one task links one sensor-motor pair, the second sensor-motor pair is linked by the other. The listing of each of these tasks is provided below.

The two tasks are identical, with the exception that DriveLeft processes LeftEye and LeftMotor, while DriveRight processes RightEye and RightMotor.

When they are started by the main task, they will both be working at the same time. These tasks are written assuming ipsilateral connections between sensors and motors; however, if one changes the cabling to produce contralateral connections, *the code does not have to be altered* to keep the robot working.

To understand these two tasks, let us consider DriveLeft alone. The first line of this task states a while (true) loop. This construction initiates an infinite loop, so that the task repeatedly carries out any commands that are in between the { and } of the loop's syntax.

The operations in the infinite loop of DriveLeft work as follows: First, OnFwd is a command that turns a motor on, rotating in a forward direction. This command needs to specify which motor, and which speed. The motor that this command affects is the LeftMotor. The speed of this motor is going to be sent as a percentage (where 100% would be full speed, and 0% would be full stop). This percentage is determined by reading the light sensor. The Sensor(LeftEye)command reads the light sensed by the LeftEye as a percentage; it is this percentage that is used as the speed of the motor. The brighter the light detected by this sensor, the faster the motor; the motor will slow down as less light is detected by this sensor.

The DriveRight task proceeds in exactly the same manner, using the other input port to determine the percentage speed associated with the motor attached to the other output port.

```
//The next two tasks run in parallel and constantly feed the values of each
//eye into the respective motor as a speed.
task DriveLeft(){
   while(true){ // Run the LeftMotor at the LeftEye's speed
      OnFwd(LeftMotor, Sensor(LeftEye));
   } //  end while loop
} //  end task
task DriveRight(){
   while(true){          // Run the RightMotor at the RightEye's speed
      OnFwd(RightMotor, Sensor(RightEye));
   } //  end while loop
} // end task
```

4.19 A COMPLETE PROGRAM

The listing below is a complete example program for our Vehicle 2, and is available from the website that supports this book (http://www.bcp.psych. ualberta.ca/~mike/BricksToBrains/). The complete program is simply a concatenation of all of the components that we have been describing in the preceding sections. This program would exist on a computer as a text file, and could then be downloaded to the NXT brick by a utility such as BricxCC (available at http://bricxcc.sourceforge.net/). Once the utility compiles and downloads the program into the robot that we assembled, the sense–act cycles that define Vehicle 2 will come to life, and the robot should be capable of demonstrating the behaviours that Braitenberg imagined in his 1984 thought experiment.

```
//Lego NXT Braitenberg Vehicle 2 code - Brian Dupuis 2008
//Plain English definitions.
//Note that these are arbitrary; LeftMotor may
//actually be wired to the right input port.
#define LeftEye S1
#define RightEye S4
#define LeftMotor OUT_B
#define RightMotor OUT_C
//The next two tasks run in parallel and constantly feed the values of each
//eye into the respective motor as a speed.
task DriveLeft(){
   while(true){ // Run the LeftMotor at the LeftEye's speed
      OnFwd(LeftMotor, Sensor(LeftEye));
   } // end while loop
} // end task
task DriveRight(){
   while(true){          // Run the RightMotor at the RightEye's speed
      OnFwd(RightMotor, Sensor(RightEye));
   } // end while loop
} // end task
//Main task. Turn on the eyes (but not their LEDs) and start the tasks.
task main(){
  SetSensorType(LeftEye, SENSOR_TYPE_LIGHT_INACTIVE);
  SetSensorType(RightEye, SENSOR_TYPE_LIGHT_INACTIVE);
  SetSensorMode(LeftEye, SENSOR_MODE_PERCENT);
  SetSensorMode(RightEye, SENSOR_MODE_PERCENT);
  start DriveLeft;
  start DriveRight;
} // end task
```

4.20 EXPLORING VEHICLE 2 BEHAVIOUR
4.20.1 Three Test Environments

With Vehicle 2 constructed and programmed, all that remains to do is to explore the behaviour of this robot. Vehicle 2 is most apt for moving about a flat world with few obstacles, because it has no sensors to detect obstacles, or reflexes to avoid them. Vehicle 2's only sensors are light detectors; it is ideally suited for environments in which light sources are present.

4.20.2 A Simple World

From the preceding instructional pages, it is obvious that Vehicle 2 is a very simple robot. When it is placed in a simple environment, its behaviour is also quite simple. This is demonstrated in the early segments of Video4-1.mpg, available from the website for this book (http://www.bcp.psych.ualberta.ca/~mike/BricksToBrains/). When the robot's world consists of a single light, ipsilateral motor connections cause the light to be avoided, and contralateral connections cause the robot to move toward the light. A slight change of embodiment in this simple world produces an interesting array of behaviours. For instance, the light sensors can be angled so that their receptive fields overlap. With this embodiment — and contralateral connections — the robot avoids lights when they are far away, but attacks them when they are nearby. Overlapping receptive fields also result in Vehicle 2 spiralling toward a light over a period of time.

4.20.3 A More Complex World

The parable of the ant (Simon, 1969), and the law of uphill analysis and downhill synthesis (Braitenberg, 1984), claim that simple devices can generate complex behaviour. Their ability to do so is contingent upon being situated in a world, and also depends upon the complexity of that world. One can increase the complexity of a robot's behaviour by increasing the complexity of its environment, without manipulating the robot at all.

The final segments of Video4-1.mpg illustrate this principle. The video illustrates a number of ways in which the environment was made more complicated, producing complex behaviours that did not require a different program to be created for the robot. Vehicle 2 demonstrates colour preferences and the ability to follow another machine — provided that the second machine has a light source mounted on it. In a dark room, the robots move slowly, and come to a stop underneath a hanging light source. They are "awakened" when the room is illuminated, and actively explore their world. If overhanging lights begin to swing, then the robots move away. Multiple robots appear to compete for resources. Imagine having to explain all of these behaviours analytically, without having direct knowledge of the robots' embodiment or programming. Would the theory that resulted be as simple as the synthetic theory represented by the instructions on the previous pages?

4.20.4 Complexities via Embodiment

A number of computer simulations of Braitenberg vehicles are available on the internet, such as Thornton's POPBUGS package, available at http://www.cogs.susx.ac.uk/users/christ/popbugs/braitenbergs.html. Why, then, would we go to the trouble of embodying Vehicle 2 as a LEGO artifact? The physical structure of the robot itself is another source of complexity. Computer simulations of Braitenberg vehicles are idealizations in which all motors and sensors work perfectly. This is impossible in a physically realized robot. Slight manufacturing differences will mean that one motor may not be as powerful as another, or that one sensor may be less sensitive than another. Such differences will affect robot behaviour. These imperfections are another important source of behavioural complexity, but are absent when such vehicles are created in simulated and idealized worlds.

4.21 FURTHER AVENUES FOR *BRICOLEURS*
4.21.1 Exploring Embodiment

The preceding instructions define one possible Vehicle 2. Of course, many alternative versions of this robot can be explored. For instance, alternative Vehicle 2 robots can be created by exploring alternative robot embodiments. One could start with minor changes of the existing robot: how does it behave when light sensors are slid to different positions? What happens when the light sensors are angled in different directions?

More elaborate *bricolage* involves redesigning some of the robot's structure. What occurs when the front slider is replaced with a balanced wheel, or a wheel that isn't completely balanced, or with a wheel that is not able to rotate a full 360°? What is the result of using different sensors, such as temperature sensors?

4.21.2 Manipulating Environments

One of the lessons of Vehicle 2 is that changing the robot is but one avenue to changing its behaviour. One can also manipulate behaviour by modifying the environment, while leaving the robot alone. The robot's environment can be explored by changing the number and location of lights, or by adding mirrors. The type of light can also be manipulated: the light sensors used in our version of Vehicle 2 are highly sensitive to infrared light, and therefore can process signals from remote controls used for televisions! Recognizing that an embodied robot is part of its world, light sources could be attached to the robot chassis, and more than one robot run at the same time.

4.21.3 Modifying Code

Yet another avenue for robot development would be to modify the robot's program. The program reads the light sensors in PERCENT mode. What might be the effect of reading the sensors in RAW mode, and then doing some sort of processing of these readings that does not involve percentages? For example, one could compare raw sensor readings to each other, or to some standard value, or to an average light reading that is updated by the robot. As well, the robot's light sensors could be initialized to be active; they would then become additional sources of light that could affect behaviour, particularly if the environment contained surfaces that reflected the LED emissions.

4.21.4 *Bricolage,* Not Design

All of the avenues mentioned above consider robot exploration from a synthetic perspective. That is, robot *bricoleurs* take some available components of interest, use them to modify or elaborate a machine, and then observe the result in order to understand what the robot can or cannot do within an environment that is also being manipulated. The robot is created first, and produces data of interest. This is in contrast to the more analytic approach, where a theory (e.g., a robot) is constructed on the basis of existing data (e.g., the robot path introduced at the start of the chapter in Figure 4-0).

The synthetic approach could be criticized in the sense that it is not goal directed. However, the success of the synthetic approach is derived from the components that are explored. As the minimalist composers found with their music, if the effort is made to begin with an interesting set of component mechanisms, then the resulting product should provide interesting or surprising results. The next few chapters will attempt to illustrate this by providing accounts of a variety of different robots constructed from LEGO components.

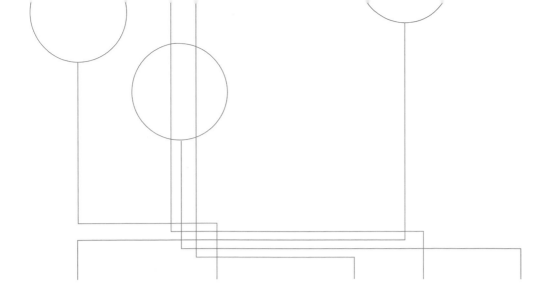

Chapter 5
Thoughtless Walkers

5.0 CHAPTER OVERVIEW

The preceding chapters have contrasted classical cognitive science with embodied cognitive science, and have also contrasted the analytic practices of the former with the synthetic methodologies of the latter. The discussion to this point might be interpreted as an argument to abandon classical cognitive science, and to replace analytic methods with synthetic ones. The point of this chapter is to prevent the reader from coming to this conclusion. This chapter considers a very general phenomenon, walking, from both analytic and synthetic perspectives. It attempts to demonstrate that analytic and synthetic methods complement one another. Each approach has its own strengths and weaknesses, and when used in combination they provide a powerful arsenal for conducting cognitive science. This theme is explored in detail by introducing two simple walking robots that do not require programming, because neither uses the NXT brick to function. One is a LEGO version of a passive dynamic walker. The other is a LEGO version of a walking sculpture created by Theo Jansen (2007). The relationship between analytic and synthetic approaches is also explored using a third LEGO robot, which moves by using worm-like movements. This robot requires a small amount of programming to permit motors to sense rotation in their axles and to react against it. This simple exploitation of feedback permits an interesting robot to be constructed and explored — without requiring multitudes of mathematical analyses of snake-like movement that are often used to inspire such robots!

5.1 ANALYSIS VS. SYNTHESIS

5.1.1 Synthetic Methodology

Previous chapters have introduced the methods of synthetic psychology (Braitenberg, 1984; Dawson, 2004; Pfeifer & Scheier, 1999). Pfeifer and Scheier (1999, p. 22) aptly describe the synthetic approach as "understanding by building." That is, the synthetic approach begins with the construction of a system from a set of interesting components. The behaviour of the system (in an environment of interest) is then observed.

Typically there exist non-linear relationships between system components, and between the system and its environment. Because of this, synthetic researchers can take advantage of *emergence* (Holland, 1998; Johnson, 2001; Sawyer, 2002). That is, the non-linear relationships governing the synthesized system are likely to produce more complex behaviour than one might expect by considering the properties of the system's components in their own right. Emergent behaviours are therefore surprises; the power of the synthetic approach is that simpler theories (i.e., a description of some components and how they are organized) can be provided to explain behavioural complexities (i.e. the surprising behaviours that emerge when the system is observed).

One of the properties of the synthetic approach is that a model is created *prior* to collecting data (Dawson, 2004; Dawson & Zimmerman, 2003). This is because the data of interest *is the system's behaviour.* In other words, the synthetic approach examines what kinds of behaviours can (and cannot) emerge from a synthesized system; it does not primarily aim to reproduce the results of a previous experiment (Pfeifer & Scheier, 1999).

5.1.2 Analytic Methodology

The synthetic approach is usually (Braitenberg, 1984; Dawson, 2004; Pfeifer & Scheier, 1999) described by contrasting it with analytic methodologies, which are "universally applied in all empirical sciences" (Pfeifer & Scheier, 1999, p. 21). In the less traditional synthetic approach, models precede data. In the more common analytic approach, data precede models. If the synthetic approach is "understanding by building," then the analytic approach is "understanding by taking apart."

The analytic approach begins with a researcher being confronted with an intact, behaving system. The system is a black box—because it was not constructed by the researcher, its internal mechanisms are unknown. The researcher collects data from which the system's internal mechanisms are to be inferred. Frequently, the analytic approach produces models of internal mechanisms. That is, a researcher makes a hypothesis about internal mechanisms, converts this hypothesis into a model, and then tests the adequacy of the hypothesized model. Such models can be in a variety of forms — models of data, mathematical models, or computer simulations (Dawson, 2004).

Regardless of the type of model, analytic researchers evaluate them in terms of their capability to generate data similar to that obtained from observing the intact system. That is, analytic researchers evaluate models by determining their fit to data that has already been collected (Dawson, 2004; Pfeifer & Scheier, 1999).

5.1.3 Complementary Methodologies

It is not uncommon to see analytic and synthetic methodologies contrasted with one another, and this can lead to a sense that these two approaches are competitors. For instance, the law of uphill analysis and downhill synthesis (Braitenberg, 1984) is a statement of why the synthetic approach is to be preferred over the analytic one. In the preceding chapters, the analytic approach seems to be exclusively associated with classical cognitive science, while the synthetic approach has been tied fundamentally to embodied cognitive science. However, it is important to recognize that these two approaches are complementary (Pfeifer & Scheier, 1999). The complementary relationship between analytic and synthetic methodologies will be explored in more detail in the current chapter, using a particular topic: the biomimetic study of walking.

5.2 BIOMIMETICS AND ANALYSIS
5.2.1 Natural Technology

One modern approach to design, *biomimetics*, involves studying Nature's solutions to problems with the aim of incorporating similar solutions into human artifacts (Bar-Cohen, 2006). "Nature's capabilities are far superior in many areas to human capabilities, and adapting many of its features and characteristics can significantly improve our technology" (p. P1). Example adaptations include making self-cleaning windows by exploiting the rough microstructure of the leaf of the lotus, developing adhesives inspired by the bristles of the gecko's foot, and inventing deployable structures by mimicking the folding of leaves (Forbes, 2006).

One area that has received particular attention in biomimetics is locomotion. In particular, researchers believe that robots that use legs will be able to move about a much greater variety of environments than wheeled robots. However, it is much more difficult to build robots that walk with legs.

For instance, Honda began work on a humanoid bipedal robot in 1986; it took until 1993 to build a set of legs that could stably carry a torso;

public unveiling of a walking robot, Asimo, did not occur until 2003 (Hirose & Ogawa, 2007). Another bipedal robot, H7, also took years to develop, and requires complex software to plan and execute movements using motors that control the angles of 30 different joints (Nishiwaki, Kuffner, Kagami, Inaba, & Inoue, 2007). It has been noted that robots like Asimo and H7 "imitate a human walk quite well, but require complex, fast, precise control mechanisms, and use far more energy than a walking human would" (Alexander, 2005, p. 58).

One hope is that biomimetics can provide insights into walking that in turn will result in simpler, more efficient, walking robots. Usually, this first involves the careful analysis of movements of biological systems that already possess legged locomotion.

5.2.2 Early Analysis of Locomotion

The analysis of movement has a long history (Andriacchi & Alexander, 2000), and has depended critically upon available technology. The earliest photographic recordings of moving animals were obtained by Eadweard Muybridge (Clegg, 2007; Solnit, 2003). In a long shed, Muybridge placed twelve stereoscopic cameras at intervals of 21 inches. He invented a shutter that, when triggered, moved a slit quickly across a camera's lens, providing an exposure of 1/1000th of a second.

Muybridge used electromagnets to trigger the camera shutters. Wires were laid from the electromagnets across a racetrack in front of the shed. When a moving horse pulled a sulky along the track, the wheels of the sulky completed a circuit, triggering the shutter mechanism. The twelve cameras were triggered in rapid sequence by the moving horse, producing a record of its movement in a series of static images. Muybridge used such photographs to provide evidence of many different horse gaits (see Section 5.21.1), and to show that in many of these gaits there were moments during which none of the horse's hooves were in contact with the ground (Muybridge, 1887/1957).

Later, Muybridge developed new clockwork shutter mechanisms that permitted him to photograph the motion of many different organisms, and permitted photographs from multiple perspectives. Muybridge's photographs revealed that many artistic renderings of horse movements are incorrect (Muybridge, 1887/1957). "Clearly, the eye was not capable of capturing the sequence of rapid limb movements of horses in motion" (Andriacchi & Alexander, 2000, p. 1217).

The introduction to the 1957 edition of *Animals in Motion* notes that Muybridge's photographs "are still the basic authorities on the movements and gaits natural to most animals, particularly to man and the horse. Despite the moving-picture and slow-motion cameras we now possess, little has been learned that Muybridge did not discover" (p. 9). Indeed, modern programs developed to analyze movement have been tested by examining their ability to reconstruct movements represented in Muybridge's photographs (Bregler, Malik, & Pullen, 2004).

5.3 FROM MOTION ANALYSIS TO WALKING ROBOTS
5.3.1 Modern Motion Analysis

Much modern research on locomotion follows the spirit of Muybridge's photographic efforts. The movement of animals is recorded using technology that is far more sensitive than the naked eye. However, modern technology incorporates records that are not purely photographic.

For instance, in addition to the recording of the movement of the cockroach *Periplaneta americana* with high-speed video (Full & Tu, 1991), its locomotion can also be studied using a force platform (Bartsch, Federle, Full, & Kenny, 2007; Full & Tu, 1991). Such a platform is capable of measuring horizontal, vertical, and lateral forces exerted by an insect's legs as it moves over a surface. This permits detailed analyses of the mechanical power being generated by the cockroach, as well as of the mechanical energy required to move the insect's centre of mass over particular distances.

Why are biomechanical measurements critical to modern analyses of animal locomotion? One reason is that biomimetics might not best proceed by merely attempting to copy an existing biological system (Forbes, 2006). Instead, it might be more likely to succeed by analyzing existing systems to discover general principles, and by using these general principles to inspire new technologies. Biomechanical measurements provide one source of information that can reveal such principles for locomotion.

For example, many insects move with their legs in a sprawled posture. These legs generate substantial lateral forces that are not in the direction of motion (Dickinson et al., 2000). Analyses of these forces suggest that elastic energy storage and recovery may occur within the horizontal plane. "By pushing laterally, legs create a more robust gate that can be passively self-stabilizing as the animal changes speed, moves over uneven ground, or is knocked askew by uneven terrain, a gust of wind, or a would-be predator" (p. 101). Similar analyses have revealed that the structural properties of insect muscles themselves permit them to stabilize a moving agent (Nishikawa et al., 2007). As well, the mechanical properties of isolated cockroach legs suggest that they might serve as springs capable of storing and releasing energy (Dudek & Full, 2006).

Another reason for biomechanical measurements is that these can be used to create elegant mathematical models that describe the fundamental characteristics of animal locomotion. A walking organism can be described as an inverted pendulum in which a rigid leg is analogous to a pendulum's cable attached to a mass (Dickinson et al., 2000). During slow walking, the mass vaults over the leg, which involves first converting kinetic energy into potential energy, and then recovering the potential energy as kinetic energy. Alternatively, when an organism runs, the rigid leg of the inverted pendulum is better viewed as a spring. Kinetic and potential energies are stored as elastic energy, and the system bounces as if it were on a Pogo stick. These two metaphors can be applied to mathematically describe the locomotion of a wide variety of bipedal, quadrupedal, and polypedal organisms (Blickhan & Full, 1993).

5.3.2 Biologically Inspired Robots

Unlike Muybridge's pioneering work, modern analyses have also been used to design walking, legged robots. One example is the hexapod robot Rhex (Altendorfer, Moore et al., 2001; Altendorfer, Saranli et al., 2001; Koditschek, Full, & Buehler, 2004). Rhex is designed to reflect the pendulum and spring models of walking or running agents (Blickhan & Full, 1993), and is in essence an idealized cockroach. It can move over badly broken terrain, and is five times faster than previous legged robots. More recent work involves developing legged robots that can also use their legs to climb vertical surfaces (Spenko et al., 2008). It has been observed (Delcomyn, 2004) that "there is little doubt that incorporating elements of biological systems into the design and control of a legged robot can confer on that robot a more sophisticated level of performance than has so far been possible without such elements" (p. 61).

5.4 ANALYSIS THAT CONSTRAINS SYNTHESIS
5.4.1 Passive Dynamic Walking

"The obvious way to make a humanlike robot walk is to provide it with motors to drive every joint, and a computer to control them" (Alexander, 2005, p. 58). Examples of this approach are bipedal robot successes like Asimo (Hirose & Ogawa, 2007) and H7 (Nishiwaki et al., 2007). Such robots are *active dynamic walkers* because their gaits are actively controlled by computers (McGeer, 1990a). However, such walking requires complex control mechanisms, and is very energy-inefficient (Alexander, 2005). Mathematical analyses of locomotion have pointed the way to an alternative form of walking that addresses these two problems.

The movement of a variety of animals can be mathematically

summarized as a single rigid leg that is used to vault or spring a centre of mass forward (Blickhan & Full, 1993). This model recognizes that legs are not analogous to wheels, and that wheel-like locomotion is rarely seen in the animal kingdom (Full, Earls, Wong, & Caldwell, 1993).

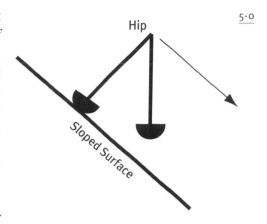

Nonetheless, some researchers have explored models of walking by translating wheels into walkers (McGeer, 1990a). McGeer imagined splitting the rim of a wagon wheel halfway between each spoke, and then removing all of the spokes but two (Figure 5-0). "Could the dynamics be such that while one leg is rolling along the ground, the other swings forward in just the right way to pick up the motion where the first leg leaves off?" (McGeer, 1990a, p. 66).

McGeer (1990a) proceeded to provide an affirmative answer to this question, and to build a working model that walked down a ramp. This model is a *passive dynamic walker* because active control is not required to generate its gait. "Gravity and inertia alone generate the locomotion pattern" (p. 63). Passive dynamic walkers are of interest they "show us that bipedal robots far simpler than their predecessors work as effectively and far more economically" (Alexander, 2005, p. 59). They may also be easily modified to add motors that eliminate the need for ramps (McGeer, 1990a; Ohta, Yamakita, & Furuta, 2001). A number of new and more advanced passive dynamic walkers have also been constructed (Collins, Ruina, Tedrake, & Wisse, 2005). These include walkers with flat feet (Wu & Sabet, 2004), with knees (Collins, Wisse, & Ruina, 2001; McGeer, 1990b), with torsos (Wisse, Hobbelen, & Schwab, 2007; Wisse, Schwab, & van der Helm, 2004), and with the ability move down stairs (Safa, Saadat, & Naraghi, 2007).

5.4.2 Search and Construct

Mathematical analyses demonstrated the possibility of passive dynamic walking. However, the complexity of these equations often leads researchers to avoid them when faced with the task of actually constructing a passive dynamic walker. Instead, researchers proceed by tinkering with plausible physical components. For example, Collins et al. (2001, p. 612) "decided to forgo three-dimensional analytic modeling" and instead worked directly with physical components, using "trial, error, and correction to minimize three-dimensional effects" (p. 612). This illustrates one way in which analysis can inform synthesis. The mathematical analysis of passive dynamic walking revealed the possibility that such a device could work, and provided some general guidelines about the device's nature. These guidelines help limit or constrain a developer's search for a working physical configuration. That is, some physical arrangements of parts will not be explored because they are completely at odds with the mathematics. As a result, the process of synthesizing a working device can be guided by prior analysis.

5.5 A LEGO PASSIVE DYNAMIC WALKER
5.5.1 Synthesis after Analysis

5-1

McGeer's (1990a) mathematical analyses indicated that one should be able to build a working, straight-legged, passive dynamic walker. He proceeded to convert "in principle" into "in practice" by constructing a simple demonstration machine. Given its simplicity, can McGeer's straight-legged walker be constructed from LEGO parts? The next few pages provide instructions for building the passive dynamic walking system illustrated in Figure 5-1. If the reader would prefer to use wordless, LEGO-style instructions, they are available as a pdf file from the website that supports this book (http://www.bcp.psych.ualberta. ca/~mike/BricksToBrains/).

5.5.2 Parts and Foraging

The passive dynamic walker is comprised of two modules: the walker itself and the ramp it walks on. You will need to assemble both components to observe its behaviour. The parts used to construct the system in Figure 5-1 are illustrated in Figure 5-2. Treat the Figure 5-2 parts list as a *suggestion*. The exact bricks used, particularly on the ramp, matter very little to the overall behaviour of the system. Only the final proportions truly matter.

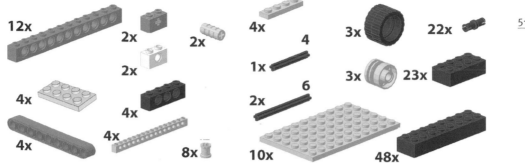

5.6 BUILDING A STRAIGHT-LEGGED HINGE

5.6.1 Centre Post

The passive dynamic walker is essentially a hinge with weights, and is simple to construct. Since the weight distribution on the walker matters, take care to assemble the robot symmetrically.

The hinge consists of a centre post and an outer support. Begin by constructing the centre post, as shown in Step 1 of Figure 5-3. Next, run a length-4 axle through the holes in the two connected 1 × 2 beams, attaching an axle joiner to each side to hold the axle in place (but permitting it to rotate in the hole). A length-6 axle holding three bushes is attached to each axle joiner (Step 2 in Figure 5-3), providing the basis of the walker's hinge joint.

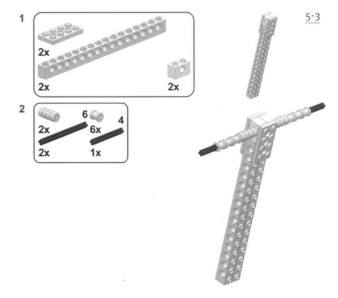

5.6.2 Support Legs

The second part of the walking "hinge" is a pair of outer legs that swing together, and permit the walker to stand upright as a tripod. A support leg (Step 3, Figure 5-4) is constructed as follows: Lay a 1 × 16 Technic beam with holes, and a 1 × 2 Technic beam with an axle hole, end to end. Use two 1 × 4 plates to join these two beams together. Two of these support legs must be constructed. Note that each support leg is half as wide as the centre post, because the centre post is constructed from pairs of beams, while only single beams are used in the support legs.

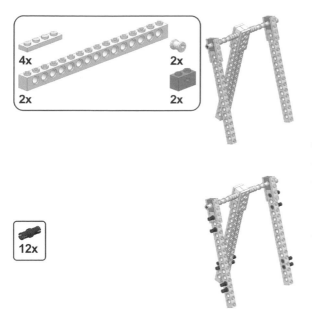

The support legs can now be attached to the centre post by inserting a leg on each free axle end of the centre post. A bush is then used to hold a support leg in place. Black Technic pins (i.e., those without friction) can now be added to all three legs, as shown in Step 4 of Figure 5-4. These pins will be used to connect additional beams that will serve as weights.

5.7 WEIGHTING THE WALKER

5.7.1 The Need for Weights

McGeer's (1990a) formal analyses of passive dynamic walking indicated that one crucial element was the weight distribution along the robot's legs. In order for gravity to pull the walker down a slope with legs swinging, weights are required at both the top and the bottom of the hinged structure that was built in Section 5.6

5.7.2 LEGO Weights

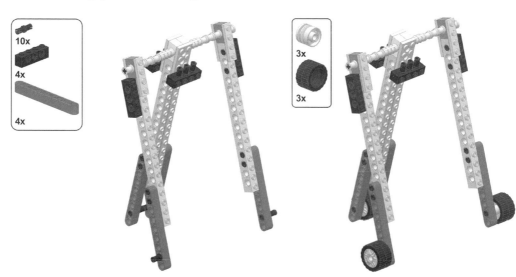

Weights are added to the appropriate locations of the LEGO passive dynamic walker by attaching beams and wheels to serve as heavy feet. The first step is to attach beams and liftarms to all three legs using the pins that were added in Figure 5-4. The locations of these new pieces are illustrated in Figure 5-5. Note that several new black Technic pins are then inserted into these components. Be certain to attach extra pins to the 1×4 beams on the centre leg and to the liftarms; these serve both as extra weight adjustment and as mounting points for the feet.

Finally, mount three 30.4×14 wheels on the pins in the liftarms, as shown in Figure 5-6. These wheels will not spin—instead, they serve as rounded, heavy, high-traction feet. This completes the passive dynamic walker.

5.8 A SPECIALIZED ENVIRONMENT

5.8.1 The Need for Holes

The completed walker (Figure 5-9) will not walk down an ordinary inclined ramp. This is because the walker has no knees, and thus no general capability of swinging its feet free of a ramp's surface. In order for the system to walk down a ramp, gaps in the ramp's surface must be strategically placed to permit the legs to swing freely.

5.8.2 Building a Ramp with Gaps

Begin by assembling four struts, of final dimensions 2×52, two bricks thick. Be sure to prevent the brick seams from overlapping the two layers in each strut. These struts are separated by enough distance to fit a length-6 beam between them, forming three "channels," as shown in Figure 5-7.

Ten 6×8 plates will be attached to these struts in a staggered formation. Attach the first plate to the two struts in the centre of Figure 5-8 (the leftmost plate in the figure). Only use two pips of the plate to attach it to the struts. This will leave four pips on the plate free; other parts will be attached to these free pips later. The next two plates will be beside each

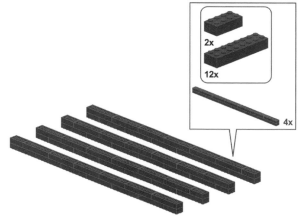

5-7

other, but shifted along the struts down from the first plate that was connected. One of these next plates is connected to the two struts on the left of Figure 5-8; the other is connected to the two struts on the right. They are shifted down the struts so that there are three empty pips (i.e., pips to which no plates are attached) on the centre struts between the first plate and the next two. Next, a single plate is attached across the centre channel, again shifted along so that there are three empty pips between it and the two plates that were previously added to the left of it in Figure 5-8.

5-8

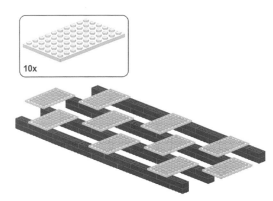

The steps described above (add one plate to the centre channel, then advance and add two plates to the side channels) are repeated until all ten plates are attached, each time advancing a plate's position to leave three blank pips between it and the preceding plates. The final plate is only attached by one pip for now (see the right end of Figure 5-14, which shows the ramp structure viewed from the top). The result is a "checkerboard" pattern of plates attached to the struts. The holes or gaps in the "checkerboard" will permit the walker's straight legs to swing when it moves down this ramp.

5.9 RAISING THE RAMP

5.9.1 Reinforced Ends

In order to use the platform constructed in Section 5.8 as a ramp, its ends must be reinforced, and a support must be constructed to elevate one end of the ramp.

The ends of the platform are reinforced with LEGO beams. Build an extension to one end of the ramp as illustrated in Step 3 of Figure 5-9.

5-9

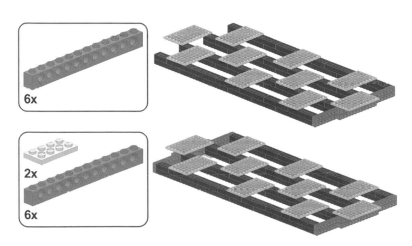

Mirror this assembly on the other end of the platform (Step 4 in Figure 5-9). Then use two 2 × 4 plates to reinforce the mirrored end piece by attaching it to the two centre struts side, supporting it with plates. This end needs this reinforcement because of the longer extension of the last plate beyond the struts.

5.9.2 Elevating the Platform

The final step in building the platform is to create a support that will raise one of its ends to convert it into an inclined ramp. Assemble a support "arch" six bricks high. The exact formation of this arch does not matter; its height is what is important. Attach this arch to the extended 6 × 8 plate on the ramp, as shown in Figure 5-10. Note that the upside-down plates provide a relatively smooth surface for the walker to step onto, and again the gaps between plates provide space for the walker's feet to swing mid-step.

Finally, add bricks to the opposite end of the ramp, as is also shown in Figure 5-10. It is important to use LEGO bricks (and not some smaller or larger LEGO pieces) and to mount them exactly as shown. This is because these bricks serve as the "feet" of the ramp, and their placement will affect the platform's angle of elevation.

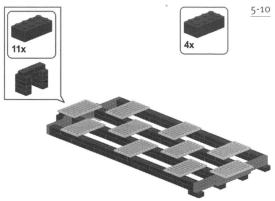

5-10

Adjusting the height of the elevating arch (by adding or remov- ing bricks, or substituting plates in place of bricks) can produce radically different behaviour in the walker; experimentation is encouraged, as the 6-brick height here was arrived at by trial-and-error and is in all likelihood not the optimal height.

5.10 FROM TALKING THE TALK TO WALKING THE WALK
5.10.1 Passive Dynamic Walking

Mathematical analyses revealed the in-principle possibility of passive dynamic walking (McGeer, 1990a). We have already seen that these analyses led to practical successes with the development of a number of different passive dynamic walkers. One question remains: is our LEGO system capable of joining this club?

If the LEGO passive dynamic walker is successful, then it should be capable of taking a succession of steps and walk from the top of the ramp to its end. One way to test this with the LEGO walker is to place its centre support leg on the top centre plate, gently swing its outer legs downward, and release the walker. Examples of this system working as a passive dynamic walker can be seen in Video5-1.mpg, available from the website for this book (http://www.bcp.psych.ualberta.ca/~mike/BricksToBrains/). A careful viewing of this video will reveal other approaches to starting the walker on its downhill journey.

5.10.2 Implications

One implication of the LEGO passive dynamic walker is that it provides an example of prior analysis informing later synthesis. That is, McGeer's (1990a) mathematical analysis of passive dynamic walking in general, and of such walking in two-dimensional straight-legged systems in particular, guided the construction of our LEGO device. For example, the prior analysis helped constrain design decisions about the length of the legs, the position and amount of weights on the walker, the size and slope of the ramp, and so on.

A second implication of this system is that it illustrates the need to consider an agent in the context of its environment. In earlier chapters, we saw that classical cognitive science adopted the idea of a disembodied mind that had descended from Descartes' claim *cogito ergo sum*. Embodied cognitive scientists have been strongly influenced by reactions against Descartes, such as Martin Heidegger's *Being and Time* (Heidegger, 1962). Heidegger was extremely critical of Cartesianism, noting that Descartes had adopted many of the terms of older philosophies, but had failed to recognize a critical element: "The ancient way of interpreting the Being of entities is oriented towards the 'world' or 'Nature' in the widest sense" (p. 47). Heidegger attempted to correct this flaw by arguing that a primary mode of existence was *Being-in-the-world*. Being-in-the-world was not just being spatially located in an environment, but instead was a mode of existence in which an agent was *engaged* with entities in the world.

Being-in-the-world is related to the concept of affordances developed by psychologist James J. Gibson (Gibson, 1979). "The *affordances* of the environment are what it *offers* the animal, what it *provides* or *furnishes*, either for good or ill" (p. 127). Gibson stressed that what the environment afforded an agent depended upon the agent's abilities or dispositions. "Note that the four properties listed—horizontal, flat, extended, and rigid—would be *physical* properties of a surface if they were measured with the scales and standard units used in physics. As an affordance of support for a species of animal, however, they have to be measured *relative to the animal*. They are unique for that animal. They are not just abstract physical properties" (p. 127). For this reason, Gibson argued that agents and environments were inseparable. "It is often neglected that the words *animal* and *environment* make an inseparable pair" (p. 8). The LEGO passive dynamic walker illustrates this inseparability of agent and environment. The straight-legged hinge that was constructed in Section 5.7 has the disposition to walk, but requires a specialized environment—a particular set of affordances—to have this disposition realized. These affordances are provided by the slope, surfaces, and gaps of the ramp constructed in Section 5.9. The LEGO passive dynamic walker will only walk when it encounters the affordances of the ramp. Clearly, the walking is not in the hinge, or in the ramp, but in the interaction between the two. Passive dynamic walking is not a characteristic of a device, but is instead a characteristic of a device being in a particular world.

5.11 SYNTHESIS IN AID OF ANALYSIS

5.11.1 The Opposite Direction

Up to this point, we have discussed how prior analysis can guide later synthesis. However, the complementary nature of these two approaches can also work in the opposite direction. At this point in Chapter 5 we will change direction and explore how prior synthesis can support later analysis.

5.11.2 Analytic Intractability

It has long been believed that advances in the "soft sciences" require the development of specialized formalisms. Such formalisms emerged with formal information theory (Khinchin, 1957; Shannon, 1948; Wiener, 1948). "There now exists a well developed logic of pure mechanism, rigorous as geometry, and likely to play the same fundamental part, in our understanding of the complex systems of biology, that geometry does in astronomy" (Ashby, 1960, p. v).

These techniques have evolved into the diverse formal methods now employed at the so-called computational level of analysis (Dawson, 1998; Marr, 1982; Pylyshyn, 1984). Researchers can prove what types of information processing problems a system can—and cannot—solve. "The power of this type of analysis resides in the fact that the discovery of valid, sufficiently universal constraints leads to conclusions ... that have the same permanence as conclusions in other branches of science" (Marr, 1982, p. 331).

Cognitive formalisms are frequently non-linear. In a non-linear system, the whole is not merely the sum of its parts (Luce, 1999). Instead, there are complex interactions between parts; the behaviour that emerges from a non-linear system is more complicated than one would predict from knowing the properties of its constituents.

There are many different sources of non-linearity in cognitive theories. These include explicitly non-linear equations (Dawson, 2004), iteration (Rescorla & Wagner, 1972; Stone, 1986), recursion (Hofstadter, 1979; Pylyshyn, 1984), or feedback (Ashby, 1956, 1960; Bateson, 1972; Wiener, 1948). The advantage of incorporating such non-linearity, as already noted, is that it permits more complex behaviour to emerge from the interactions of components (Holland, 1998; Johnson, 2001; Sawyer, 2002).

The problem with non-linearity, though, is that it makes system analysis extraordinarily complex, sometimes to the point of making it impossible. Consider, for example, the study of feedback to machines (Ashby, 1956, 1960). For Ashby, a machine was simply a device whose output can be completely predicted from knowing a) its current internal state, and b) its current input. With this general definition, machine inputs and outputs can be defined as being numerical, and so in principle machine behaviour can be explored analytically using mathematics. Ashby was interested in using equations to analyze the behaviour of systems that were constructed by having several different machines interacting with one another in feedback relationships.

However, Ashby (1956, p. 54) realized that even a four-component feedback system could not be completely understood analytically. "When there are only two parts joined so that each affects the other, the properties of the feedback give important and useful information about the properties of the whole. But when the parts rise to even as few as four, if every one affects the other three, then twenty circuits can be traced through them; and knowing the properties of all the twenty circuits does *not* give complete information about the system."

How did Ashby solve this problem? Ashby replaced the analytic analysis of his system with a synthetic one. He built a system that was comprised of four identical machines, incorporated mutual feedback, and observed the resulting behaviour. "A better demonstration can be given by a machine, built so that we know its nature exactly and on which we can observe what will happen in various conditions" (Ashby, 1960, p. 99). The next section briefly describes Ashby's synthetic approach.

5.12 ASHBY'S HOMEOSTAT
5.12.1 Homeostat Design

The machine that Ashby (1956, 1960) synthesized to study feedback was called the *Homeostat*. The Homeostat was a system of four identical component machines; their inputs and outputs were electrical currents. The purpose of each component was to transform the input current into the output current. This was accomplished by using the input current to change the position of a pivoted magnet (attached to an observable needle) mounted on the top of the component. All things being equal, a large input current would cause a large deflection of the magnet (and needle), which in turn would result in a proportionately large current being output from the component.

The electrical current that was input to one unit was the sum of the electrical currents that were output by each of the other three units, after each of these three currents was weighted using a potentiometer. The result was a dynamic system that was subject to feedback. "As soon as the system is switched on, the magnets are moved by the currents from the other units, but these movements change the currents, which modify the movements, and so on" (Ashby, 1960, p. 102).

The Homeostat automatically moderated the currents produced by its components. Each unit was equipped with a 25-valued uniselector or stepping switch, where each value was a randomly assigned resistance that mediated current. If the output current was below a predetermined threshold level, the uniselector did not activate, and resistance was unchanged. However, if the output current exceeded the threshold, the uniselector activated, advanced, and changed the resistance value.

In general, then, the Homeostat was a device that monitored its own internal stability (i.e., the amount of current being generated by each of its four component machines). If subjected to external forces, such as an experimenter moving one of its four needles by hand, then this internal stability was disrupted and the Homeostat was moved into a higher energy state. When this happened, the Homeostat would modify the internal connections between its component units by advancing one or more of its uniselectors, returning it to a lower energy state.

5.12.2 Behaviour of the Homeostat

Ashby (1960) tested the Homeostat by placing some of its components under his direct control, manipulating these components, and observing the changes in the system as a whole. Many surprising emergent behaviours were observed. Ashby found that the system was capable of learning when the needle of one component was used to "punish" the Homeostat for an incorrect response (i.e., for moving one of its needles in the incorrect direction). Over time, the Homeostat adapted so that it moved the goal needle in the desired direction. Ashby also found that the Homeostat could adapt to two different environments that were alternated from trial to trial.

The Homeostat is one of the earliest examples of the synthetic methodology. It permitted dynamic feedback to be studied at a time when mathematical accounts were intractable, and also at a time when computer simulations of feedback were not feasible. The large number of different internal states that were available to a working Homeostat provided the machine with many degrees of freedom with which to produce a low energy state. These same degrees of freedom made it difficult for the analytic approach to be applied to it. "Although the machine is man-made, the experimenter cannot tell at any moment exactly what the machine's circuit is without 'killing' it and dissecting out the 'nervous system'" (Grey Walter, 1963, p. 124). In other words, it was much easier to produce interesting behaviour in the Homeostat than it was to analytically explain it. However, the fact that the Homeostat produced interesting behaviour indicated that future analytic accounts of it were, in principle, *possible*.

5.13 THE GREAT PRETENDER
5.13.1 Synthesis and Scaling Up

Because of the inherent complexity, Ashby (1960) did not analyze systems whose few components were involved in non-linear interactions. Instead, he used a synthetic approach — the construction of the Homeostat to observe what would be produced by four interacting sub-units — to reveal that such a system could adapt to a variety of environments. Via synthesis, Ashby observed what the Homeostat could and could not do, and argued that the principles that governed the Homeostat also governed adaptation in living organisms.

However, the link between the Homeostat and biological agents was strained by a difference in scale: "The Homeostat is, of course, grossly different from the brain in many respects, one of the most obvious being that while the brain has a very great number of component parts the Homeostat has, effectively, only four" (Ashby, 1960, p. 148). Ashby went on to speculate on what would be required for the efficient functioning of a Homeostat built from a large number of component units. He realized that for it to work efficiently, functional connections between components would have to be limited, and independently discovered stigmergy (see our earlier discussion beginning with Section 1.10.2): "Coordination between parts can take place through the environment; communication within the nervous system is not always necessary" (p. 222).

Ashby did not go on to actually build a larger-scale Homeostat. However, there are many examples available in which the properties of scaled-up systems are explored synthetically. Let us consider one such example related to a topic central to the current chapter: walking. In particular, let us explore the many-legged walking machines, Strandbeests, created by sculptor Theo Jansen (Jansen, 2007).

5.13.2 Strandbeest

Theo Jansen is a sculptor famous for his large installations that resemble animal skeletons. These walking skeletal systems are called Strandbeests, and are intended to be a new species of organism that live on, and adapt to, conditions of a seaside beach. Jansen's artistic aim is given on his website (www.strandbeest.com): "I make skeletons that are able to walk on the wind, so they don't have to eat. Over time, these skeletons have become increasingly better at surviving the elements such as storms and water and eventually I want to put these animals out in herds on the beaches, so they will live their own lives."

The Strandbeests reflect Jansen's long-time interest in evolutionary theory and genetic algorithms (Jansen, 2007). Prior to 1990, he used genetic algorithms to program creatures that existed as lines on a computer's screen, and whose likelihood of reproducing depended upon how quickly they moved. He then transferred this interest to dynamic artifacts, building moving machines from plastic tubing. These machines have become wind-powered devices that walk along beaches, capable of anchoring themselves against storms, storing air, gathering sand, avoiding water, and using pneumatic nerve cells. Jansen continues to use genetic algorithms by racing Strandbeests against one another, and using parts of the losers to make (possibly imperfect) copies of the winners. He is currently exploring how to make his Strandbeests self-replicate.

Jansen's (2007) work is of interest to us because his Strandbeests have been developed synthetically. That is, he has discovered how to make these large walking systems not by analyzing movement, but by letting his basic building blocks (plastic tubes) guide his discovery of multilegged walking structures. Rather than analyzing what makes a gait "lifelike," Jansen explored, from the ground up, configurations of tubing that resulted in amazingly lifelike, many-legged, walking sculptures. In the sections that follow, we will gain some understanding of Jansen's synthetic discoveries by constructing our own Strandbeest out of LEGO components.

5.14 A LEGO STRANDBEEST
5.14.1 Alternative Material

Jansen's (2007) Strandbeest originated from the goal of building creatures from particular material. "I want to make everything out of plastic tubing. Just as nature as we know it consists largely of protein, I want to make my own life-forms from a single material" (Jansen, 2007, p. 35).

One reason that Jansen began his inventions by choosing their basic material is that he has found that limitations imposed by this choice guide (or force) his later creative thinking. "Remarkably, chance is more likely to play a role when there are restrictions. Financial restrictions, for example, may mean that drawers in the workplace stay closed. This necessitates looking for other possibilities elsewhere. During this search new ideas automatically emerge, ideas that are often better than the ones you first had. Again, the restrictions of the plastic tubing oblige you to look for technical solutions that are less than obvious" (p. 37).

5-11

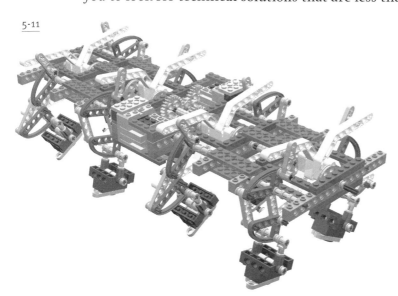

In following sections, we will explore building a machine that is inspired by the Strandbeest, but is composed of yet another "material": LEGO parts. The machine that we will create is illustrated in Figure 5-11. Because we are choosing a material that differs from Jansen's, we will be forced — by our material — to make some creative decisions in order to create a device that behaves in a similar fashion to those that Jansen has created and described. The primary goal of the device that we will build is to walk; when it walks we will see that it does so in a fashion that is remarkably lifelike, as is the case for Jansen's self-locomoting devices.

5.15 SEGMENTED DESIGN

5-12

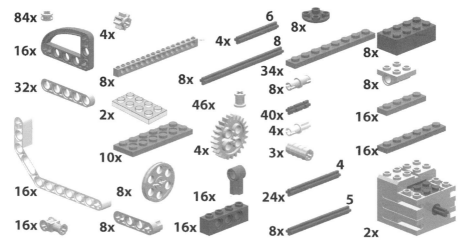

5.15.1 Parts and Foraging

If the reader would prefer to use wordless, LEGO-style instructions, they are available as a pdf file from the website that supports this book (http://www.bcp.psych.ualberta.ca/~mike/BricksToBrains/). The components in Figure 5-12 are used to construct the LEGO Strandbeest. This machine's gait is dependent completely on the particular proportions of its leg components: if certain parts are not available, creative substitutes must be developed that maintain the proportions that will be described in more detail in Section 5.18. This is one way in which the LEGO parts can drive the creative process, just as the plastic tubing guided Jansen (2007, p. 37): "What is handy about the artist's method is that you yourself don't have to devise or invent anything. The material does that for you. So it was the plastic tubes that put the idea of a new nature into my head."

5.16 FROM THE GROUND UP
5.16.1 Ankles and Feet

Unlike Vehicle 2 (and many other robots in this book), the Strandbeest is capable of moving on soft surfaces such as carpets. It accomplishes this by having feet with ankles, which are controlled solely by gravity and geometry. Naturally, you will need to build one foot for every leg you want, but like the walker itself the foot design is repeatable (that is, every leg, regardless of position or orientation, uses the same foot design).

To assemble these feet, begin by snapping two 1 × 5 liftarms to the top of a 2 × 4 LEGO brick, as shown in Figure 5-13. This can be quite tight, so press firmly. Attach a 1 × 4 inverted dish to the centre of the liftarms once they connect.

Next, flip the Figure 5-13 assembly over. A length-5 axle is used to hold two type 1 connectors and a 2 × 2 plate (with axle hole) in place with half-bushes, as shown. Once this is connected to the brick, slide two 2-axles into the connectors, as in Figure 5-14.

Finally, add two 1 × 4 beams to the brick, and a connector (with axle hole) to each 2-axle. The beams serve as stoppers to prevent the foot from flipping over. The resulting assembly is shown in Figure 5-15.

5-13

5-14

5-15

5.16.2 Feet vs. Wheels

The feet described in Section 5.16.2 are to be attached to robot legs of unique design. Among roboticists there has been considerable debate concerning whether autonomous robots should use wheels or legs. Wheels have the advantage of being very efficient and stable, because their axle does not move up and down as the wheel rotates. Largely for this reason, successful autonomous robots sent to Mars (such as the rovers Spirit and Opportunity, which were designed to last 90 Martian days, but at the time of this writing are still working well over 1,500 Martian days past their "expiry date") have used wheels instead of legs.

In principle, however, legs have many advantages over wheels. Because legs do not have to be continuously in contact with the ground, they can step over small obstacles in uneven terrain, and therefore are more adaptive to a wider variety of circumstances than are wheels (Jansen, 2007). The problem with legs is their potential for instability, because they are not wheel-like — the hips to which they are attached usually move up and down with each step, unlike the axle of a wheel.

Jansen (2007, p. 51) has discovered how to construct legs to solve this problem; like McGeer (1990a), he discovered how to convert walking into wheel-like locomotion: "The upper and lower leg parts move relative to one another in such a way that the hip joint (at the juncture with the upper leg) remains at a constant height, just as with the axle of a wheel." Let us now turn to constructing such special legs.

5.17 A STRANDBEEST LEG
5.17.1 Precise Proportions

The Strandbeest translates rotational motion along a fixed axis into horizontal motion, much like a wheel. Furthermore, the precise proportions of the different parts of the leg convert that motion into a smooth walking gait.

Begin by laying out two 3 × 5 L-shaped liftarms with quarter ovals (see Figure 5-16), and connecting them to four length-2 axles, a length-8 axle, and a length-4 axle in the positions shown.

5-16

5-17

5-18

Next, slide two 1 × 5 liftarms onto the pegs as shown in Figure 5-17, securing one of them with half-bushes. The other is left unsecured for now.

Next, slide two LEGO bushes onto the length-8 axle on the 1 × 5 liftarm's side, and three on the other side of the same axle, as illustrated in Figure 5-18.

Next, on the upper remaining length-2 axle, slide a half-bush, followed by a 1 × 4 liftarm as shown in Figure 5-19. Finally, the foot can be attached using both ends of the axle at the bottom of Figure 5-17. The positioning of the attached foot is also shown in Figure 5-19.

5-19

5.18 LEGO LEGS AND HOLY NUMBERS

5.18.1 Completing a Leg

Leg construction is completed by attaching two double-bent liftarms to the assembly by their second pinhole, as in Figure 5-20. The upper one uses a low-friction pin (grey, not black) to connect to the 1 × 4 liftarm, while the lower one is secured to the length-4 axle using a half-bush. To complete the Strandbeest, many copies of this leg must be assembled. For now, **just build two of them.**

5-20

5.18.2 The Holy Numbers

The lifelike motion of Jansen's (2007) Strandbeest depends crucially on the proportions of various leg parts. Jansen used genetic algorithms to choose proportions that optimized the shape of the leg's walking curve (to maximize hip stability) and the duration that the leg was in the air (which was minimized). The resulting "holy numbers" are provided in Figure 5-21, which depicts Jansen's leg design. Each black line in the figure represents one plastic tube. The numbers give tube lengths in arbitrary units. "It is thanks to these numbers that the animals walk the way they do" (Jansen, 2007, p. 57).

A leg of the LEGO Strandbeest (Figure 5-20) is an attempt to implement these "holy numbers" as accurately as possible. Figure 5-22 shows a side view of the Figure 5-20 leg superimposed by a Jansen leg. The units in this figure are LEGO studs; this configuration of lengths is our best approximation of the optimal lengths in Figure 5-21. Thus, the structure of the LEGO leg — the actual parts, and their relative placement — is the direct result of seeking this match while being restricted to different materials. The exact structure of a LEGO leg is dictated by the constraints imposed by LEGO parts.

5-21
5-22

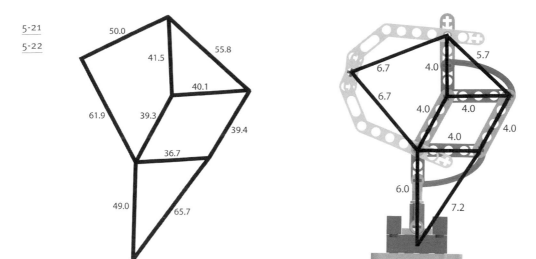

5.19 REINVENTING THE WHEEL
5.19.1 Pairing Legs into a Module

After assembling two legs (Figure 5-28), create a leg module. First, position both legs as shown in Figure 5-23. Run a length-6 axle through the first pinhole on each liftarm, joining the four liftarms together.

Next, slide a wedge belt wheel onto each side of the length-6 axle that connects the four liftarms. Pin each wheel in place with half-bushes (Figure 5-24). **The length-6 axle must pass through the same hole on both wedge belt wheels (see below).** Insert a length-4 axle through the axle hole in the centre of each wedge belt wheel, followed by a half-bush.

5-23
5-24

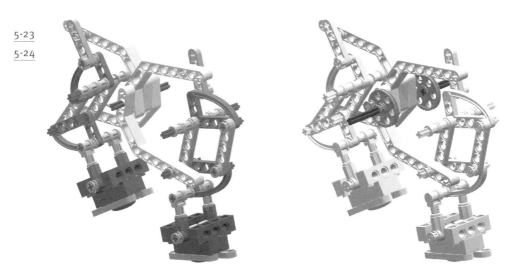

When wedge belt wheels are attached (Figure 5-32), note that there are six different holes that could be used (Figure 5-25). Within a leg module, the length-6 axle must be inserted in the *same hole* of both wedge belt wheels. If this is not done, then the legs will not move properly.

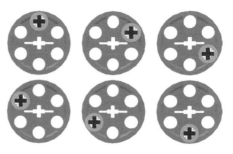

Next, place a 1 × 16 Technic beam on each side of the assembly. The length-8 axles from each leg pass through the third pinhole from the end, while the length-4 axles you just placed go through the central hole. Pin the beam in place with a half-bush on each length-4 axle (Figure 5-26).

Brace the two beams with the 1-width plates, as demonstrated in Figure 5-27. The 6-length plates brace the 1 × 8 plates crossing the assembly, reducing the slipping one would expect from 1-width bracings.

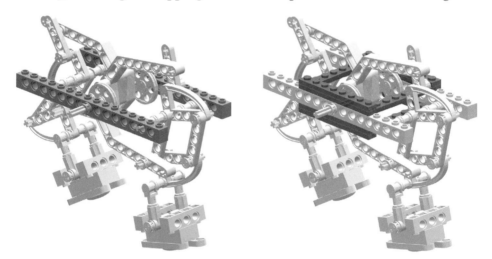

The legs can now be moved by rotating the central axle by hand. Movement might be restricted because of the tightness of various connections. Fine-tune the connections, making sure that all axles are flush with their connectors, that the half-bushes holding the beams in place are snug, and that the legs move smoothly.

5.20 QUADRUPED
5.20.1 Mounting the Modules

Using the instructions from Sections 5.16 through 5.19, build four legs, and use these legs to create two leg modules. Once this is done, the simplest version of a walking Strandbeest — a quadrupedal version — can be assembled.

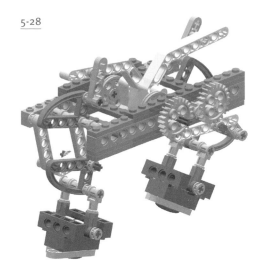

5-28

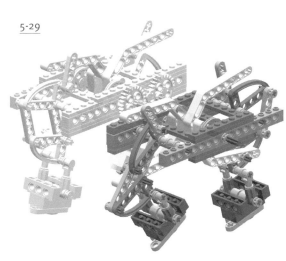

5-29

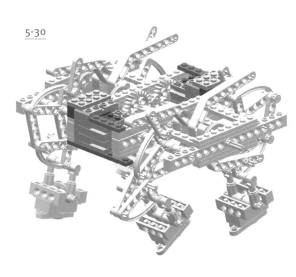

5-30

Begin by replacing the central half-bush on one leg module with a 24-tooth gear. Using axle pins, mount an 8-tooth gear and another 24-tooth gear to the right of the axle, followed by seven 1 × 4 plates in the configuration shown in Figure 5-28.

Repeat this step on the second leg module to create two identical assemblies that can be connected using a Technic axle joiner on the central axle such that the gear trains face each other, as in Figure 5-29.

Using a pair of 1 × 4 plates, 1 × 8 plates, 1 × 4 beams and 2 × 4 plates, attach a pair of 9V mini-motors to the joined leg modules. The positions of the motors are shown in Figure 5-30. Note that the two leg modules are separated by seven LEGO pips, so the 1 × 8 plates will have some overhang, and the axle joiner (from the above step) should have some play. The motors should hold 8-tooth gears, each of which should mesh with a gear that is part of a leg module.

5.20.2 Gait Exploration

With the motors attached, a quadruped Strandbeest is finished. Its walking gaits can now be explored. Attach both motors to the same LEGO battery pack (using two wires from the battery pack's output) and turn it on. If the robot seizes up, the wires are not connected in the same direction; if they are connected correctly the motors should turn smoothly. Later in this chapter, we will present a video that illustrates the behaviour of this quadrupedal Strandbeest, as well as of one that involves more than two leg modules.

5.21 MANIPULATING QUADRUPED GAITS
5.21.1 Quadruped Gaits

As was noted in Section 5.3.2, early photographic analysis of animal locomotion revealed a number of distinct animal gaits (Muybridge, 1887/1957). Quadruped gaits that Muybridge reported included the *walk*, in which a horse supported itself first with 3 legs, and then with 2 legs, in alternation. The particular feet involved in support during the walk alternated in a distinct sequence; there were eight different support stances in the sequence before it repeated itself.

The *amble* was identified as the evolution of the walk into a faster form of locomotion. The horse supported itself first with 2 legs, then with 1 leg, in an alternating sequence of eight different support stances before the entire sequence of the amble began again.

"The *trot* is a system of progress in which each pair of diagonal feet are alternately lifted with more or less synchronism, thrust forward, and again placed on the ground; the body of the animal making a transit, without support, twice during each stride" (Muybridge, 1887/1957, p. 41). The *rack*, or the *pace*, was identical to the trot with the exception that lateral leg pairs were alternated instead of diagonal leg pairs.

The *canter* is a more complex gait than the previous four. It involves the same sequence of footfalls as the walk, but at irregular intervals. It comprises a spring from a forefoot, followed by a landing on the diagonal hind foot, with the body being supported by 1, 2, or 3 legs in a complex sequence.

Finally, the *gallop* is the gait that produces the most rapid quadrupedal movements. Each foot impacts the ground in a sequence. One sequence is known as a transverse gallop, while an alternative sequence is known as a rotational gallop. During galloping, a horse's body is supported by 0, 1, or 2 feet at any given moment.

5.21.2 Exploring Strandbeest Gaits

The quadruped LEGO Strandbeest is also capable of producing a variety of different gaits, which are determined by the phase of the legs in one module relative to the legs in the other. One can manipulate relative phase in increments of 90° as follows: detach the axle that joins adjacent leg modules together (as seen earlier in Figure 5-36). Rotate the 40-tooth gear of module to shift its legs, and then reattach the axle. (The fact that the axle and axle joiner are cross-shaped constrains this phase manipulation. Phase shifts that are not multiples of 90° will not permit the axle to be reattached.)

A second variable that affects phase involves the wedge belt wheels that were attached in Figure 5-32. Within a module, both wedge belt wheels must be attached using the same hole (Figure 5-33). However, across modules, different holes can be used. In certain instances, when different modules use different wedge belt wheel holes, different phase manipulations become possible.

Imagine that in one leg module, the hole that is used is the one at the top left of Figure 5-33. If this hole, or the hole 180° from it (bottom left Figure 5-33), is used in a second module, then the phase between the modules can only be manipulated in 90° increments when the modules are disconnected and manipulated as described earlier in this section. However, if the second module uses any of the other four possible holes illustrated in Figure 5-33, then the phase difference between the modules can be manipulated in increments of 60°.

One of the revelations provided by the early analysis of animal movement was that in some instances all four legs might be off the ground at the same time (Muybridge, 1887/1957). The slow movement of the quadruped Strandbeest prevents this from ever happening. However, you will observe that by manipulating the phase of adjacent modules, different gaits can be produced. How many of these gaits were observed by Muybridge? Can this machine produce any gaits that Muybridge did not observe in his analytic studies of locomotion?

5.22 AN OCTAPEDAL STRANDBEEST
5.22.1 Additional Legs

One of the advantages of Jansen's (2007) modular leg design is that it is fairly easy to scale Strandbeests up and explore their behaviour. For instance, what gaits might be seen in an octapedal Strandbeest? To answer this question, one builds additional leg modules, adds them to the quadrupedal Strandbeest, and observes the result—while at the same time manipulating phases between adjacent leg modules.

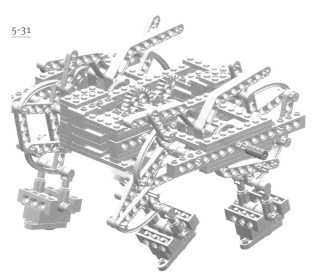

5-31

To continue exploring the Strandbeest, you will need to assemble two additional leg modules (Section 5.19). Once these extra leg modules are assembled, mounting them to the Strandbeest is extremely simple. Begin by sliding a Technic axle joiner to the extended length-4 axles on either side of the walker, as shown in Figure 5-31.

Next, slide one of the new leg modules into the joiner on each side of Figure 5-31, as illustrated in Figure 5-32. Be careful when doing this to line up the legs in adjacent modules so that their phase relationship is what you desire. For instance, to set adjacent modules 180° out of phase, ensure that the legs in one module are in "up" position, while the legs in the adjacent module are in a "down" position. This should be the case for each of the four adjacent leg modules that are illustrated in Figure 5-32.

To complete the octapedal Strandbeest, the leg modules added in Figure 5-32 must be reinforced. To do this, snap on five 2 × 6 Technic plates between each of the new segments and the existing ones: two on the ends, and one in the centre, as shown in Figure 5-33.

5.22.2 Walking with Eight Legs

The walker will now operate as an octapod, with eight legs working in unison. Connect the motors to a single LEGO battery pack and turn it on just as before. Observe the walking behaviour. How does it compare to that of the quadrupedal Strandbeest? When the machine walks forward, how are its legs coordinated? What happens when you reverse the direction of the motor, or manipulate leg phases? What happens when the two motors are run from separate battery packs? What happens when the central axle joiner is disconnected, producing a walker that uses two independent quadrupedal modules? How does this thoughtless machine coordinate its legs so that it moves forward?

5.23 STRANDBEESTS IN ACTION
5.23.1 Observing Strandbeest Gaits

When the octapedal Strandbeest is completed, we are in a position to begin to explore the dynamics of Strandbeest locomotion. Video05-2. mpg, available from the website for this book (http://www.bcp.psych. ualberta.ca/~mike/BricksToBrains/), provides some examples of walking in both the quadrapedal and the octapedal versions of the Strandbeest. It also illustrates how changing the phases of the leg modules—by 90° increments in the video—clearly affects Strandbeest gaits. More importantly, the effect on gait interacts with the number of leg modules in use. That is, when identical phase manipulations are performed on the quadrupod and the octapod, different gaits are observed.

The video also demonstrates another advantage of adopting the synthetic methodology—that is, of building the Strandbeest from the ground up without an extensive pre-analysis of its structure or behaviour. The advantage is that the working model can be exposed to environments that were not originally considered when the machine was being built. For instance, the video shows how the Strandbeest's legs, and the positions of its ankles, smoothly adapt to changes in slope.

Another advantage of the synthetic approach is that one can quickly explore how the machine reacts to changes in its embodiment. Two such changes are explored in the video. In one, the two central motors in the octapedal Strandbeest are run in opposite directions. The result is that the machine executes a smooth turn. In the second, the axle joiner connecting the two central leg modules of the octapedal Strandbeest is disconnected. As a result, its left and right sets of legs are independent machines. When it is run, these two independent sets of legs become desynchronized, and the machine begins to turn. However, in spite of their independence, the two sets of legs later resynchronize, and the machine stops turning.

How is it possible for the independent legs to resynchronize? This phenomenon is a reminder that the Strandbeest is an embodied agent that exists in an actual environment, and that its behaviour (like that of the LEGO passive dynamic walker) must be considered with the environment kept firmly in mind. While the Strandbeest is not strongly situated in its environment, because (unlike Vehicle 2) it does not take measurements with sensors, it is at least *weakly situated* in its world (see also Dawson, 2004, Chapter 6). That is, the exact position of a leg not only depends upon the current position of its axles and wedge belt wheels, but also upon the floor that is pushing against the leg. This in turn will have an effect on the positions of other parts of the robot. So, if desynchronization produces an awkward stance for the robot, this in turn can slow the movement of some legs down (e.g., increased force might strain one motor, slowing it down). This can change the position of one set of legs relative to the other set, and permit resynchronization to occur. In short, just as was the case in Ashby's (1960) account of a large-scale Homeostat, the disconnected legs can communicate with one another because they are embodied, and weakly situated, in a common environment.

5.23.2 Exploiting Stronger Situation

Above, the coherent walking behaviour of the Strandbeest was explained by appealing to feedback between legs that are embodied and weakly situated. A similar account is provided by Dawson's (2004, Chapter 6) analyses of the different gaits produced by multi-legged walkers constructed from K'Nex components.

These two examples lead to an obvious next step: the development of a robot that propels itself by taking advantage of feedback between components that are more strongly situated. To end this chapter, we will discuss one such example of a straightforward LEGO robot. This robot is a snake-like or worm-like machine that is created by chaining four identical motors together in a sequence. Coherent movement can be produced by having motors react to their own rotation, which can be caused by the motor itself, or by the motor's axle being influenced by the rotation of other motor axles in the robot's body.

5.24 ALTERNATIVE GAITS AND ROBOTIC SNAKES
5.24.1 Snake-like Movement

The two LEGO robots that have been described in this chapter illustrated the complementary nature of analytic and synthetic approaches to locomotion. Both were extremely simple—so simple, in fact, that they did not need an NXT brick to control their behaviour. The final robot described in this chapter, called the Wormeostat, is more advanced in the sense that it uses NXT bricks to control motors. However, it too is a very simple device, and shows how local feedback can be used to control globally coherent actions that are capable of producing movement. The Wormeostat also illustrates a synthetic alternative to a great deal of analytic research that has been used to develop robots that move in the world by emulating the locomotion of such beasts as snakes and inchworms.

Japanese roboticist Shigeo Hirose began research on robots that employ snake-like locomotion in the early 1970s, and has continued to be a leader in this field of research (Endo, Togawa, & Hirose, 1999; Hirose, 1993; Hirose, Fukushima, & Tsukagoshi, 1995; Hirose & Mori, 2004; Hirose & Morishima, 1990). When Hirose began his research, he was interested in two different questions. First, at the time it was not clear how snakes were able to move forward without the use of legs, and Hirose was interested in using engineering to explore this issue. Second, Hirose believed that a robot that was snake-like in form would be extremely versatile, capable of moving across a variety of terrains, and able to solve a variety of problems. This is because a snake's body can be used in many different ways: as "legs" when locomoting, as "arms" when climbing, and as "fingers" when grasping. Indeed, snake-like and inchworm-like robots are being applied to such diverse tasks as search and rescue (Ito & Murai, 2008; Matsuno, 2002) and for surgery (Phee et al., 2002).

Snakes can propel themselves using one of four different kinds of gaits: serpentine, side-winding, concertina, and rectilinear (Saito, Fukaya, & Iwasaki, 2002). For instance, when snakes use the serpentine gait, their body takes a particular form called the serpenoid curve. When in this shape, there is greater friction tangential to the snake's body and less friction in the direction parallel to its body. As a result, movement of body parts causes the entire body of the snake to move forward instead of sideways (Hirose, 1993).

Hirose exploited such principles when developing a serpentine robot, which he called the active cord mechanism. The original version was about 2 metres long, and had 20 different joints that linked modular servo-mechanisms. The mechanism at each segment of this robot enabled a particular joint to bend to the left or to the right. Casters were attached to each segment to manipulate friction — a segment could easily be pushed in the direction normal to the robot's body, but could not be so easily moved in the direction orthogonal to the body. Serpentine was achieved by bending the robot's body at a particular angle at the front of the machine, and then sending this angle back toward the rear of the robot as a travelling wave.

5.24.2 Analyzing Snake Locomotion

Hirose's robots were inspired by a biomechanical analysis of the locomotion of living snakes (Hirose, 1993). Not surprisingly, the literature on snake-like movement for robots provides numerous mathematical models of the dynamic forces that are involved in propelling snakes forward in two or three dimensions (Bayraktaroglu, 2009; Chirikjian & Burdick, 1995; Hirose & Morishima, 1990; Ito & Murai, 2008; Lobontiu, Goldfarb, & Garcia, 2001; Saito et al., 2002; Shan & Koren, 1993; Skonieczny & D'Eleuterio, 2008; Transeth, Leine, Glocker, & Pettersen, 2008; Transeth, Leine, Glocker, Pettersen, & Liljeback, 2008). These contributions are similar to the detailed analytic research that inspired the passive dynamic walkers discussed earlier in the current chapter. However, it is also possible to synthetically produce such devices. In the next section, we introduce a LEGO robot that can move like a snake, and which was inspired by tinkering with some of the basic principles of the Homeostat (Ashby, 1956, 1960).

5.25 THE WORMEOSTAT: A SYNTHETIC SNAKE OR WORM
5.25.1 Feedback and Motion

The Wormeostat is a simple robot that consists of a chain of four NXT motors, as shown in Figure 5-34. These motors are linked in such a way that when one of them moves (i.e., when one of them rotates its axle), it can physically cause other motors in the robot to move too by contorting the robot's body.

Each of the motors in the Wormeostat's body is analogous to one of the four electrical components in the Homeostat (Ashby, 1956, 1960). That is, in addition to being able to affect the other components, each motor reacts to the effects of the other motors and itself. If a motor rotates above a threshold level — either by running, or by being rotated by the action of other motors — then the motor attempts to rotate in the opposite direction in order to return to its original position. If the motor's rotation is below threshold, then it does not react.

When such feedback is linked in the body of the Wormeostat, the result is that the body contorts in waves of motion up and down, like a caterpillar or worm. (The motors can be mounted sideways so that the contortions are side to side, analogous to the movement of a snake.) These contortions cause the robot to move forward, and even permit it to climb over obstacles.

5.25.2 Motion from Friction

The reason that the contortions of the Wormeostat's body cause it to move forward is because it has a number of friction points that provide resistance to movement in one direction, but do not resist movement in the opposite direction. This is similar to Hirose's use of differential friction in his robot snakes (Hirose, 1993).

5.26 FORAGING FOR WORMEOSTAT PARTS

5.26.1 Building the Wormeostat

The next few pages provide instructions for building the Wormeostat that was illustrated in Figure 5-34. If the reader would prefer to use wordless, LEGO-style instructions, they are available as a pdf file from the website that supports this book (http://www.bcp.psych.ualberta. ca/~mike/BricksToBrains/).

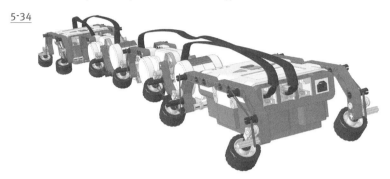

5-34

5.26.2 Parts and Modules

The complete Wormeostat is comprised of two modules. Each module consists of one NXT brick and two motor components, all linked together in a flexible assembly. When the two modules are connected together, the Wormeostat is finished. All of the parts that are used to construct this robot are illustrated below in Figure 5-35.

However, it is important to realize that there are subtle differences between the two modules. The instructional images that follow illustrate how to build each module, and how to link them together. Be careful to look at the details of these instructional images, though, so that the two modules that you will construct are not identical.

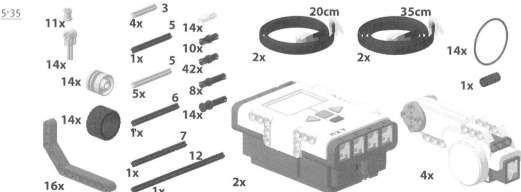

5-35

5.27 MOTOR AND TIRE ASSEMBLIES
5.27.1 Motor Modules

Much of the Wormeostat's body is composed of four assemblies, each of which is an NXT motor to which additional parts will be attached. The first step in building the Wormeostat is to begin to build these four components, as illustrated in Figure 5-36. An axle that is 5 studs in length, four pins, and two double-bent liftarms, are attached to each NXT motor as in the figure.

5.27.2 Tire Assemblies

Other components that will eventually be attached to the motors are a number of tire assemblies. These assemblies are not designed to work as wheels, but will instead provide points of friction that will prevent the robot from sliding backward when force is applied. Each of these components is very simple, and 14 of them must be constructed. The instructions for tire assembly construction are also provided in Figure 5-36. They will be attached later in the robot's construction.

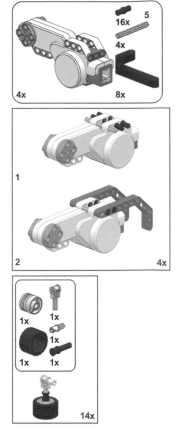

5-36

5.28 PREPARING TWO NXT BRICKS
5.28.1 Control and Friction

The Wormeostat uses two NXT bricks to control the feedback amongst motors. One brick controls the two motors in the front half of the robot, while the other brick controls the two motors in the back half of the robot. In addition to holding the tasks that run the motors, each brick also holds a set of tire assemblies (Figure 5-36) that serve as points of friction that prevent the robot from sliding backward when it squirms. In this step, two NXT bricks are prepared to hold these tire assemblies using pins and double bent liftarms, as is illustrated in Figure 5-37.

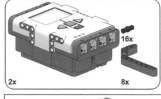

5-37

5.29 FRONT END FRICTION

5.29.1 Motor Friction

In this step, friction-producing tires are added to a motor that will be part of the front end of the robot. Pins are used to prevent the tire assemblies from pushing too far forward, and elastics are used to pull the tire assemblies back into place. Examining the motor assembly that is the top part of Figure 5-38 provides the logic of this design.

If this motor were pushed to the right of this figure, then the tire assemblies would be pushed into the surface that the robot was on, would provide friction, and would prevent this motion. In contrast, if this motor were pushed to the left of this figure, then the tire assemblies would be pushed up, and would therefore not provide friction, and motion to the left would not be inhibited. When the motor shifted position (e.g., if it twisted forward), the tire assemblies would be pulled back into position by the elastics, and would be ready to provide friction again if required.

5.29.2 Brick Friction

5-38

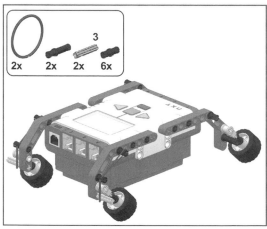

To complete this step, tire assemblies identical to those added to the motor are added to the NXT brick as well.

With a casual glance, the front half of the Wormeostat appears to be identical to its back half (e.g., Figure 5-34). However, this is not the case — a careful inspection reveals subtle but important differences. One example of such a difference can be seen in the front NXT brick. Notice the direction in which the tire assemblies are pointing, and note that the buttons on the NXT brick are pointing in the same direction. Note that the assembly involving the second NXT brick (Figure 5-42, presented later this chapter) is a mirror image of Figure 5-38 — the two NXT brick assemblies are not exactly the same!

5.30 A SECOND FRONT END MOTOR
5.30.1 Reflected Construction

The front half of the Wormeostat includes two motors. The tire assemblies are attached to the second motor as illustrated below in Figure 5-39. Note that this motor assembly is different from the other front-end motor that was illustrated in Figure 5-38: the motor has a reflected orientation, so that the tire assemblies that are attached to it have a different position relative to the double-bent liftarms.

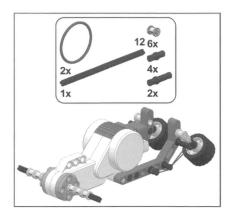

5-39

5.31 COMPLETING THE FRONT HALF
5.31.1 Connecting Three Components

The front half of the Wormeostat is created by joining the three modules that have been described in the preceding pages. The motor assembly illustrated in Figure 5-39 is attached to the front NXT brick assembly as illustrated in Figure 5-40. Note that the rotating axle of this motor is what links the motor to the brick. This is important; it means that this motor can lift the front brick upward, permitting the Wormeostat to climb over obstacles. Note too that elastics are added to the tire assemblies at the connection between the brick and the motor.

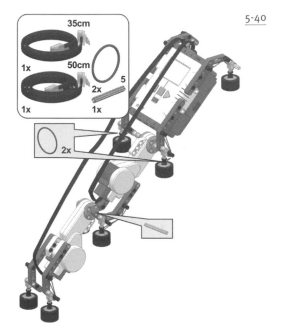

5-40

A 5-stud–length axle is then used to connect the motor illustrated in Figure 5-38 to the motor that is attached to the NXT brick, making a chain of three major components. Take note of the orientation of the motors to one another, and of the fact that all of the friction-producing tire assemblies are pointing in the same direction.

Construction of the front half of the Wormeostat is completed by attaching cables between the motors and Output Ports A and B of the NXT brick.

5.32 MODULES FOR THE REAR HALF
5.32.1 Replicating Components

The rear half of the Wormeostat is also created from three major components, two involving motors and one involving an NXT brick. The instructions for building the two motor modules are provided in Figure 5-41. Notice an important difference between these instructions and those for the front motors. The lower motor in Figure 5-41 is the one that will be attached to the rear NXT brick. It will be attached using the axles that are inserted into the double-bent liftarms attached to the motor. This means that the rotating axle of this motor will *not* be directly connected to the NXT brick.

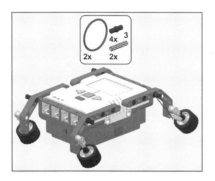

The rear NXT brick module is completed as shown in Figure 5-42. Note that while it is similar to the front NXT brick module, the two are not identical, as was discussed earlier in Section 5.29. This brick assembly is a mirror image of the one that was provided earlier in Figure 5-38.

5.33 COMPLETING THE REAR HALF
5.33.1 A Second Chain

The rear half of the Wormeostat is created by linking the three components described in Section 5.32 together in a chain. This is illustrated below in Figure 5-43. Again, pay attention to subtle differences between this figure and Figure 5-40. The rotating axles of the motors in Figure 5-43 point away from the NXT brick, not toward it. Also note the orientation of the motors relative to one another. Once the components are united as shown, this step is completed by attaching cables between the motors and Output Ports A and B of the NXT brick.

5.34 THE TOTAL WORMEOSTAT
5.34.1 Linking the Halves

The Wormeostat can now be completed by attaching its two halves together with a 5-stud–length axle, as illustrated in Figure5-44. Note that

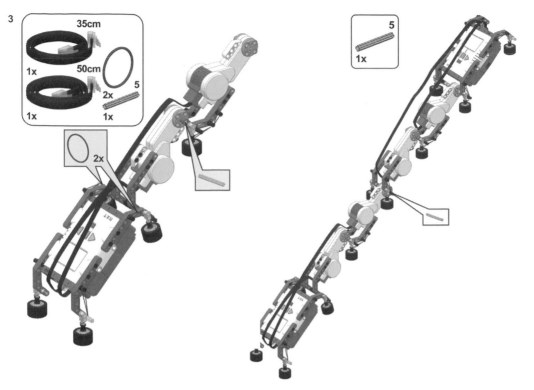

when this step is completed, the four motor components have alternating orientations. Also note that all of the tire assemblies point in the same direction.

5.34.2 Programming Feedback

Now that the Wormeostat has been constructed, NXC code must be written to create the feedback between motors that will lead to the robot's movement. The pages that follow describe the program.

5.35 WORMEOSTAT CODE FOR MOTOR 1
5.35.1 Motor Behaviour

In general, the Wormeostat's movement is produced by having a motor sense how much its axle has rotated in a brief period of time. If this rotation is below a threshold value, then the motor does nothing. However, if the rotation is above this threshold, then the motor turns on and attempts to rotate the axle back to its original position.

The NXC code below provides three tasks that accomplish this. Motor1SenseRotation senses motor rotation. Motor1ComputeRotation computes a response to this sensed rotation. Motor1React uses the computed response to run the motor, but only if the sensed rotation is above a threshold.

```
//Wormeostat_v1 -- Run this program on both NXT bricks simultaneously.
//The motors feed back to themselves electrically and to each other physically.
#define Motor1 OUT_A
#define Motor2 OUT_B
const int  RefreshDelay = 50; //time between sensing rotation and running motor
const int  Strength = 80;  //motor strength in percent
const int Threshold = 8;  // Used to decide to run motor, or not
const int reverse = -1, forward = 1; //a few basic constants
//Motor1========================================================
int Motor1Response, Motor1Activation, Motor1Sense, Motor1Lag, Motor1Direction;
task Motor1SenseRotation(){
        //This task senses the amount that Motor1 has rotated. It records this
        //value in a variable (Motor1Sense). This permits more than one task
        //to have access to this value at the same time.
        while (true){
                Motor1Sense = MotorRotationCount(Motor1);
        }
}
//Compute movement: Compute rotation required to return to initial position.
task Motor1ComputeRotation(){
        while (true){
                Motor1Response = (-1 * Motor1Sense); //Response opposite to sensed rotation
                //Hold computed value until it is reached or until motor jams
                until ((Motor1Response == Motor1Sense) || MotorOverload(Motor1));
                Wait(Motor1Lag); // short delay before recomputing
        }
}
//React by running the motor as computed by Motor1ComputeRotation.
// Only run the motor if sensed rotation is greater than a threshold
task Motor1React(){
        while (true) {
                //Set the direction in which to run the motor
                if(Motor1Sense > Motor1Response) Motor1Direction = reverse;
                else Motor1Direction = forward;
                //If sensed rotation is above threshold, run the motor to react
                if (abs(Motor1Sense - Motor1Response) > Motor1Activation){
                        OnFwdReg (Motor1, Motor1Direction * Strength, OUT_REGMODE_SPEED);
                }
                else Float(Motor1); //If below threshold, do not run the motor
        }
}
```

5.36 WORMEOSTAT CODE FOR MOTOR 2
5.36.1 Second Verse Same as the First

Each NXT brick is used to control the two motors nearest the brick. All of these motors are controlled in exactly the same way, because each motor is an identical component that simply reacts to movement by trying to reverse the motion that it has sensed. The NXC code listed below is the code used by one NXT brick to control the second motor that it is responsible for. Note that this code is identical to the code that was described in the previous section, with the exception that all variable names refer to Motor2 instead of Motor1.

```
//Motor2 mirrors Motor1 in all ways.=====================
int Motor2Response, Motor2Activation, Motor2Sense, Motor2Lag, Motor2Direction;
task Motor2SenseRotation(){
        //This task senses the amount that Motor2 has rotated. It records this
        //value in a variable (Motor2Sense). This permits more than one task
        //to have access to this value at the same time.
        while (true){
                Motor2Sense = MotorRotationCount(Motor2);
        }
}
//Compute movement: Compute rotation required to return to initial position.
task Motor2ComputeRotation(){
        while (true){
                Motor2Response = (-1 * Motor2Sense); //Response opposite to sensed rotation
                //Hold computed value until it is reached or until motor jams
                until ((Motor2Response == Motor2Sense) || MotorOverload(Motor2));
                Wait(Motor2Lag); // short delay before recomputing
        }
}
//React by running the motor as computed by Motor2ComputeRotation.
// Only run the motor if sensed rotation is greater than a threshold
task Motor2React(){
        while (true) {
                //Set the direction in which to run the motor
                if(Motor2Sense > Motor2Response) Motor2Direction = reverse;
                else Motor2Direction = forward;
                //If sensed rotation is above threshold, run the motor to react
                if(abs(Motor2Sense - Motor2Response) > Motor2Activation){
                        OnFwdReg (Motor2, Motor2Direction * Strength, OUT_REGMODE_SPEED);
                }
                else Float(Motor2); //If below threshold, do not run the motor
        }
}
```

5.37 WORMEOSTAT MAIN TASK
5.37.1 The Main Task

The main task is used to initialize the variables that control the responses of each motor under an NXT brick's control, and then to activate the three tasks that run each motor. The NXC code for the main task is provided below. The complete Wormeostat program is available from the website that supports this book (http://www.bcp.psych.ualberta.ca/~mike/BricksToBrains/).

5.37.2 Modular Duplication

As is apparent from the instructions for building it, the Wormeostat is a highly modular machine constructed from identical functional components (four motor assemblies and two NXT bricks all linked together in a flexible structure). All four motors run exactly the same three procedures (SenseRotation, ComputeRotation, React), but can influence one another, producing interesting behaviour as discussed in the next section. Each NXT brick is responsible for controlling two of these motor modules. Thus, each NXT brick will run exactly the same program: the main task listed below, and the two sets of motor routines described in the preceding two sections.

```
//Main===========================================
task main()
{
// Set Initial Activation Levels and Lag
  Motor1Activation = Threshold;
  Motor1Lag = RefreshDelay;
  Motor2Activation = Threshold;
  Motor2Lag = RefreshDelay;
//Activate Motor1 Modules
  start Motor1SenseRotation;
  start Motor1ComputeRotation;
  start Motor1React;
//Activate Motor2 Modules
  start Motor2SenseRotation;
  start Motor2ComputeRotation;
  start Motor2React;
}
```

5.38 THE WORMEOSTAT'S BEHAVIOUR
5.38.1 Fireside Dogs

The Homeostat (Ashby, 1956, 1960), and autoassociative artificial neural networks (Hopfield, 1982, 1984; Hopfield & Tank, 1985), are examples of systems a) that are constructed from multiple components that are identical in nature, and b) in which the activity of these components depends upon feedback amongst them. In general, these systems manipulate the signals transmitted from component to component (e.g., by altering the activity of each component) until a stable state is reached. When this state is reached, component activities stabilize, and the entire system is inert. It remains in this state until some external influence disrupts the stability that has been achieved. The Homeostat was "like a fireside cat or dog which only stirs when disturbed, and then methodically finds a comfortable position and goes to sleep again" (Grey Walter, 1963, p. 123). The Wormeostat is another example of such a system.

5.38.2 Wormeostat Movement

The Wormeostat is able to convert the feedback amongst its four motors into a coherent movement that propels it forward, with a fairly jerky gait, over a variety of terrains. One of the reasons that it is capable of movement is the variable friction that is provided by the tires connected to the underside of the robot. In one position, these tires come in contact with the terrain and provide friction that causes forces to be transferred through the body of the robot and prevent backward movement. In another position, the tires are no longer in contact, permitting a segment of the robot's body to be slid forward in the direction of the forces that are being applied. Video 5-3 presents some examples of Wormeostat's movement; this video is available from the website for this book (http://www.bcp.psych.ualberta.ca/~mike/BricksToBrains/),

Of particular interest is that the "fireside dog" nature of the Homeostat was preserved in the Wormeostat. Usually the programs in the two NXT bricks are started, successively, with the entire robot's body resting on the floor. Because the body is at rest, none of the motors detect rotation, so there is no attempt by any motor to move. As a result, the motor programs are all running, but there is no movement. However if the robot is disturbed, it suddenly erupts into movement that it attempts to stabilize. For instance, if one segment of the robot is pulled up from the floor, this launches a complex feedback process that produces contortions that are converted into locomotion, such as those illustrated in Figure 5-45.

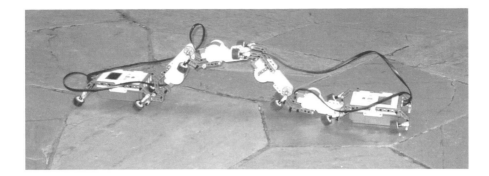

If someone comes along, though, and restrains the robot so that it cannot move as is being done in Figure 5-46, then after a short period the rotations detected by the motors fall below threshold, and Wormeostat reverts to its "sleeping" state. As shown in Video 5-3, a relatively small poke of the robot will awake it from this state and make it move again.

5.39 IMPLICATIONS

5.39.1 Two Cultures

It is generally acknowledged that there exists a rift between science and the humanities, a rift that Charles Percy Snow explored in his famous essay "The Two Cultures" (Snow, 1969).

This same rift would appear to separate those who apply analytic methods from those who apply synthetic ones. For instance, Jansen (2007) identifies the former as being "engineers," and the latter as being "artists" (see also Section 5.15.1). "Given the restrictions of this material [plastic tubing] I was forced to seek out escape routes that were neither logical nor obvious. The strategy I followed to assemble the animals is in fact the complete opposite of that taken by an engineer" (Jansen, 2007, p. 35). Jansen built his Strandbeests from the ground up, being informed by the constraints of his parts, and not beginning with a carefully analyzed design. In contrast, he feels that engineers build "devices that are first thought out and then assembled. That's how engineers work. They have ideas and then they make these ideas happen" (p. 35).

Within cognitive science, there is a definite tension between analysis and synthesis. We have seen arguments that analytic theories will be more complicated than synthetic ones, because they usually ignore the frame of reference problem (Braitenberg, 1984; Pfeifer & Scheier, 1999). It has also been argued that models created from analytic methods are evaluated by their fit to existing data, and therefore are less able to provide the surprising insights that often accompany models created via synthetic methods (Dawson, 2004; Dawson, Boechler, & Valsangkar-Smyth, 2000; Dawson & Zimmerman, 2003).

A famous anecdote (Josephson, 1961) nicely illustrates the tension between the analytic and synthetic cultures. In 1878, Francis Upton was hired to work as a specialist in mathematical physics in Thomas Edison's Menlo Park laboratory. Upton was a Princeton graduate, who had studied for a year in Germany under Helmholtz. As one of his first tasks, Upton was asked by Edison to calculate the volume in cubic centimetres of a pear-shaped glass bulb used in the lab for experiments on electric lighting. With his mathematical training, Upton adopted an analytic approach to solve this problem. "Upton drew the shape of the bulb exactly on paper, and got the equation of its lines, with which he was going to calculate its contents" (Josephson, 1961, p. 193). After an hour, Edison asked Upton for the results, and was told that the mathematician was only halfway done and needed more time. Edison responded that a synthetic approach would have been more efficient: "'Why', said Edison, 'I would simply take that bulb, fill it with a liquid, and measure its volume directly'."

5.39.2 Mending the Rift

The Upton anecdote was told many times, because it illustrated "the contrast between the practical, 'Edisonian' rule-of-thumb method and the mathematical scientists' different mode of attack on the same problem" (Josephson, 1961, p. 193). However, what the anecdote leaves out is that Edison realized that Upton's approach complemented his own, and that he needed Upton's analytical skills to translate his intuitions into rigorous equations. Menlo Park succeeded because Edison combined analytic and synthetic methods. Jansen (2007) also recognizes that the two methods can be combined. His use of genetic algorithms to determine his holy numbers is one example of such integration.

The complementary nature of the two approaches is also illustrated in the walking machines that we have been discussing in Chapter 5. Detailed analyses (McGeer, 1990a) laid the groundwork for the creation of simple passive dynamic walkers, such as the LEGO model that we built. However, as was illustrated with the Homeostat, analysis may become difficult or impossible when non-linear systems are scaled up. In this case, synthetic models such as the Strandbeest and the Wormeostat can be created in the absence of analysis. Their existence can then be used to constrain later analysis that will ground the "practical demonstration" in rigorous mathematics.

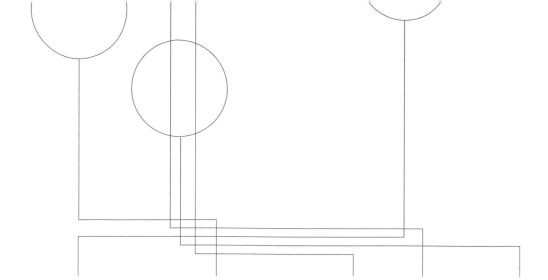

Chapter 6
Machina Speculatrix

6.0 CHAPTER OVERVIEW

The first autonomous robots were created by William Grey Walter in the late 1940s in Bristol, England. Grey Walter was a master publicist for his research, and the feats of these machines received a fair amount of coverage in the British press. These descriptions make clear that these machines, called *Machina speculatrix* by their inventor, were capable of generating a variety of complex behaviours. However, at the root of this complexity was an extremely straightforward and elegant design. As we have seen with other machines described in previous chapters, *Machina speculatrix* was a fairly simple robot that performed its complex behaviours because it was situated in an interesting environment. One purpose of this chapter is to provide a brief historical introduction to this particular type of robot. The second purpose is to provide detailed instructions about how to build a functionally similar robot out of LEGO Mindstorms components in order to explore its behaviour.

The LEGO Tortoise is a more complicated device than the LEGO robots that have been described in preceding chapters. Part of this complexity comes from the fact that the motors that are used to control two general behaviours (moving forward and steering) are under the influence of two different sensory systems (light detection and obstacle detection). We have found that the programming of such robots is facilitated by adopting an approach called the subsumption architecture (Brooks, 1999). An account of this general approach, and a subsumption architecture for the LEGO Tortoise, is the topic of Chapter 7.

6.1 WILLIAM GREY WALTER

6.1.1 Biographical Highlights

Born in Kansas City, Missouri, in 1910, William Grey Walter moved with his family to England when he was seven years old. His secondary education was at Westminster School. He then entered King's College at Cambridge, where he took a First in Natural Sciences in 1931 (Bladin, 2006). He studied nerve physiology as a postgraduate, and received his MA in 1934. Unfortunately this work did not result in his being awarded a fellowship at King's. "Cambridge still could not accept the brain as a proper study for the physiologist" (Grey Walter, 1963, p. 55).

Frederick Golla led the Central Pathological Laboratory at Maudsley Hospital in London, and had been interested in the clinical possibilities of electroencephalography (EEG) since at least 1929 (Grey Walter, 1963, p. 55). Inspired by a presentation of William Lennox at the 1935 London International Neurology Conference, Golla recruited Grey Walter for the task of developing an EEG machine.

Beginning in 1935, Grey Walter himself designed, assembled, and tested the initial EEG equipment. His tests on neurological patients confirmed the utility of EEG for studying epilepsy that had been reported by Lennox, but Grey Walter also demonstrated that his machine could be used to locate brain tumours (Bladin, 2006). This pioneering work included the discovery and naming of delta and theta rhythms. In 1947 Cambridge awarded him his ScD in recognition of his enormous scientific contributions.

In 1939, Golla moved his research group from London to the Burden Neurology Institute in Bristol. Grey Walter became head of this institute, holding this position for thirty years. He died of a heart attack in 1977. One obituary (Cooper, 1977) listed his many important practical contributions, which included developing the first British machine for electroconvulsive therapy, creating the first portable electroencephalograph, designing the first automatic low-frequency wave analyzer, and making electrodes for recording deep in the brain during neurosurgery. A second obituary noted that "few scientific disciplines owe as much to one man as electroencephalography and clinical neurophysiology owe to Grey" (Shipton, 1977).

6.1.2 A Very Public Robot

Though famous for his discoveries in EEG, Grey Walter is also celebrated for his pioneering work in cybernetics and robotics. He created the first biologically inspired robots (Holland, 2001, 2003a, 2003b); they are the subject of this chapter.

Grey Walter was an unconventional man for his time. "His popular and academic reputation encompassed a heterogeneous series of roles ranging from robotics pioneer, home guard explosive experts, wife swapper, t.v.-pundit, experimental drugs user, and skin-diver to anarcho-syndicalist champion of leucotomy and electro-convulsive therapy." (Hayward, 2001, p. 616). Part of his unconventionality was reflected in his desire and ability to generate publicity about his research.

For instance, the existence of Grey Walter's robots, which he labelled *Machina speculatrix* but were more commonly called Tortoises, was first revealed in a story in the *Daily Mail* in 1949, and the BBC created a television newsreel film about these robots in 1950 (Holland, 2003a). The robots were displayed at the 1951 Festival of Britain, and in 1955 he released them to roam about the audience at a meeting of the British Association. "The tortoises, with their in-built attraction towards light, moved towards the pale stockings of the female delegates whilst avoiding the darker legs of the betrousered males" (Hayward, 2001, p. 624).

The purpose of this chapter is to explore *Machina speculatrix* in more detail, first by examining Grey Walter's accounts of their behaviours and inner workings, and second by constructing our own version of this robot from LEGO components.

6.2 THE TORTOISE

6.2.1 Appearance

William Grey Walter constructed his first Tortoise, Elmer (for Electro-Mechanical Robot) in 1948; a second machine, Elsie (for Electro Light Sensitive Internal External) was built several months later (de Latil, 1956). Both were comprised of wartime surplus components. "The first model of this species was furnished with pinions from old clocks and gas meters" (Grey Walter, 1963, p. 244). They were eventually replaced by six new devices built in 1951 by 'Bunny' Warren (Holland, 2003a), two of which are currently displayed in museums.

Grey Walter (1963, p. 113) classified the machines "*Machina speculatrix*, inevitable name of the species for the discerning, though 'tortoise' to the profane." The latter name arose because each machine was covered by a tortoise-like shell. In the 1949 *Daily Mail* article, they were described as "toys containing an electric brain" (Holland, 2003a, p. 2090).

There was also an enclosure that was used to recharge robot batteries. "On the floor, in a corner of the room, was a sort of hutch in a portable box illuminated by a very strong lamp inside it" (de Latil, 1956, p. 212). Figures that illustrate the Tortoise entering its hutch, as well as other figures that illustrate Tortoise behaviour, are available at (http://www.ias.uwe.ac.uk/Robots/gwonline/gwarkive.html).

While other robots predated the Tortoises, such as the "electric dog" created in 1912, and "philidog," exhibited in Paris in 1929, Grey Walter's robots represented an advance because they were autonomous (Sharkey & Sharkey, 2009). This resulted in behaviour that was much more life-like than either "robotic dog," because the feedback exploited by the Tortoises caused them to explore their environment.

6.2.2 Behaviour

The general press was provided a fairly eclectic description of the behaviours of the Tortoises. The Daily Mail reported that "the toys possess the senses of sight, hunger, touch, and memory. They can walk about the room avoiding obstacles, stroll round the garden, climb stairs, and feed themselves by automatically recharging six-volt accumulators from the light in the room. And they can dance a jig, go to sleep when tired, and give an electric shock if disturbed when they are not playful" (Holland, 2003a, p. 2090).

The general public received similar written accounts from Grey Walter himself in his popular book *The Living Brain* (Grey Walter, 1963). For instance, he described his invention as "an electro-mechanical creature which behaves so much like an animal that it has been known to drive a not usually timid lady upstairs to lock herself in her bedroom, an interesting blend of magic and science" (pp. 124–25).

Of course, Grey Walter had particular scientific interests in his machines, and used behavioural observations to support a general theoretical position that played an important role in cybernetics (Grey Walter, 1950a, 1950b, 1951). Grey Walter (1963, p. 112) described the Tortoise as an example of "a free goal-seeking mechanism." The purpose of the Tortoise was to "demonstrate the first of several principles exemplified in the mechanisms of most living creatures" (p. 113). He generated a list of nine such principles, and chose particular Tortoise behaviours to illustrate these principles in action (Grey Walter, 1950a). Some of these behaviours are described in the pages that follow.

Grey Walter's synthetic use of robots was intended to explore how basic biological and neurological principles might affect behaviour, and this tradition of research remains an important influence on many modern researchers (Reeve & Webb, 2003). Grey Walter's "ingenious devices were seriously intended as working models for understanding biology: a 'mirror for the brain' that could both generally enrich our understanding of principles of behaviour (such as the complex outcome of combining simple tropisms) and be used to test specific hypotheses (such as Hebbian learning)" (Reeve & Webb, 2003, p. 2245). For this reason alone, they are worthy of review; let us begin by considering their diverse behaviours.

6.3 SPECULATION AND POSITIVE TROPISMS
6.3.1 Exploration as Speculation

Grey Walter was intrigued by the idea that new developments in machines, beginning with James Watt's steam engine governor, challenged traditional distinctions between the living and the non-living (Grey Walter, 1950a). The notion that a living organism's behaviour could be controlled by negative feedback was commonplace in biology and physiology. The new machines of interest to Grey Walter also appeared to be governed by this principle, a discovery that was fundamental to the then-new field of cybernetics (Ashby, 1956; Wiener, 1948). He was interested in the extent to which fairly simple machines could generate interesting, active, lifelike behaviour in virtue of receiving negative feedback. He developed *Machina speculatrix* to explore this issue.

He gave these robots this "mock-biological name" because "they illustrate particularly the exploratory, speculative behavior that is so characteristic of most animals" (Grey Walter, 1950b, p. 43). De Latil (1956, p. 209) provided a first-hand account of this behaviour: "Elsie moved to and fro just like a real animal. A kind of head at the end of a long neck towered over the shell, like a lighthouse on a promontory and, like a lighthouse; it veered round and round continuously." Grey Walter argued that the Tortoises were different from other computing devices because of their constant search for "problems to solve" (Grey Walter, 1963, p. 126), a search that would cover several hundred square feet in an hour.

6.3.2 Phototropism

What, though, did the Tortoises search for? Grey Walter designed his machines to seek light of moderate intensity. When this stimulus was not visible, the Tortoise would engage in active exploration, "like the restless creatures in a drop of pond water, it bustles around in a series of swooping curves" (1963, p. 126). The "head at the end of a long neck" mentioned by de Latil was a photocell mounted on the robot that could be used to detect light. When moderate light was detected, exploration would stop, and the robot would approach the light source.

De Latil (1956) visited Grey Walter's house to see a demonstration of the Tortoises. It was a "dull dark day," and when Grey Walter switched on a lamp the attention of the robot was immediately drawn. "She continued on her way towards the attraction of the light source. Even so, she interrupted her direct course towards the light by one or two 'hesitation waltz' steps. At last, however, her course became firm and direct. Her head no longer turned, but remained fixed in the direction of the light as if fascinated by it" (p. 210).

Grey Walter described this behaviour as being "analogous to the reflex behavior known as 'positive tropism', such as is exhibited by a moth flying into a candle" (1950b, p. 43). He frequently used biological and psychological terms to describe his robots because of his aim to illustrate that cybernetic principles were capable of imparting very life-like qualities to machines.

6.3.3 Inferring Internal Mechanisms

One way to distinguish analytic from synthetic approaches to explaining behaviour is to consider the kinds of theories that would be proposed by external observers of agents, and to compare these with those of the agents' creator (Braitenberg, 1984). Braitenberg's position was that the analytic account generated by the external observer would be obtained with greater difficulty, and would likely also be more complicated than necessary.

To this point we have been provided some general notions of the appearance of Grey Walter's robots, as well as an account of how these machines behave in darkness, and in moderate light. We are thus in a position to adopt the analytic approach, and to begin to generate a theory that explains why the Tortoises act as they do. What mechanisms can you propose that would produce the sorts of behaviour that Grey Walter demonstrated to de Latil? Would these mechanisms be capable of generating other behaviours if the robot's environment was changed?

6.4 NOT ALL LIGHTS ARE THE SAME
6.4.1 A Negative Phototropism

Importantly, the Tortoises were designed to be attracted to light, but not to *any* light. That is, not all lights produced the same behaviour in *Machina speculatrix*. While Elsie would be attracted to moderate light, she would be repelled by light that was too bright or intense. The detection of bright light elicits a negative tropism in which "the creature abruptly sheers away and seeks a more gentle climate" (Grey Walter, 1950b, p. 44).

One problem that is encountered when considering the behaviours exhibited by the Tortoises is that there is little information available. There are a few written descriptions of a standard set of theoretically interesting behaviours (de Latil, 1956; Grey Walter 1950a, 1950b, 1951, 1963).

The BBC film of these robots is still in existence, and is probably the only moving picture of *Machina speculatrix* in action. Holland (2003a) provides a written transcript of this film, but it is not available for general viewing.

Fortunately, Grey Walter left an important static record of robot behaviour that foreshadowed Herbert Simon's parable of the ant (Simon, 1969). For a series of experiments that were conducted at his home, Grey Walter mounted candles on the robots. He then took extended exposure photographs of his machines. As a result, the candles produced bright lines in these photographs that traced the path taken by the robots over an extended period of time. Some of these photographs can be viewed at http://www.ias.uwe.ac.uk/Robots/gwonline/gwarkive.html.

Figure 6-1 is an example of a path that could be produced by the robot when it encountered an environment in which there was a single light source (a lamp) placed on the floor in front of it. In general terms, the robot circles the light at a moderate distance, as is roughly indicated by the arrows in Figure 6-1. However, the actual path of the robot is much more complicated than this, as is indicated by the solid dark line in the figure. For another example of this behaviour, see Figure 8 in Holland (2003b).

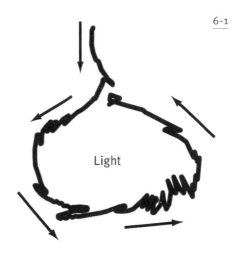

Light

Holland (2003a) also discovered a document at the archives of the Burden Neurology Institute entitled "Accomplishments of an artifact," which was likely written by Grey Walter and provides descriptions of several of Grey Walter's pictures of Tortoise behaviour. The text that accompanies the photograph of Elsie circling the light was, "Attracted at first by a distant bright light the creature reaches the zone of brilliant illumination where it is repelled by the excessive brilliance of the light and circles round it at a respectful distance, exhibiting a search for optima rather than maxima—the idea of moderation of the classical philosophers" (Holland, 2003a, p. 2107).

6.4.2 Analysis of Behaviour

The behaviour illustrated in Figure 6-1 indicates that *Machina speculatrix* will produce a trajectory in which it "circles" a light at a safe distance, although the path taken by the robot is much more complicated than mere circling. "The machine circles around [the light] in a complex path of advance and withdrawal" (Grey Walter, 1950b, p.44). Presumably this path is the result of some combination of its positive and negative phototropisms. What mechanisms would be required to generate such behaviour in the robot? To what extent could these putative mechanisms be used to predict the precise, complex path of the robot?

6.5 CHOICE

6.5.1 Buridan's Ass

In *On The Heavens* (Aristotle, 1953/350 BC), Aristotle (Book 2, Section 13.III, p. 27) describes a particularly drastic state of indecision: "the man who, though exceedingly hungry and thirsty, and both equally, yet being equidistant from food and drink, is therefore bound to stay where he is."

Since mediaeval times, this situation has become known as a paradox called "Buridan's ass." In this paradox, two equally attractive stacks of hay are placed at identical distances from a donkey. The poor animal starves to death because the stacks of hay, being identical in every respect, do not provide the information that enables the donkey to choose one over the other. The paradox is named for French philosopher Jean Buridan (c. 1300–1358). He himself did not use the paradox; instead, it is a parody of his theory of free will, in which one must choose an action that produces the greatest good.

Presumably the force of this parody is that a living organism would never fall prey to it. We have already seen that one of Grey Walter's interests was in using cybernetic principles to generate lifelike behaviour in machines. Indeed, one aspect of human psychology and animal behaviour that *Machina speculatrix* was intended to demonstrate was "the uncertainty, randomness, free will or independence so strikingly absent in most well-designed machines" (Grey Walter, 1950b, p. 43).

6.5.2 Complicating the Environment

What would a Tortoise do if it found itself in the position of Buridan's ass? To answer this question, Grey Walter complicated the robot's environment by adding a second light source. An example of the behaviour that would be expected is illustrated in Figure 6-2. The dark lines indicate paths taken around the lights, and the arrows indicate the direction of motion. A photograph demonstrating the behaviour, and taken by Grey Walter, can be seen as Figure 9 in Holland (2003b).

Holland (2003a, p. 2108) reports that Grey Walter described the behaviour that it depicts as follows: "The solution of the dilemma of Buridan's ass. The photoelectric cell which functions as the creature's eye scans the horizon continuously until a light signal is picked up;

the scanning stops, and the creature is directed towards the goal. This mechanism converts a spatial situation into a temporal one and in this process the dilemma of two symmetrical attractions is automatically solved, so that by the scholastic definition the creature appears endowed with 'free-will'. It approaches and investigates first one goal and then abandons this to investigate the other one, circling between the two until some other stimulus appears or it perishes for want of nourishment." Holland notes that this demonstration was viewed by Grey Walter as one of the Tortoise's most impressive achievements.

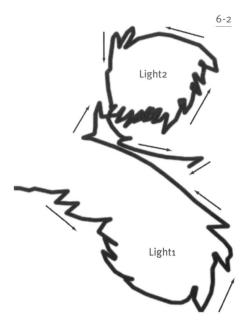

It is clear that one method for making the behaviour of the Tortoise more complicated is to increase the complexity of its environment. The robot that generates the example trajectories in Figures 6-1 and 6-2 is the same; the only difference is the number of light sources. Are the possible mechanisms that you have been inferring for *Machine speculatrix* sufficient to account for this result, not to mention for the robot's ability to choose?

6.6 ADDITIONAL NEGATIVE TROPISMS
6.6.1 Avoiding Obstacles

Because of their constant exploration of their environment, the Tortoises were likely to encounter physical objects that would serve as barriers between them and their goal (intermediate light). To cope with this, Grey Walter built in a second negative tropism toward "material objects."

Figure 6-3 provides an example path produced by this negative tropism. The robot can sense the light through the legs of a stool, but encounters the stool as a physical obstacle that must be avoided before the journey to the light can continue.

Another example is a photograph (see Holland, 2003b, Figure 7) — labelled "discernment" by Grey Walter — which he described (Holland, 2003a, pp. 2106-2107) as follows: "Presented with a remote goal (seen at the top of the slide) the creature encounters a solid obstacle that it cannot move, and although it can still see the candle it devotes itself to circumventing the obstacle (of which it retains a short memory) before it circles round in an orbit and reaches the objective."

Grey Walter viewed this as evidence of "discernment" because, he

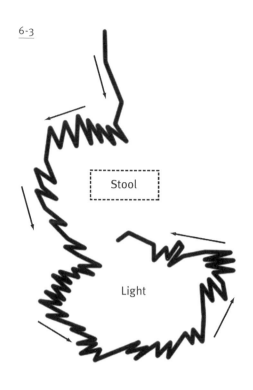

Stool

Light

argued (e.g., 1950a, p. 210), the obstacle-avoiding responses that were induced by encountering the barrier in the photo did more than provide a means for escaping this problem. In addition, this behaviour eliminated the attractiveness of the light. As long as the obstacle was being detected, the tortoise had no interest in seeking the light. "The search for lights and attraction to them when found is not resumed for a second or so after a material conflict." Grey Walter explained this by stating that the Tortoise had a brief memory of the obstacle.

De Latil (1956) provided several accounts of how the Tortoises behaved when obstacles were encountered. "I enclosed Elsie in a barricade of furniture, but by banging herself and reversing and knocking herself and backing and turning again, she managed to find her way out" (p. 209). When Grey Walter placed a box in Elsie's way, "she got a shock, seemed to hesitate, and no longer continued her way towards the light, although she could see it quite well shining above the level of the box" (p. 210). The ensuing discernment behaviour, in which the light was ignored while the box was avoided, and light seeking was delayed, particularly interested de Latil: "Now came the surprising turn. Elsie acted as if she remembered her shock."

What mechanisms are required to create this negative tropism for material objects? How should these mechanisms interface with those responsible for Elsie's other phototropisms, and for obstacle memory?

6.6.2 Avoiding Slopes

Machina speculatrix was also given a negative tropism toward going up or going down steep slopes. Whenever the robot finds itself moving along a slope that is too steep it generates behaviours similar to those seen when it bumps into objects. What mechanisms would you propose to explain such behaviour in this machine?

6.7 DYNAMIC TROPISMS
6.7.1 Toys vs. Tools

Humankind has always had a fascination with the possibility of creating machines that mimic living creatures, such as the duck and flute-player automata created by de Vaucanson in the eighteenth century (Wood, 2002), or the infamous chess-playing Turk that was constructed by von Kempelen in the eighteenth century, and was in and out of the public eye until its destruction by fire in 1854 (Standage, 2002). In more modern times, this fascination has been fueled by the many depictions of robots and cyborgs in popular fiction and movies (Asimov, 2004; Caudill, 1992; Ichbiah, 2005; Levin, 2002; Menzel, D'Aluisio, & Mann, 2000).

The life-mimicking automata of previous centuries were only partially successful in the sense that they exhibited a predetermined and predictable set of behaviours, and failed to change or adapt their behaviour over time. This lack of learning or of the ability to adapt has long been used to differentiate living organisms from machines (Grey Walter, 1950a; Wiener, 1964). Automata were toys, not tools.

Grey Walter viewed his robots as "electro-mechanical animals" that differed from most other machines because of their dynamic and unpredictable behaviours. One compelling type of evidence that could be marshalled in support of this position was the fact that the phototropisms changed over time, and were not static characteristics of *Machina speculatrix*.

6.7.2 Changing Light Sensitivity

To be more precise, the Tortoises were first and foremost electrical devices. This was a key theoretical position adopted by Grey Walter, because his EEG research had convinced him that the "nervous mechanism underlying even such elaborate functions as original thought and imagination may someday be definable in electric terms (1950a, p. 208). As well, the reason that he built a positive tropism for moderate light into his machines was because he interpreted this tropism as signalling the robot's need for the electrical energy that it required.

We have previously noted that the hutch was an enclosure that was designed to recharge the Tortoise batteries, and that also contained a very bright light. When the fully charged Tortoise encounters the bright light in the hutch, it is repelled by it. The Tortoise withdraws from the hutch, and begins to explore elsewhere in its environment. However, after a period of exploration, the Tortoise's negative phototropism appears to have changed in magnitude. This is because the robot is no longer repelled by the hutch light. Instead, the robot is attracted to it, and has a tendency to re-enter the hutch on its own to permit recharging (see Holland, 2003a, Figures 10 and 11). Grey Walter labelled the photograph that Holland reproduced as his Figure 10 "simple goal seeking," and described it as follows: "Started in the dark the creature finds its way into a beam of light and homes on the beam into its feeding hutch" (Holland, 2003a, p. 2110).

Importantly, different instances of this simple goal seeking can have the same general result, but at a fine level of detail the particular trajectory taken by the machine will vary as a function of the robot's location, its electrical state, and other environmental factors. For instance, the two photographs that demonstrate this behaviour in Holland (2003a) record the robot's movement from two different starting points—one at the left of the hutch, the other at the right. The trajectories in the two photographs are strikingly different, though the end result is the same (i.e., entering the hutch).

Holland (2003a) points out that in practice the success of returning to the hutch was rather limited. He observes that when the original hutch caught fire from its lamp and was destroyed, the replacement hutch had a lamp—but did not include a charging system! Nonetheless, Holland agrees that one of the Tortoise's characteristics was the dynamic nature of its phototropisms.

How would the mechanisms inferred by an external analysis of behaviour be modified in order to account for this dynamic dimension of robot behaviour?

6.8 SELF-RECOGNITION
6.8.1 Self, Not Machine

In Section 6.7 it was argued that adaptability was one characteristic that separated machines from living beings. What other characteristics might also serve this role? One possibility is that living beings—and some would argue only human beings—are the only agents capable of self-consciousness or self-awareness (Dawkins, 1993; Dennett, 1991, 2005; Varela et al., 1991).

Mirrors have had an important role in the study of self-consciousness since they were used to provide evidence that chimpanzees could recognize themselves, and therefore demonstrate a form of self-consciousness (Gallup, 1970). Gallup allowed chimpanzees to play with mirrors for 10 days. Then, when they were anaesthetized, a mark was placed on their forehead. They were unaware of the mark until they were given the mirror, which they used to guide their hands to touch the mark on their own head (and not the mark reflected in the mirror). Monkeys did not behave in this way; machines, too, should not.

6.8.2 The Mirror Dance

Nevertheless, two decades before Gallup's (1970) publication Grey Walter was using mirrors to demonstrate that *Machina speculatrix* was also capable of self-recognition. In the robots, a small pilot light was connected to the front of the robot, and turned on when the machine was turning, but not when it was moving straight. This light was bright enough to elicit positive phototropism when the pilot light was reflected in a mirror. However, as soon as the robot headed toward this reflection, the light would turn off. Of course, this in turn produced further exploration, turning the pilot light on again. "The creature therefore lingers before a mirror, flickering, twittering and jigging like a clumsy Narcissus" (Grey Walter, 1963, p. 128). Grey Walter described this famous mirror dance as evidence of self-recognition.

Figure 6-4 illustrates an example path that the Tortoise could take when encountering its reflection in a mirror. Grey Walter's photographic evidence of this behaviour is reproduced as Figure 6 in Holland (2003a).

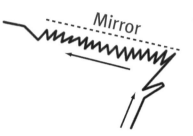

6-4

Does self-recognition emerge from the mechanisms proposed to this point to explain the various behaviours of Grey Walter's robots?

6.9 MUTUAL RECOGNITION
6.9.1 The Relative World

To this point we have seen illustrations of a number of different Tortoise behaviours. In most cases Grey Walter increased the complexity of the observed behaviour by modifying the robots' environments, and not the robots themselves. The simplest environment contained a single light. The environment was made more complicated by adding a second light, or by having a light and one or more physical obstacles.

The mirrored environment was more complicated still, because the mirror served as a barrier that reflected the light source, which was one of the robots. As a result, in the mirrored environment the light source was very complicated, in the sense that it changed position when the robot changed position.

The dynamic nature of the mirrored environment illustrates a key concept that is central to Grey Walter's robots: that robot behaviour is the result of the interaction between the robot and its sensed environment, and is not merely a function of the internal workings of the robot alone.

De Latil (1956, p. 237) was particularly impressed by this concept: "The object that I see, that I sense, cannot be thought of as an isolated entity; if I sense it, it is part of my system. Such a relativistic view of external reality is of course commonplace in many philosophies." These observations from de Latil could easily be part of more modern, more anti-representational, accounts of cognition (Braitenberg, 1984; Brooks, 1999, 2002; Clark, 1997; Pfeifer & Scheier, 1999; Varela et al., 1991; Winograd & Flores, 1987).

6.9.2 Social Environments

Grey Walter created an even more complicated, dynamic environment when he activated two of the Tortoises at the same time. In this case the candles mounted on the two machines, as well as their flickering pilot lights, could fuel the phototropisms. However, they provide added complexity because the light sources are constantly moving, and their movement is influenced by feedback relations between each robot.

Grey Walter entitled Figure 6-5 below "social organization," and provided the following description (Holland, 2003a, p. 2104):

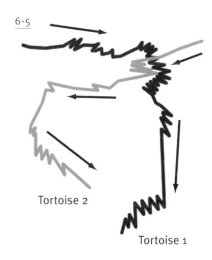

6-5

Tortoise 2

Tortoise 1

The formation of a cooperative and a competitive society. When the two creatures are released at the same time in the dark, each is attracted by the other's headlight but each in being attracted extinguishes the source of attraction to the other. The result is a stately circulating movement of minuet-like character; whenever the creatures touch they become obstacles and withdraw but are attracted again in rhythmic fashion. While this evolution was in progress the light in the feeding hutch was turned on; the common goal disrupted the cooperative organization and transformed it into a ruthless competition, in which both creatures jostled for entrance to the source of nourishment.

Are the mechanisms that the analytic observer has proposed to this point powerful enough to explain this activity, or must additional mechanisms be hypothesized?

6.10 INTERNAL STABILITY
6.10.1 Feedback and Cybernetics

At the time of the Second World War, it was recognized that feedback was a fundamental mechanism of control (Ashby, 1956; Wiener, 1948).

Feedback occurs when information about an action's effect on the world is used to inform the progress of that action. For example, "when we desire a motion to follow a given pattern the difference between this pattern and the actually performed motion is used as a new input to cause the part regulated to move in such a way as to bring its motion closer to that given by the pattern" (Wiener, 1948, p. 6).

Wiener also realized that processes like feedback were central to a core of problems involving communication, control, and statistical mechanics. He provided a mathematical framework for studying these problems, and this framework defined a new discipline that Wiener called cybernetics, which was derived from the Greek word for "steersman" or "governor." "In choosing this term, we wish to recognize that the first significant paper on feedback mechanisms is an article on governors, which was published by Clerk Maxwell in 1868" (1948, p. 11).

6.10.2 Cybernetics and Simulation

The complexity of feedback relationships is mathematically challenging. As we saw in Section 5.12.1, another cybernetics pioneer, W. Ross Ashby, realized that insights into feedback required alternative methods of study. We saw that his synthetic approach was to build and observe the Homeostat (Ashby, 1960).

Modern studies have used the synthetic approach to explore the properties of dynamic feedback. Rather than adopting Ashby's approach, though, and building physical devices, these systems are usually created and studied using computer simulations.

One example is provided by autoassociative artificial neural networks, such as the Hopfield network (Hopfield, 1982, 1984). In its simplest format, a Hopfield network is a collection of simple processing units that can either be "on" or "off." A unit's state is determined by the total signal that is being sent to it by other units. If that signal is above the unit's threshold, then it turns on; otherwise the unit turns off. A Hopfield network is massively parallel, which means that every processing unit is connected to every other processing unit in the system. Thus, when one unit changes state, this (via feedback sent through the connections) can affect the rest of the network.

The weight of each connection is determined by using a simple learning rule so that the network remembers a small number of patterns that are presented to it by turning some of the units on and the others off. The task of the network is to retrieve one of these remembered patterns when it later receives a signal (i.e., some pattern that it may not have encountered before) from the environment.

When a signal is received, one of the many processing units in the network is selected at random. Then the signal to that unit is computed, and it is turned on or off as a result. Over time, all of the network's units are repeatedly selected and updated. Eventually, no matter what unit is selected, when the updating rule is applied the unit will not change its state. At this point, the network is said to have converged, and its processing units represent the information that was recalled as a result of the original stimulation. This is useful, because (for example) if one presents a "noisy" example of a previously learned stimulus, then the network responds by removing the noise.

Importantly, Hopfield network dynamics are identical to those of the Homeostat. That is, one can define a metric of the overall "energy" in a Hopfield network at any given time (Hopfield, 1982). Hopfield proved that whenever a unit in the network changed its state, this meant that total network energy decreased. Thus, as one watches a Hopfield network dynamically reacting to some stimulus, one sees a system changing its units — and the feedback signals that they send to other units in the network — in such a way as to seek a lower-energy, more stable state.

6.11 PARSIMONY
6.11.1 Two Approaches to Stability

Grey Walter (e.g., 1950a) felt that the Homeostat, in its failure to generate actions, could only be used to model the maintenance of *internal* stability. A second concern of Grey Walter's was that the Homeostat's ability to maintain internal stability was based upon a large number of internal connections. One of the themes that motivated Grey Walter's robots was parsimony (Grey Walter, 1950a). He deliberately restricted the numbers of components in the Tortoises to "two functional elements: two miniature radio tubes, two sense organs, one for light and the other for touch, and two effectors or motors, one for crawling and the other for steering" (Grey Walter, 1950b, p. 43). To him, the complexity of the behaviour of his robots indicated that the interconnectedness of elements (to themselves, and to the sensed world) was more important than the sheer number of internal building blocks in terms of accounting "for our subjective conviction of freedom of will."

The Tortoise can be viewed as a response that was motivated by these concerns, where the fundamental goal was to generate complex behaviour from much simpler componentry and from far fewer connections (Boden, 2006). Grey Walter modelled behaviour "as economically as he could — in both the financial and the theoretical sense. Not only did he want to save money [...], but he was determined to wield Occam's razor. That is, he aimed to posit as simple a mechanism as possible to explain apparently complex behavior. And simple, here, meant simple" (Boden, 2006, p. 224).

6.11.2 A Simple Machine

Why does Boden (2006) emphasize the extreme simplicity of Grey Walter's design? It is because the behaviour of the tortoises is rooted in the interactions between ridiculously small sets of components. The robot used two motors, one to move the robot using front wheel drive, the other motor to steer the front wheel. The motors were controlled by two different sensing devices. The first was a photoelectric cell mounted on the front of the steering column. The other was an electrical contact that served as a touch sensor. This contact was closed whenever the transparent shell that surrounded the rest of the robot was displaced.

Of a Tortoise's two reflexes, the light-sensitive one was the more complex. In conditions of low light or darkness, the machine was wired in such a way that its drive motor would propel the robot forward while the steering motor slowly turned the front wheel. When moderate light was detected, the steering motor stopped. As a result, the robot moved forward, approaching the source of the light. However, if the light source were too bright, then the steering motor would be turned on again at twice its normal speed. De Latil (1956) explains how modifying the thresholds associated with these reflexes could impact the behaviour of the robots.

The touch reflex was such that when it was activated, any signal from the photoelectric cell was ignored. When the Tortoise's shell encountered an obstacle, an oscillating signal was generated that first caused the robot to drive fast while slowly turning, then to drive slowly while quickly turning. Object avoidance was achieved by toggling between these two states. When the machine moved along a steep slope, the shell would change position, and avoidance behaviour would also result.

These reflexes were responsible for all of the behaviour that has been described in the preceding pages. Would an analytical observer be likely to invent such a simple theory about *Machina speculatrix*? Consider the example paths that have been illustrated in Figures 6-1 through 6-5 from the perspective of the parable of the ant (Simon, 1969). What kind of analytic theory might be proposed to explain these paths? How complicated would an analytic theory assume that the internal mechanisms of the Tortoise must be to produce such complex data?

6.12 A LEGO TORTOISE

6.12.1 A New Generation

It was argued earlier that it is unfortunate that Grey Walter's robots are not available for current study. If one is interested in exploring the behaviour of these machines, then one is limited to the record that we have been considering in the earlier pages of this chapter.

Of course, another possible course of action is to build machines in the spirit of Grey Walter's Tortoises. If we used LEGO Mindstorms components to do this, then we could build new machines, explore their behaviours, and carry on with Grey Walter's research program.

The purpose of the remaining pages of this chapter is to provide instructions for carrying this research project out. They detail the construction of a robot that we will call the Tortoise, and which is illustrated in the figure below. If the reader would prefer to use wordless, LEGO-style instructions, they are available as a pdf file from the website that supports this book (http://www.bcp.psych.ualberta.ca/~mike/BricksToBrains/).

6.12.2 Variations of Design

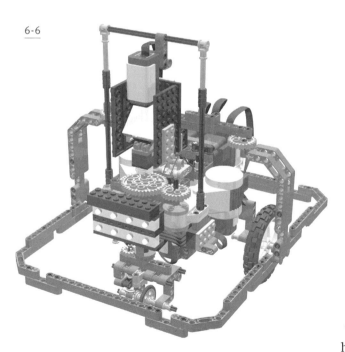

6-6

It should be clear from Figure 6-6 that our machine is not called Tortoise because of its appearance. Rather than building an exact replica of a Grey Walter robot, we have elected to develop a machine that shares the name of its ancestors because of its functional similarities. The robot that we will build has two motors, one to power a front wheel drive, the other for steering the front wheel. A light sensor is used to implement the phototropisms that Grey Walter studied. Rather than building a shell, we have constructed a surrounding bumper that permits the machine to change its motor behaviours when it detects a material obstacle or when it ventures onto steep slopes.

6.13 PARTS FOR A MODULAR DESIGN
6.13.1 Sophistication from Tweaking

The LEGO Tortoise that we will uses a light sensor not only to sense varying degrees of ambient light, but also to implement a "stick in ring" method for detecting obstacles. It is far more complex than previous robots that we have described: it uses a wider variety of LEGO pieces, and we will see in Chapter 7 that it is slightly more complicated to program. It also generates more interesting behaviour. Figure 6-7 shows the parts that should be foraged before building can begin!

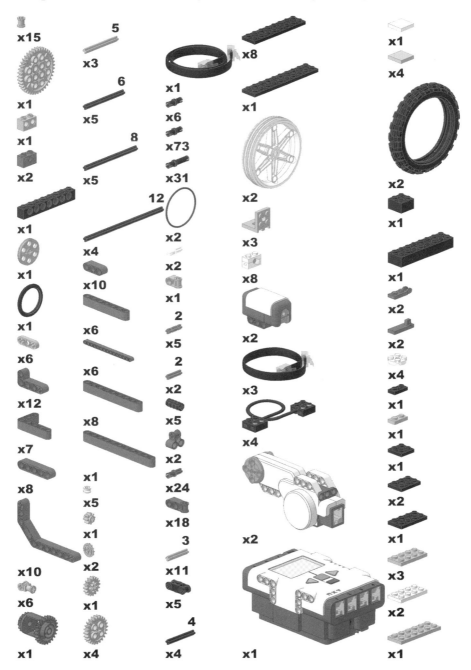

6.14 THE "SPINE" OF THE CHASSIS
6.14.1 Building a Spine

One of the first steps in the construction of Vehicle 2 was the creation of an interior "spine," which supported other parts that were attached around it (Section 4.4). The LEGO Tortoise begins with a similar approach. Steps 1 through 5 in Figure 6-8 illustrate how pins are used to connect a variety of beams together to fashion a spine that serves as the supporting core of the machine.

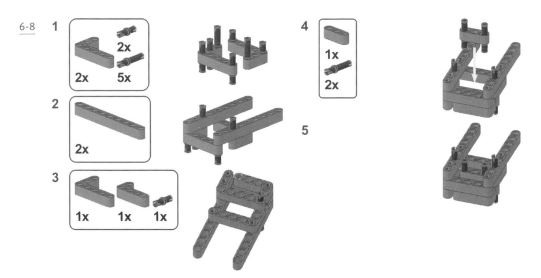

6-8

6.15 MIRRORED MOTOR ASSEMBLIES
6.15.1 Two Motor Assemblies

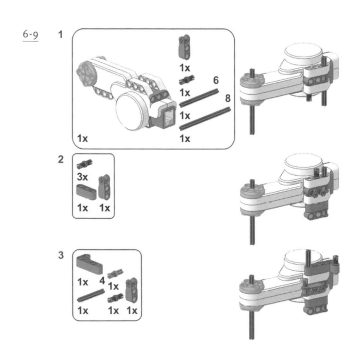

6-9

The next stage in constructing the LEGO Tortoise is to construct two separate motor assemblies. The instructions for one are shown in Figure 6-9, and the instructions for the other are shown in Figure 6-10. For the most part, the motor assemblies are mirror images of each other, and are created by attaching a small number of components to an NXT motor.

However, note that the long axle in Figure 6-9 is extended toward the bottom of the figure, while the analogous axle in Figure 6-10 is extended up toward the top of the figure (e.g., Step 3 in both images). This is because the two assemblies, though (mostly) mirror images, have very different functions. A gear attached to the underside of the motor in Figure 6-9 will be used to power the robot's front wheel drive, while another gear attached above the motor in Figure 6-10 will be used to provide power steering to the machine.

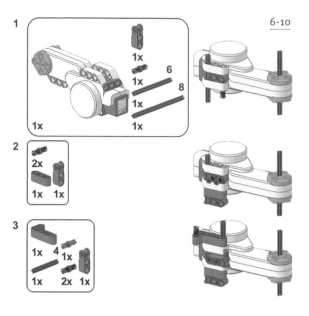

6.16 ATTACHING MOTORS TO THE CHASSIS
6.16.1 Motors and Steering Gear

With the motors constructed, they can now be attached directly to the chassis that was constructed in Section 6.14, as is shown below in Figure 6-11. Note that when the motors are connected, the main axle in the motor on the left of the figure (see Step 8) extends downward, while the main axle in the motor on the right of the figure extends upward. A supporting beam and a gear are added to this latter axle in Step 9. This gear, when rotated by the motor, will rotate a front axle in order to steer the LEGO Tortoise.

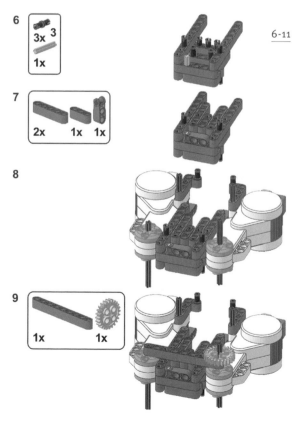

6.17 A SMALL STICK FOR BIG OBSTACLES
6.17.1 Stick-In-Ring Detector

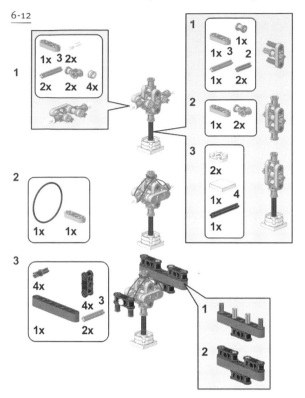

6-12

One of Grey Walter's innovations was to attach the Tortoise's shell to a stick-in-ring switch. This permitted the shell to be sensitive to objects that struck the shell from any direction. In our LEGO Tortoise, a similar stick-in-ring switch will be able to move in any direction, permitting 360° sensitivity to obstacles. However, the signal sent by the switch will not be processed by a touch sensor. Instead, a white plate is attached to the bottom of the stick (see Figure 6-12). When the shell attached to the stick causes the stick to move, the plate will move, and this movement can be detected by a light sensor. Figure 6-12 illustrates how to construct this version of Grey Walter's switch.

6.18 ADDING A DRIVE GEAR AND STICK-IN-RING SWITCH
6.18.1 Front Wheel Drive

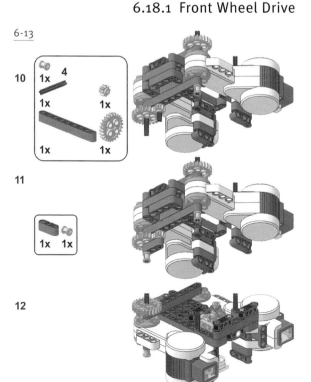

6-13

Steps 10 and 11 in Figure 6-13 demonstrate how gears are attached to the left motor (shown in the figure from below the construction to this point) to drive the robot forward. In Step 10, a small gear gang is attached to the motor axle pointing toward the bottom of the figure. In Step 11, a short beam and a bush are used to keep the gears of this gang in place.

6.18.2 Stick In the Stick-In-Ring

As shown in Step 12 of Figure 6-13, the stick-in-ring switch can now be attached to the chassis, between the rears of the two motors.

6.19 A VERTICAL FRONT AXLE
6.19.1 Front Axle Gears

After inserting the stick-in-ring switch, two bushes and elastic are used to hold it in place, as shown in Step 13 of Figure 6-14. A vertical front axle can then be constructed, as shown in Step 14 of the same figure. The vertical axis is created by joining a length-5 and length-12 axles together with an axle joiner; note (as shown in the subassembly illustrated in Step 14) that these two axles are inserted through a differential gear, and that the axle joiner is placed inside the differential.

The top of this axle is then inserted upward through the centre of the front of the chassis, and a large gear is placed at the top of this axle so that it meshes with the smaller gear that was added in Step 9 of Figure 6-14.

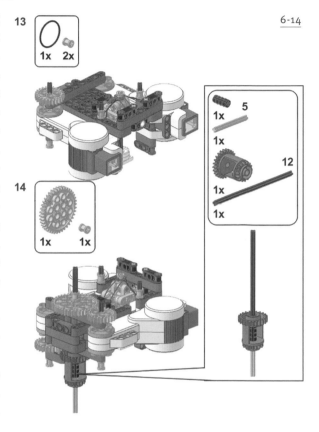

The large gear is then held in place using a bush. These meshed gears now permit the motor on the left of Step 14 of Figure 6-14 to turn this axle, which will cause the robot to turn left or right.

6.20 PREPARING THE NXT BRICK
6.20.1 Readying the Brick

The next component that will be added to the chassis is the NXT brick itself. Before this is done, six pins are first inserted into the brick as shown in Step 1 of Figure 6-15. Two beams, and two additional pins, are then inserted as shown in Step 2 of the same figure.

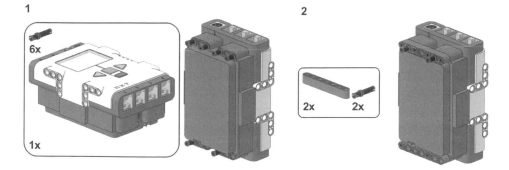

6.20.2 Stick-In-Ring Detector

When objects affect the stick-in-ring switch that was constructed in Section 6.17, the stick moves, and a white tile attached to the bottom of the switch shifts. This shifting is detected by a light sensor that is mounted directly to the NXT brick, as illustrated in Figure 6-16. When the brick is attached to the chassis, this light sensor will be directly below the stick-in-ring switch that was constructed in Section 6.17 and mounted to the chassis in Section 6.18.

6.21 SUPPORTING REAR WHEELS
6.21.1 Rear Wheel Supports

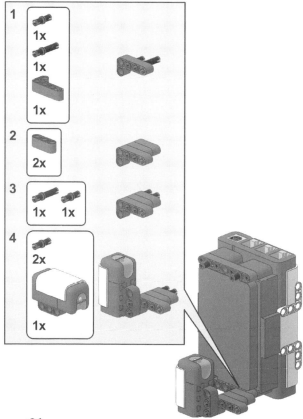

While the LEGO Tortoise will move via a front wheel drive, it will be supported by two large wheels mounted on either side of the rear of the robot. These wheels will freely rotate on axles that are in turn supported by frames that attach each wheel to the NXT brick. Figures 6-17 and 6-18 illustrate how each frame is constructed. Note that the two frames are mirror images of each other.

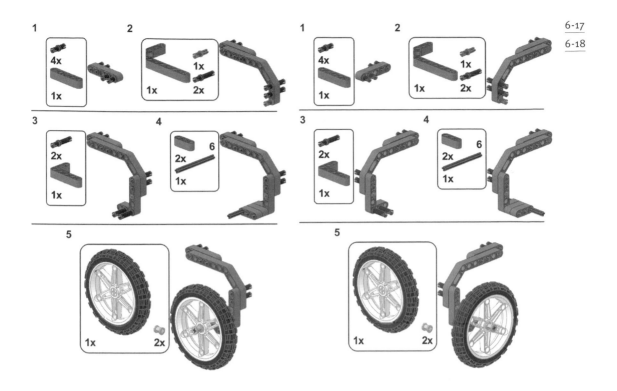

6.21.2 Brick and Wheel Attachment

Figure 6-19 illustrates the next steps in constructing the LEGO Tortoise. First, the NXT brick is attached to the back of the chassis (Step 15). Second, the two rear wheels are then attached to the NXT brick (Step 16). Note that when the NXT brick is added, the light sensor that was mounted on it will be directly below the stick-in-ring switch.

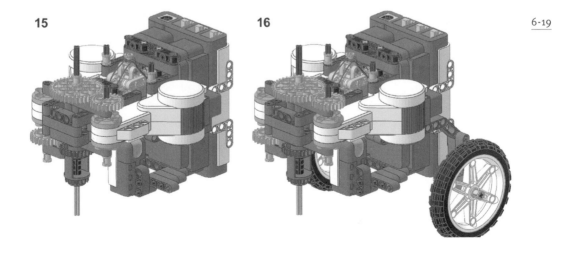

6-19

6.22 FRONT WHEEL ASSEMBLY AND ATTACHMENT
6.22.1 Front Wheel Gear Gang

The front wheel of the LEGO Tortoise is not a passive "caster," as was the case in Chapter 4's Vehicle 2, but instead is the wheel that moves the robot about. As a result, the wheel must be built into an assembly that includes a number of gears that will ultimately be driven by the motor that was earlier depicted on the left of Figure 6-11.

The steps for constructing this assembly begin in Figure 6-20. This figure illustrates the building of the side of the assembly that contains all the gears. A crown-tooth gear is attached to the end of the vertical axle inserted in Step 2 of that figure; it will rotate the horizontal axle inserted in Step 3 after being meshed with a second crown-tooth gear that is attached in Step 4. Most of the other parts are used to add reinforcement to this wheel structure.

6-20

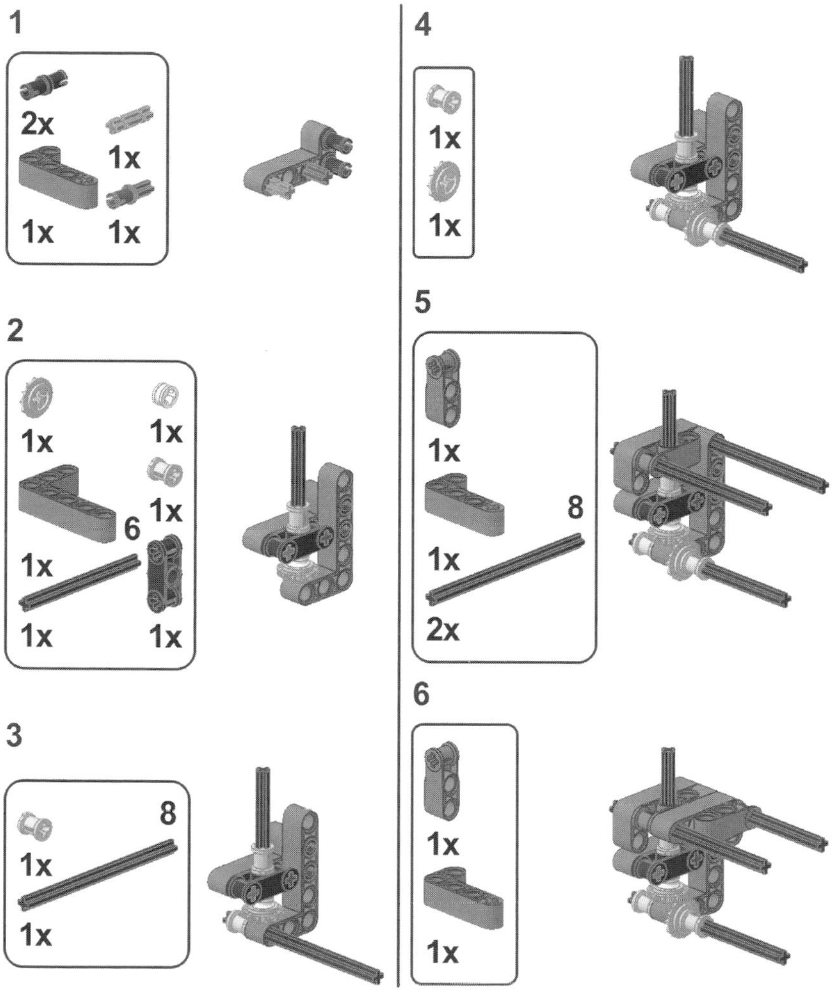

Figure 6-21 illustrates the construction of the remainder of the front wheel, which essentially involves adding the wheel and securing it with more reinforcing parts. The 24-tooth gear added in Step 10 of that figure will be driven by the motor to power the wheel and propel the robot.

The front wheel assembly can now be attached (Figure 6-22) to the vertical front axle that was built in Section 6.19. The assembly attaches to the lower axle; note that the uppermost gear of the front wheel assembly will mesh with the lower gear on the differential.

6.23 PILOT LIGHT ASSEMBLY
6.23.1 A LEGO Pilot Light

Much of the interesting behaviour of Grey Walter's original Tortoise — such as the famous mirror dance — resulted when the robot's light sensor reacted to a "pilot light" that was mounted on the robot. The pilot light would turn on when the steering mechanism was active, but would turn off when the robot moved in a straight line (i.e., when the steering motor was off as well).

Figure 6-23 illustrates how to build our version of the pilot light, which is realized as an array of 8 LEGO light bulbs. An NXT light sensor will be able to detect the reflection of this many lights in a mirror. The electricity for these lights is provided by old-style RCX cables. As a result, an RCX to NXT adapter cable must also be used in order for the NXT brick to power this pilot light.

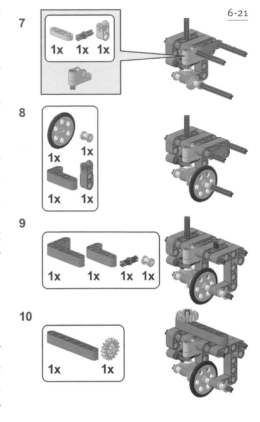

6-21

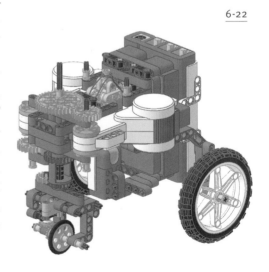

6-22

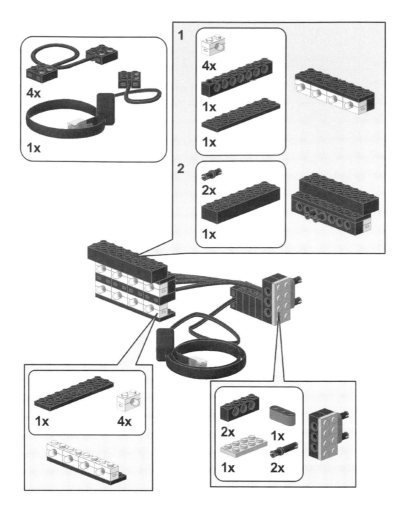

6.24 ATTACHING PILOT LIGHTS AND CONNECTING WIRES
6.24.1 Pilot Light Wiring

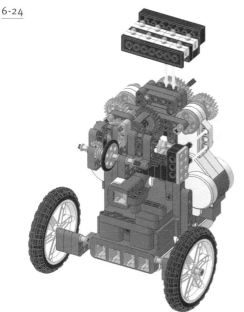

The pilot light is inserted into the front of the chassis "spine," as illustrated in Figure 6-24. The brick assembly that holds its many wires together is then attached to the motor on the right of Figure 6-24.

Now the cable that powers the pilot lights can be attached. It is thread between the stick-in-ring light sensor and the NXT brick, underneath the rear wheels (Figure 6-25), and around the back of the robot to be inserted into Output Port B (Figure 6-26).

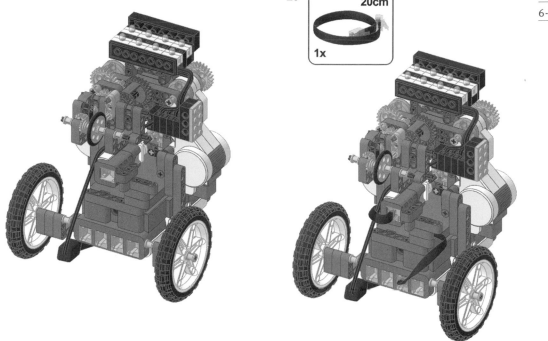

We can now connect the stick-in-ring light sensor to Input Port 2 (Figure 6-27). A short cable can be thread between the light sensor and the NXT brick, around the side of the robot (the right side of the figure) and inserted into the port.

6-27

6.25 A PERISCOPE MIRROR
6.25.1 A 360° Rotating Mirror

6-28

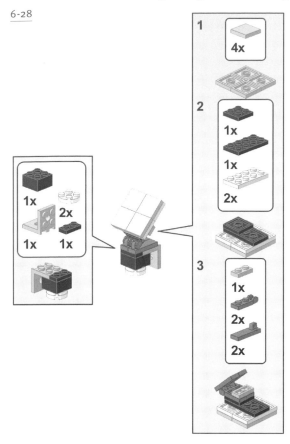

The electric eye in the original Tortoise was mounted on top of the steering axle and could be rotated a full 360°. This is not possible to do with a LEGO light sensor, because of the cable that attaches it to the NXT brick. If a LEGO light sensor were mounted on top of an axle that can be rotated 360°, the light sensor's cable would be wound around the axle until it was pulled off the NXT brick.

To solve this problem, a light sensor was mounted on a fixed mount above the axle, so that the light sensor was stationary. It looked down upon an angled mirror, which served as a "periscope," reflecting light up into the light sensor. This "periscope mirror" was mounted directly on a vertical, rotating, front axle, and could rotate freely. The mirror itself was a square plate. As a result, it was directional: when pointed toward a light, it would reflect light upward to the light sensor, but when turned away from the light, it would reflect less light upward. As a result, this approach to light sensing was functionally equivalent to the one used by Grey Walter.

Figure 6-28 illustrates the construction of the mirror component of this periscope. We have found that if the surfaces of the plates that act as the mirror (the surface constructed in Steps 1 through 3 of the figure) are covered with

6-29

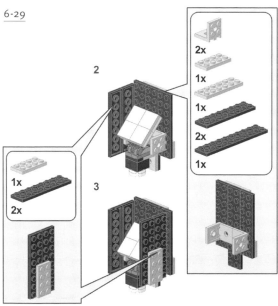

aluminum foil, then its action as a mirror will be satisfactory. After it is built, LEGO plates are used to create a "blinder" that surrounds the mirror on three sides, as shown in Figure 6-29.

6.26 SENSING LIGHT FROM THE PERISCOPE
6.26.1 Attaching the Periscope

After construction, the periscope is attached rectly to the top of the vertical front axle (Figure 6-30). Now, it will rotate with this axle, and — because it has no cables — can rotate a full 360° without a problem.

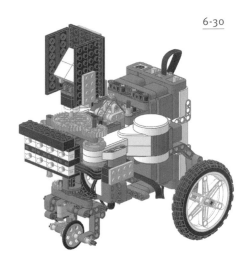

6-30

6.26.2 Sensing Periscope Light

In order to work as a directional light detector, the light reflected from the periscope must be directed into a light sensor. In order to accomplish this, a light sensor must be mounted directly above the rotating mirror. Figure 6-31 illustrates the construction of a frame that is used to suspend a light sensor above the periscope. Each bottom end of this frame is inserted into one of the motors, as is illustrated in Figure 6-32.

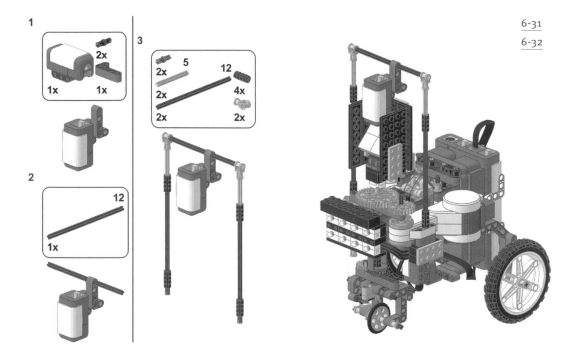

6-31
6-32

6.27 ADDING MORE CABLES
6.27.1 Periscope Wiring

The light sensor above the periscope can now be cabled into Input Port 1 (Figure 6-33). To keep this cable from tangling with the rear wheel, a short beam and two axle pins are used to hold it against the NXT brick.

6.27.2 Motor Wiring

As well, the two motors can now be wired to the NXT bricks, as shown in Figure 6-34. The motor on the robot's right is connected to Output Port C, and the motor on the robot's left is connected to Output Port A.

6-33
6-34

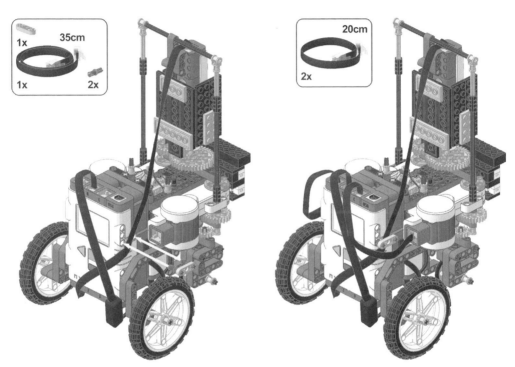

6.28 A SURROUNDING SHELL
6.28.1 Shell Design

The original Tortoise was surrounded by a shell that triggered a reaction when bumped. This shell was attached to a metal stick that was suspended in the centre of a metal ring. When the shell was bumped, the stick and ring would come in contact, completing a circuit and triggering a relay. We have constructed a LEGO version of the stick-in-ring switch (Section 6-17), but must still add a shell that will surround the robot and later be attached to our switch. Figures 6-35 through 6-38 illustrate the substeps that are required to construct this surrounding shell.

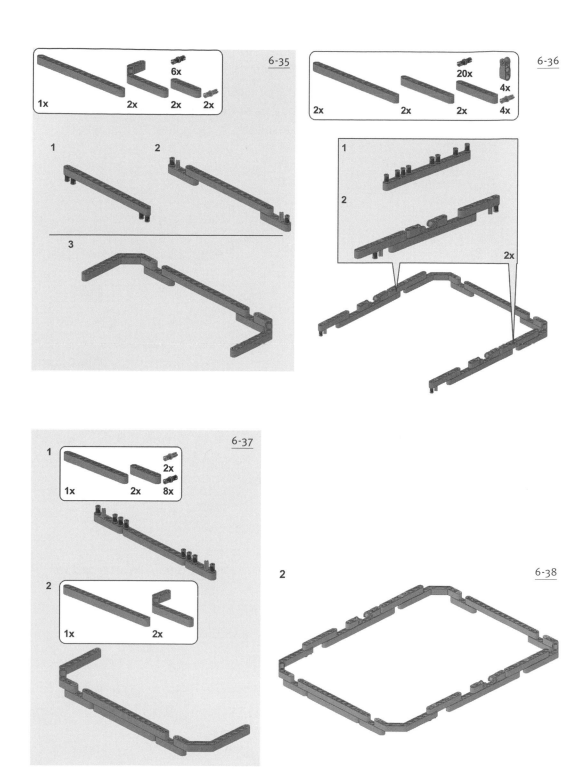

6-35

6-36

6-37

6-38

6.29 SUSPENDING THE SHELL

6.29.1 The Suspension System

A rigid connector must be created to be attached to the shell constructed in Section 6.28. This is because any movement of the suspended shell must be converted into movement of the stick-in-ring switch. Figure 6-39 shows how to build the two sides of this suspension system.

Figure 6-40 illustrates the construction of a central piece that will connect the two side pieces.

6-39
6-40

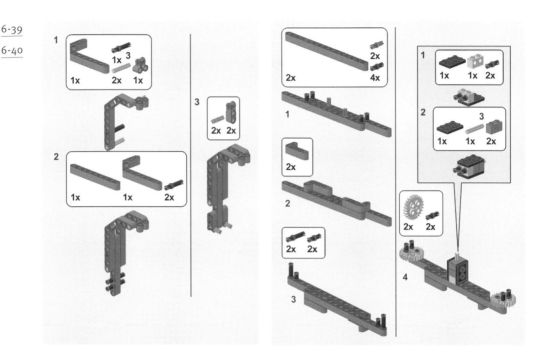

The two side pieces and the central "bridge" can now be attached to the shell as shown in Figure 6-41.

6-41

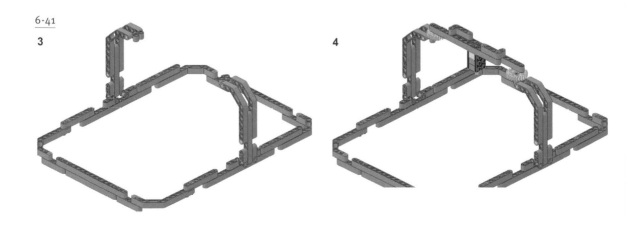

6.30 COMPLETING THE TORTOISE

6.30.1 Attaching the Shell

To complete the construction of the LEGO Tortoise, the shell must be attached. This is accomplished by connecting the central "bridge" across the shell directly to the top of the stick-in-ring switch (Figure 6-42).

6-42

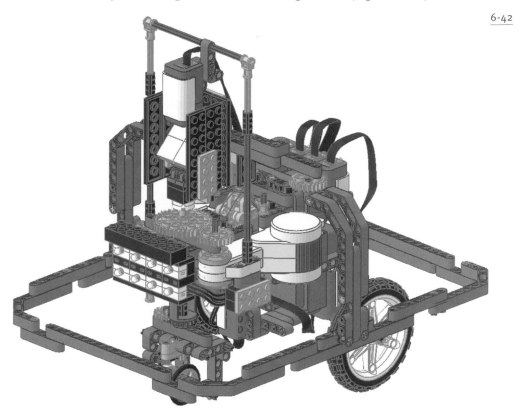

6.30.2 Next: Tortoise Programming

The preceding pages have described how to embody our modern version of the Tortoise using LEGO parts. However, in order to bring this modern robot to life a program must be created to create the simple reflexes that Grey Walter originally studied.

The next chapter provides the details of such a program. However, it also attempts to illustrate a particular programming philosophy called the *subsumption architecture* (Brooks, 1999). Brooks has argued that the subsumption architecture provides a radically different approach to the control of mobile robots, and has many advantages over more traditional approaches. Given that Grey Walter viewed *Machina speculatrix* as a collection of a small number of interacting sense–action reflexes, it is appropriate to merge his robots with Brooks' modern perspective.

6.30.3 Embodiment Issues

We will see that many interesting observations can be made by rank-ordering Tortoise reflexes in a hierarchy, and layering these reflexes on top of each other in accordance with this rank-ordering. From this perspective, many of our insights into the behaviour of the modern Tortoise will come from manipulating Tortoise software.

As was the case with Vehicle 2, this is not the only approach that is available to exploring the Tortoise. The instructions that have been provided have made a number of design decisions that have changed the modern Tortoise from the ancestral models built by Grey Walter. For instance, the relative positions, widths, and sizes of the Tortoise wheels might have profound effects on how the Tortoise moves.

However, before the impact of alternative embodiments can be studied, some sort of program needs to be added to our Tortoise. At the end of Chapter 6, the robot is embodied, but it is not really situated, because no program exists that determines how what the robot senses gets translated into movements. The next chapter details an example of the subsumption architecture for our LEGO Tortoise.

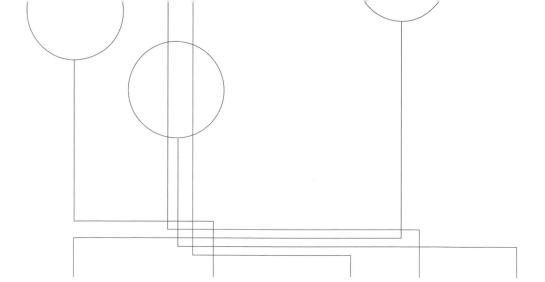

Chapter 7
The Subsumption Architecture

7.0 CHAPTER OVERVIEW

While the most specific aim of this chapter is to provide a program to be used to control the LEGO Tortoise that was constructed in Chapter 6, it has a more general aim as well: to provide a brief introduction to the subsumption architecture (Brooks, 1989, 1999, 2002), which is a successful and flexible approach that has been used to program a number of different robots. The subsumption architecture explicitly rejects the classical view that cognition's purpose is planning, and that cognition mediates perception and action (see our earlier discussion of this issue in Chapter 3). Instead, the subsumption architecture views a robot as being a stacked set of modules. Each module is a sense–act mechanism—thinking or representation is removed as much as possible. Lower levels in a subsumption architecture are modules that govern more basic or general abilities. Higher levels in a subsumption architecture are modules that take advantage of the abilities provided by lower levels and provide more complex abilities. Higher levels might provide weak control over lower levels (for instance, by inhibiting them for a moment); lower levels have no access to, or control of, modules that are higher in the hierarchy.

This chapter proceeds by describing some of the motivation for the development of the subsumption architecture, with a brief account of a famous example (Brooks, 1989). It then provides a more detailed example of this type of architecture by providing the code for the various levels that are used to control the behaviour of the LEGO Tortoise. At the end, it provides a brief description of the behaviour of both the complete Tortoise and the Tortoise when higher-level modules are re-moved. The implications of such behaviour are discussed at the end of the chapter.

The subsumption architecture is an extremely useful approach for developing our robots, particularly as they become more complicated because of employing multiple sense–act mechanisms. Additional examples of the subsumption architecture are also provided for the Lemming robot described in Chapter 8 and for the antiSLAM robot that is the topic of Chapter 9.

7.1 A SANDWICH OF VERTICAL MODULES

7.1.1 Cognitivism

The cognitivist movement arose in psychology in the late 1950s as a re-action against the behaviourist school (Gardner, 1984). A growing number of psychologists were frustrated with behaviourism's reluctance to explore internal processes that could not be directly observed, as well as with its general position that humans were passive responders to environmental stimuli.

The cognitive revolution responded to behaviourism by adopting an information-processing metaphor (Miller, 2003). Rather than passively responding to the environment, humans were described as active processors who received information, manipulated this information to make important content explicit, and then used this explicit content to guide action. This general approach has been called the *sense–think–act cycle* (Pfeifer & Scheier, 1999). The hypothesis that cognition is in-formation processing amounts to the claim that the human mind is a complex system that receives, stores, retrieves, transforms, and trans-mits information (Stillings et al., 1987). For cognitivists, explanations must include accounts of the processes that manipulate information. As a result, successful cognitive theories must include proposals about the manner in which information is represented, as well as proposals about the rules or procedures that can transform these representa-tions (Dawson, 1998). These theories are instances of what has come to known as the classical approach, as the symbolic approach, or as the representational theory of mind (Chomsky, 1980; Fodor, 1968, 1975; Newell, 1980; Pylyshyn, 1984).

7.1.2 The Classical Sandwich

There are many important consequences of adopting the classical ap-proach. Two of these involve the relationship between perception, cog-nition, and action (Hurley, 2001). First, cognition — the rule-governed manipulation of representations — is viewed as being the central characteristic of the mind. Second, perception and action are seen as being peripheral characteristics of the mind, as well as being separate from one another.

These two consequences arise because the classical view tacitly assumes that one of the primary purposes of cognition is planning (see Chapter 3). For example, it has been argued that mind emerged from the natural selection of abilities to reason about the consequences of hypothetical actions (Popper, 1978). This permitted fatally incorrect actions to be discarded before actually being performed. "While an uncritical animal may be eliminated altogether with its dogmatically held hypotheses, we may *formulate* our hypotheses, and criticize them. Let our conjectures, our theories die in our stead!" (p. 354).

Classical theories tend to take on a stereotypical form when it is assumed that the function of cognition is planning. Hurley (2001) described this form as a set of vertical modules that stand between (or that are sandwiched by) perception and action. Each vertical module is classical in nature — each involves a particular form of representation, and particular processes that modify these representations. Critically, there are no direct connections between perception and action (see Figure 3-1 in Section 3.3). That is, perception can only indirectly inform action, by sending information to be processed by the central, vertical modules, which in turn ultimately choose which action is to be performed. In her critique, Hurley called this structure the *classical sandwich.*

Many researchers now question the classical sandwich, and are considering alternative roles for cognition. Some have argued that cognition is not used to plan, but is instead used to control action (Clark, 1997; Varela et al., 1991). The classical sandwich is being disassembled, because direct links between perception and action are appearing in cognitive theories (Brooks, 1999, 2002). We earlier saw an example of this with the "leaky mind" model depicted in Figure 3-3 (Section 3.12).

7.2 THE NEW LOOK AND ITS PROBLEMS

7.2.1 The New Look in Perception

Why would the classical sandwich be challenged? One answer to this question comes from exploring an example of perceptual theory and its performance when it is imported into a behaving robot.

In the late 1940s and early 1950s, psychologist Jerome Bruner and his colleagues performed a number of experiments that led to a radically new, cognitive theory of perception (Bruner, Postman, & Rodrigues, 1951). These experiments indicated that subjects' perceptual experiences were strongly influenced by their expectations or experiences.

The implication of such results was that perception (i.e., categorizing the visual world, as opposed to sensing light) is equivalent to cognition (Bruner, 1957). "A theory of perception, we assert, needs a mechanism capable of inference and categorizing as much as one is needed in a theory of cognition" (Bruner, 1957, p. 124). The view that perception was in essence an active cognitive process became known as the *New Look*. More modern variants of this perspective have also appeared (Gregory, 1970; Rock, 1983).

For Bruner (1957), the New Look was "a general view of perception that depends upon the construction of a set of organized categories in terms of which stimulus inputs may be sorted, given identity, and given more elaborated, connotative meaning" (p. 148). This is consistent with Hurley's (2001) classical sandwich, because "perception" would merely deliver the stimulus inputs, and perceptual categorization would be accomplished by the vertical modules.

7.2.2 Shakey Implications

Beginning in 1966, the Stanford Research Institute conducted research on a robot nicknamed "Shakey" (Nilsson, 1984). Shakey plotted its own path through a controlled indoor environment, using a television camera, an optical range finder, and touch sensors. Shakey communicated what it sensed via radio signals to a central computer that updated Shakey's model of the world, and planned Shakey's next behaviour. Shakey illustrates the classical sandwich, with its vertical modules being contained in the central computer.

Shakey's model of the world was a set of predicate calculus expressions. This predicate calculus also represented Shakey's goals. A planning system (called STRIPS) would attempt to derive a sequence of actions to convert the current model to one in which the goal was accomplished. Shakey would then execute the sequence of actions to physically accomplish the goal.

Shakey was capable of performing a number of impressive tasks, many of which are illustrated in the film SHAKEY: Experimentation in Robot Learning and Planning, which is available from http://www.ai.sri.com/movies/Shakey.ram. It could plan routes through its environment, navigate around obstacles, and move obstacles (large painted blocks) to desired locations. However, Shakey also revealed two of the main problems with the classical sandwich.

First, Shakey was as successful as it was because it was placed in a carefully constructed and controlled environment that simplified visual processing and model building. "Shakey only worked because of a very careful engineering of the environment. Twenty years later, no mobile robot has been demonstrated matching all aspects of Shakey's performance in a more general environment" (Brooks, 1999, p. 61).

Second, Shakey was extremely slow. Even when placed in a tailored environment that was not complicated to sense or to model, it took several hours for the robot to complete a task (Moravec, 1999). This was because maintaining and using the world model was computationally expensive. The problem with the sense–think–act cycle in robots like Shakey is that by the time the (slow) thinking is finished, the resulting plan may fail because the world has changed in the meantime. This problem is dramatically accentuated by increasing the complexity of the world model that is maintained in the interior of the classical sandwich.

7.3 HORIZONTAL LAYERS IN THE HUMAN BRAIN
7.3.1 Evidence from Action

The classical sandwich amounts to the claim that sensing is separated from acting by a great deal of thinking, modelling, and planning. However, this claim begins to be severely challenged when researchers study the neural mechanisms that coordinate perception and action (see also Section 3.7.2).

For example, consider the examination of the patient known as DF, who suffered extensive brain damage as the result of carbon monoxide poisoning (Goodale et al., 1991). DF's brain injuries did not impair basic sensation, such as detection of colour, or the spatial resolution of images. However, higher-level perception was severely impaired. DF had severe visual form agnosia, and could not describe the orientation or shape of any visual contour, no matter what visual information was used to create it.

Amazingly, while DF could not consciously report orientation or shape information, such information was available to control some of her behaviour. In particular, when DF was asked to perform a motor activity, such as grasping an object, or inserting an object through an oriented slot, her actions were identical to controls, even to the fine details that are observed when such actions are initiated and then carried out. "At some level in normal brains the visual processing underlying 'conscious' perceptual judgments must operate separately from that underlying the 'automatic' visuomotor guidance of skilled actions of the hand and limb" (Goodale et al., 1991, p. 155).

DF's brain injuries caused visual form agnosia, but left visuomotor coordination intact. Importantly, other kinds of brain damage produce a very different pattern of deficits that support the notion of "separate operation" noted above.

Patients who suffer damage to the posterior parietal cortex can exhibit optic ataxia, in which they are unable to use visual information to reach out and grasp objects when presented in the part of the visual field affected by the brain injury. Some of the motor skills studied in patient DF have also been studied in the patient VK, who was suffering optic ataxia (Jakobson, Archibald, Carey, & Goodale, 1991). One of the main differences between VK and DF was that VK demonstrated a number of visuomotor abnormalities. For instance, the size and shape of her grasp were only weakly related to the size and the shape of the to-be-grasped object, and grasping movements took much longer to be initiated and to be executed. A second main difference was that VK, unlike DF, had no difficulty recognizing the orientation and shapes of visual contours.

In short, DF and VK illustrate the double dissociation between brain mechanisms responsible for the conscious awareness of visual form and brain mechanisms responsible for complex visually guided actions, such as grasping objects.

7.3.2 Sandwich Alternative

These results can be used to argue that the human brain is not completely structured as a "classical sandwich." On the one hand, Goodale concedes that one likely function of the visual system is the creation of a model of the external world; this is the kind of function that the classical sandwich captures, and is disrupted in DF (Goodale & Humphrey, 1998). However, a second function, revealed by the double dissociation, is the control of action. This is accomplished by converting visual information directly into motor commands. This is not part of the classical sandwich, because it assumes that there is a much more direct link between vision and action.

It has been argued that the two functions mentioned above can coexist, can interact, and can complement one another (Goodale & Humphrey, 1998). However, some would argue that the typical relationship between vision and action has much more to do with controlling action than building models. This has led to a proposed architecture that is a direct challenge to the classical architecture and to the sense–think–act cycle. We now turn to considering this alternative proposal.

7.4 HORIZONTAL LINKS BETWEEN SENSE AND ACTION
7.4.1 A Sandwich Alternative

One alternative to Hurley's (2001) classical sandwich is much more in line with the results from neuroscience that indicate that 1) an important function of vision is the control of action, and 2) that this function is mediated by pathways that are distinct from those involved in constructing models or representations of the world (Goodale, 1988, 1990, 1995; Goodale & Humphrey, 1998; Goodale et al., 1991; Jakobson et al., 1991). The alternative is to replace the classical sandwich's vertical layers that separate perception from action with horizontal layers that directly connect perception and action.

One of the primary motivations for such horizontal layers is to recast what cognition is all about. The classical approach, as illustrated with Shakey, is that cognition amounts to planning, and that this planning is required to mediate perception and action. In short form, the classical view endorses the sense–think–act cycle. The alternative view is to assume that cognition is not planning, but instead is the control of action (Clark, 1997). "The idea here is that the brain should not be seen as primarily a locus of inner *descriptions* of external states of affairs; rather, it should be seen as a locus of internal *structures* that act as operators upon the world via their role in determining actions" (p. 47). Importantly, these structures serve as links between sensing and acting, not as general processes that stand between sensing and acting.

This alternative view represents a strong reaction against the notion of central control that is fundamental to classical cognitivism. In the classical sandwich, each vertical layer is defined by a particular representational medium (i.e., symbols of a particular type) and by a set of rules that manipulate these representations. In addition, though, there must be some control mechanism that chooses which rule to apply to the symbols at any given time. In the thinking and problem-solving literature, control is usually described as deciding "what to do next," or as searching a problem space (Simon, 1969).

Much of cognitive science is inspired by the metaphor of the digital computer (Pylyshyn, 1979), or Turing's universal machine (Turing, 1936, 1950). In these devices, control is centralized, and is used to choose one rule at a time to manipulate symbols. Not surprisingly, classical cognitive theories usually appeal (either directly or indirectly) to some sort of central control mechanism.

Many of the reactions against classical cognitivism, such as the connectionist movement (Schneider, 1987) or situated cognitive science (Greeno & Moore, 1993; Touretzky & Pomerleau, 1994), point to problems with the classical view that are due to this central control. Classical systems are often described as slow, brittle, and unable to gracefully degrade (Feldman & Ballard, 1982). One solution to such problems is to decentralize control.

Historically, the first major modern proposal for decentralizing control is to take advantage of the distributed representations that are characteristic of artificial neural networks (McClelland & Rumelhart, 1986; Rumelhart & McClelland, 1986). Arising slightly later, and having a more recent impact, are the sense–act models that have been developed within behaviour-based robotics (Brooks, 1999, 2002; Pfeifer & Scheier, 1999). We will consider one of these approaches, Brooks' subsumption architecture, in more detail in following sections.

However, it is important to be aware that the possibility of decentralized control has broader implications for cognitive science. For example, arguments against the "Cartesian theatre" view of consciousness (Dennett, 1991, 2005) are completely consistent with this sort of decentralization. So too is the notion of high-level cognitive phenomena emerging from a "society of mind" (Minsky, 1985, 2006). Similarly, decentralized control makes possible the notion of the mind leaking into the environment, challenging the classical views of boundaries of the mind (Clark, 1997, 1999; Wilson, 2004).

7.5 THE SUBSUMPTION ARCHITECTURE
7.5.1 Modularity of Mind

One of the key innovations in modern cognitive science is the idea that many cognitive functions are modular (Fodor, 1983). A module can be viewed as a special purpose machine that only has access to a limited amount or type of information, which permits the machine to be fast. The module is designed to solve a particular problem, is associated with specific neural circuitry, and cannot be influenced by the contents of higher-order beliefs, desires, or expectations. A module would not be part of Bruner's (1957) New Look!

Modularity is not inconsistent with the classical sandwich. Many of the horizontal layers that separate perception from action could be modular in Fodor's (1983) sense. For example, the computational theory of vision proposed by David Marr (Marr, 1976, 1982; Marr & Hildreth, 1980; Marr & Ullman, 1981) consists of a set of modules that produce a sequence of preliminary representations of visual information (the raw primal sketch, the full primal sketch, the 2½D sketch). These modules are used as the foundation for a meaningful, useful mental representation of the world. Marr's theory is modular, and is also classical in exactly the sense that is questioned by Hurley (2001).

7.5.2 Vertical Modules

The subsumption architecture that has been proposed by roboticist Rodney Brooks (1999) is modular, but explicitly departs from the classical sandwich, and rejects the sense–think–act cycle. The subsumption architecture is a set of modules, each of which can be described as a sense–act mechanism. That is, every module can have access to sensed information, as well as to actuators. This means that modules in the subsumption architecture do not separate perception from action. Instead, each module is used to control some action on the basis of sensed information.

A second characteristic of the subsumption architecture is a hierarchical arrangement of modules into different levels. Lower levels are modules that provide more basic or more fundamental sense–act functions. Higher levels provide more complex sense–act functions, which depend upon (or take advantage of) those provided by lower-level functioning.

The hierarchical structure of the subsumption architecture reflects the generality of the sense–act function provided by each level. The functions provided by lower levels are more general, in the sense that they are required in a very broad array of situations. Higher-level functions are more specific, designed to be used in a narrower range of situations.

The hierarchical relationship amongst levels in the subsumption architecture is also reflected in how they communicate with one another. Consider the lowest level, Level 0, and a level built immediately above it, Level 1. Level 1 has access to the sense data of Level 0, and can send signals that alter Level 0 (e.g., by inhibition). However, the reverse is not true: Level 0 cannot access or influence Level 1. In other words, the hierarchy of the subsumption architecture is one of control.

Importantly, control in the subsumption architecture is not centralized. There is no central clock, and no serial processing. All of the levels in the architecture run in parallel and asynchronously.

The modular and hierarchical nature of the subsumption architecture is also reflected in how it is created. A designer decides on what the lowest level, the broadest sense–act function, should be. This level is created, and then never revisited. "We start by building a complete robot control system which achieves level 0 competence. It is debugged thoroughly. We never alter that system" (Brooks, 1999, p.10). Then this process is repeated for the next level, and continues until all of the levels in the subsumption architecture have been completed.

7.6 ADVANTAGES OF THE SUBSUMPTION ARCHITECTURE
7.6.1 Reasons for Revolution

Brooks (e.g., 1999) proposed the subsumption architecture as an explicit reaction against classical notions of sense–think–act and centralized control, particularly as these notions were realized in classical research on autonomous robots. Not surprisingly, Brooks has argued that the success of his own robots illustrates that the subsumption architecture has many demonstrable advantages over robots that are designed with the classical sandwich in mind.

7.6.2 Coping with Multiple Goals

Grey Walter's robots (Grey Walter, 1950a, 1950b) had to accomplish more than one goal: seeking moderate light, approaching such light when it was found, avoiding bright light, and avoiding physical obstacles.

A classical system that uses centralized control, such as Shakey, must carefully consider all of the robot's goals at any given time to plan an appropriate course of action. This process becomes more and more complicated as the number of goals multiplies. Furthermore, as changes in the environment occur, their impact on the robot's goals must be re-evaluated, leading to what has been called the frame problem (Pylyshyn, 1987). In the frame problem, a classical system is lost in thought, evaluating the impact of world changes on its world model, unable to perform actions in a timely manner.

Brooks (e.g., 1999) argued that the subsumption architecture can deal with multiple goals that are in conflict or change in priority. This is because different goals are associated with different levels, and each level pursues these goals in parallel. Coping with multiple goals emerges from the control structure of the architecture. For example, a higher level pursuing the goal of avoiding objects can inhibit lower levels pursuing the goal of moving, but only when an object is encountered to make the higher level's goal a more immediate priority.

7.6.3 Combining Multiple Sensors

A classical system is also challenged when the number and types of sensors used to create a model increase. This leads to a need to increase the complexity of the model, and the resources used to update the model, which is a variation of what is known as the packing problem (Ballard, 1986).

The subsumption architecture provides an elegant approach to dealing with multiple sensors, because usually each level uses only a subset of the sensors that are available. As well, increases in computational demands are kept in check because each level is converting sensed information into action, and is not updating an internal model of the world.

7.6.4 Robustness

A robust robot generates useful behaviour even when it has problems with sensors. The subsumption architecture is argued to be robust (Brooks, 1999) because of its hierarchical nature. Lower levels provide the most basic and the most generally applicable behaviours, while higher levels provide more sophisticated and specialized capabilities. If higher levels in the architecture fail, or are processing inputs too slowly, the lower levels are still operating. As a result, the robot still performs some appropriate behaviour under challenging situations.

7.6.5 Speed with No Modelling

One of the key criticisms of the classical approach is that the need to maintain an internal model of the world results in an agent that is too slow to take action under the real-time demands of the world. The subsumption architecture deals with this problem by removing the need to build internal models. Instead, the world is used as a model of itself, which the subsumption architecture senses but does not represent. "The realization was that the so-called central systems of intelligence—or core AI as it has been referred to more recently—was perhaps an unnecessary illusion, and that all the power of intelligence arose from the coupling of perception and actuation mechanisms" (Brooks, 1999, p. viii).

7.7 CONCRETE EXAMPLES
7.7.1 Walking Robots

The advantages of the subsumption architecture have been demonstrated in a number of different walking robots. One famous example is Genghis, a six-legged walking robot (Brooks, 1989). Each leg is affected by two motors, one for swinging it back or forth, and another for lifting it up or down. The subsumption architecture controls leg movement, and interesting walking behaviour emerges from interactions amongst the architecture's layers.

Level 0 for this robot is **standup**; each leg's motors are set to hold the leg in a position so that all legs together enable the robot to stand.

Level 1 is **simple walk**. This includes mechanisms to set a leg down if it is not down already, to balance the robot (so that if one leg moves forward, the other legs will move backward slightly, and to move legs up and forward. Most of these mechanisms work independently for each leg, but one mechanism totals the back-and-forth position of each leg in an attempt to coordinate the legs. The result is that different walking gaits can be produced in the robot.

Level 2 is **force balancing**. The force on each leg is monitored by measuring the force placed on the motor that raises a leg up or down. Whenever the force is too high, a signal is sent to lift the leg. This is an attempt to compensate for rough terrain.

Level 3 is **leg lifting**. By measuring the force on the motor used to swing a leg forward, this level determines if the leg is hitting an obstacle, and will send a signal to lift the leg higher if the measured force is too high.

Level 4 is **whiskers**. Two whiskers on the front of the robot are used to detect obstacles. If a whisker is depressed, the front leg of the robot on the whisker's side will be raised higher.

Level 5 is **pitch stabilization**, and is used to provide more sophisticated balancing than is provided by Level 2. Pitch stabilization senses the angle of the robot's body, and will lift either the front or the rear legs to provide better stability.

Level 6 is **prowling**. Six infrared sensors are mounted on the front of the robot, and with this level the robot will only move if it has detected something else that has moved nearby.

Level 7 is **steered prowling**. The infrared sensors are capable of noting the direction of detected movement, and this level sends signals to move legs in such a way that the robot turns in this direction.

This subsumption architecture was built into Genghis, and it produced a number of very interesting behaviours. The robot could walk over a number of different of terrain types without having to be altered from one terrain to the next. It could demonstrate more than one gait, including one in which subsets of legs supported the robot as alternating tripods, and one in which the gait ripples through the legs from the back to the front. The robot could follow moving objects, such as people, using its higher architectural levels.

7.7.2 The Tortoise

The subsumption architecture for Genghis was wired into its structure. The basis for a sense–act relation in a level was a simple machine called an augmented finite state machine, which is a simple device that has a set of registers for storing information such as machine states, a finite state machine that determines what should happen to registers given the current machine state, and wires that permit one of these machines to send signals to another. Fifty-seven such machines were wired into a network to provide Genghis' built-in subsumption architecture (Brooks, 1989).

In the pages that follow, we will explore our own subsumption architecture for the LEGO Tortoise. Instead of soldering one together, we will attempt to use the spirit of the subsumption architecture to guide our writing of tasks in NXC. The complete code for this robot is available from the website that supports this book (http://www.bcp.psych.ualberta.ca/~mike/BricksToBrains/).

7.8 LEVEL 0, BASIC MOVEMENT
7.8.1 A Fundamental Function

To begin the Tortoise program, we must first decide on the most fundamental function to include as the lowest level of a subsumption architecture. Our choice for Level 0 is *basic movement*.

By basic movement, we mean a functional layer that will cause the Tortoise to move itself forward. The function is so basic that it has no sensors. However, it does have direct access to the drive motor. The purpose of this function is to ensure that the drive motor is running, causing the front wheel to move the robot.

The NXC code for Level 0 is provided below. Note that this level is defined by a single task, called task level_0 (). This task turns on the motor that drives the front wheel (which is given the name DriveMotor in the main task), and propels it forward at DriveSpeed (which is also defined in the main task). As long as this task runs, this motor is on, because the command that turns the motor on in a forward direction (OnFwd) is contained within a while(true) loop.

If the Tortoise only used Level 0, then how would it behave? It would move in one direction, and its motor would not stop. It might move in a straight line, or it might turn; this would depend completely upon the direction that the front wheel was pointing when Level 0 started. Whatever direction it moved, this direction would not change, because Level 0 has no influence on the steering motor.

```
//Level 0: Turn the drive motor on.
int DriveSpeed;
task level_0(){
    while(true){
        OnFwd(DriveMotor, DriveSpeed);
    }
}
```

7.9 LEVEL 1, STEERING
7.9.1 Exploration

The logic of the subsumption architecture is that higher-level layers that take advantage of whatever functions are already provided by lower layers. For the Tortoise, Level 1 is **steering**. This level causes the steering motor to turn on, resulting in the front wheel turning. It does not cause the front wheel to drive the robot, which is instead accomplished by the lower Level 0.

This layer operates the steering motor at the front of the Tortoise. It is slightly more complicated than Level 0, because the front motor can be turned at different speeds: a medium speed for exploration, a fast speed when the robot is dazzled, and at zero speed — the motor is off — when the robot has sensed medium light.

The NXC code for Level 1 (task level_1 ()) is provided below. Note that this code is written to reflect the fact that later, higher levels in our architecture will affect robot turning. They will do this by sending a signal down to this level that sets the speed at which the turning motor will run.

When Level 1 is considered not by itself, but instead in the context of the pre-existing Level 0, we can see how it takes advantage of basic movement to advance robot behaviour. That is, when both of these Layers are operating, the robot will move forward and change direction; when it is exploring or avoiding it produces a distinctive "staggering" motion. This is only possible because the steering level is taking advantage of the basic movement provided by Level 0.

```
//Level 1: Turn the steering motor on.
int TurnSpeed;
task level_1(){
    while(true){
        OnRevReg(TurnMotor, TurnSpeed, OUT_REGMODE_SPEED);
    }
}
```

7.10 LEVEL 2, SENSING AMBIENT LIGHT
7.10.1 Light Affects Lower Levels

The next level in the Tortoise is Level 2, which brings the phototropisms to life. Recall that Grey Walter's robots explored in the dark, moved straight in moderate light, and then turned away when they were dazzled by bright light. The NXC code below for task level_2 () shows how such behaviour is added to the Tortoise.

Level 2 is an infinite loop that always measures light by reading the LightSensor, which is set to RAW mode by the main task. This means that it returns a large number in the dark, and a smaller number when brighter light is detected.

The Level 2 task compares the current light sensor reading to two threshold values that are defined in the main task, called dark and bright. If it is dark, nothing is done — the robot explores in the fashion defined by Levels 0, 1; because steering is initiated, the pilot light is turned on. Note that during exploration, the robot is driven at half speed. If moderate light is detected, then the front turning motor is stopped by sending a signal to the Level 1 task, the pilot light is extinguished, and the robot is driven at full speed. If bright light is detected, the turning motor is run at half speed, the pilot light is turned on, and the robot is driven at full speed.

Note that this level works by using sensed light to change the behaviours of lower levels. The signals that are sent depend upon the amount of light sensed, and manipulate the lower levels in such a way that the general light-sensitive behaviours that Grey Walter described are produced. Note, too, that what this level senses is affected by the operations of lower levels, which position the robot — and its periscope mirror — in particular positions relative to whatever light sources might be in the environment.

```
task level_2(){
    while(true){
        OnFwd(PilotLight, lightSwitch(TurnSpeed));//Pilot lights on if turning.
        if (Vision == 1) {See = Eye;}
        //Sensor in dark threshold
        if (See <= dark){
            DriveSpeed = HalfDrive;
            TurnSpeed = FullTurn;
        }
        else {
            //Sensor in moderate threshold
            if (See < bright){
                DriveSpeed = FullDrive;
                TurnSpeed = Zero;
            }
            else {
                //Sensor in bright threshold
                DriveSpeed = FullDrive;
```

```
                    TurnSpeed = HalfTurn;
                }
            }
        }
    }
//This just toggles the lights.
int lightSwitch(x){
        if (x == 0) return 0;
        else return 100;
}
```

7.11 LEVEL 3, OBSTACLE AVOIDANCE
7.11.1 Sophistication from Tweaking

With the first three levels working, the Tortoise will explore the environment by sensing ambient light. In many instances, this exploration will cause it to bump into an obstacle. Level 3 is **obstacle avoidance**, which permits the Tortoise to find its way around an obstacle when it is encountered. An obstacle is detected when the robot's shell is depressed, and one or more of the touch sensors are triggered. This level links obstacle avoiding behaviour to this situation. Importantly, the ability of Level 3 to do this depends upon the behaviours created by all of the lower levels.

Level 3 operates by constantly reading the touch sensors in front, which is done by the Bumpers expression in the NXC code below (task lvl3 ()). If the sensors return a value of 1, then an obstacle has been encountered, and it is depressing the shell. The Level 3 task then "tweaks" the robot's behaviour in an attempt to move away from the obstacle. These "tweaks" consist of sending signals that are used by, and change the behaviour controlled by, the lower levels.

If the shell is depressed, the first thing that Level 3 does is make the robot's turning insensitive to light by setting See = 0, which affects Level 2. It then sets the sensed light to dark (overriding the light sensor) for a short period of time, and then sets it to bright, again for a short period of time. The duration of these states depends on current light conditions. These states are toggled back and forth until the shell is no longer depressed. At that time, the light sensor is reactivated, and all routines — including Level 3 — go back to their usual operation.

By toggling the two sensed light conditions, Level 3 sends signals that affect the behaviour of Level 2. These signals cause changes in the speeds of the steering and drive motor, producing one state that Grey

Walter called "steer hard, push gently," and another called "steer gently, push hard." The robot is still steering via Level 1, and driving via Level 0, so the overall result is the robot changing direction in various ways that eventually cause it to move away from the obstacle that depressed the shell.

```
//Level 3: The shell can temporarily override the light sensors.
int TimeConstant, Bumped; //Reaction time constant and threshold for contact.
task level_3(){
    while(true){
        until (Shell < Bumped);
        Vision = 0;
        while (Shell < Bumped){//Flicker between dark and bright.
            See = dark ;
            Wait(TimeConstant * Eye);//Itill flicker differently if it sees light.
            See = bright;
            Wait(TimeConstant * Eye * 2);
        }
        Vision = 1;
    }
}
```

7.12 THE MAIN TASK
7.12.1 Modular Design

In the code below a number of # define commands name the input and output ports. Code for Levels 0 through 3 — which are saved in separate files — is then included. The main task initializes motor speeds, sensor settings, timing constants, and the sensor values that define dark and bright. It also starts each of the tasks that define each level of our subsumption architecture. The code is organized level by level, to reflect the nature of our architecture. By deleting a Level's # include command, and by deleting the code that initializes the level's variables in the main task, one can study how the Tortoise behaves with a different version of its architecture (i.e., a version with one or more selected levels ablated).

```
//Tortoise NXT code
//Definitions in plain English
#define DriveMotor OUT_C
#define TurnMotor OUT_A
#define EyePort S1
```

```
#define Eye SENSOR_1
#define ShellPort S2
#define Shell SENSOR_2
#define PilotLight OUT_B
task main(){
    //Set up hardware.
    SetSensorType(EyePort, SENSOR_TYPE_LIGHT_INACTIVE);
    SetSensorMode(EyePort, SENSOR_MODE_RAW);
    SetSensorType(ShellPort, SENSOR_TYPE_LIGHT_ACTIVE);
    SetSensorMode(ShellPort, SENSOR_MODE_RAW);
    //Init level 0.
    DriveSpeed = 70;
    start level_0;
    //Init level 1.
    TurnSpeed = 40;
    start level_1;
    //Init level 2.
    Zero = 0;
    HalfDrive = 40; FullDrive = 60;
    HalfTurn = 7; FullTurn = 20;
    dark = 450; bright = 700;
    Vision = 1;
    start level_2;
    //Init level 3.
    TimeConstant = 5;
    Bumped = 620;
    start level_3;
}
```

7.13 OBSERVING TORTOISE BEHAVIOUR
7.13.1 Level 0

Let us first consider LEGO Tortoise performance as levels are added one by one. This is demonstrated in Video 7-1.mpg, available from the website that supports this book (http://www.bcp.psych.ualberta.ca/~mike/BricksToBrains/). The first behavioural segment in this video shows what occurs when only Level 0 is operating. The robot moves forward, in whatever direction the front wheel was pointing when the robot was activated. The robot is insensitive to the light in its environment.

7.13.2 Level 0 + Level 1

The next behavioural segment in the video illustrates the effect of adding Level 1 to Level 0. Level 1 rotates the front axle a full 360°. When combined with Level 0, this produces a wandering movement. Although the robot approaches the light in the video, this is merely accidental — the light sensor in the robot is not active at this time. As well, the robot bumps into the light, but its shell is also not active at this time. The robot, in fact, blindly wanders into the light, and then blindly wanders away from it.

7.13.3 Level 0 + Level 1 + Level 2

The next behavioural segment illustrates the robot's behaviour when light sensitivity is added to the previous two levels of the subsumption architecture. Note, now, that the robot's behaviour is much more "light directed." First, rather than randomly wandering into the light, the robot moves directly toward it. Second, rather than randomly wandering away from the light, the robot circles the light at a respectful distance.

This version of the robot does not yet have obstacle avoidance activated. However, it appears that the robot is able to move away from the light when it is bumped. In actuality, this movement depends entirely on the front-wheel drive rotating away from the light when bright light is sensed. As well, the embodiment of the shell permits the robot to slide off an obstacle, much as the shape of a locomotive's cowcatcher permits it to move obstacles out of a train's path.

7.13.4 All Four Levels

The total LEGO Tortoise adds obstacle detection to the previous three levels. In essence, obstacle detection functions as follows: when the robot's shell is depressed, light is momentarily ignored. The lower levels are manipulated by Level 3 to produce turning behaviour that moves the robot away from the obstacle. After a brief period of time, normal operations resume. The period of time during which the robot is insensitive to light depends upon the amount of light that is currently being sensed. In short, it "remembers" obstacles for a period of time, but then forgets them and returns to being a light-sensing robot.

The combination of the four levels can produce behaviour that appears to be much more complicated than one might predict from knowing about the construction and programming of the LEGO Tortoise. For instance, during the final behavioural segment of the video, a shoe is tossed at the robot, activating its shell. From an analytic perspective, the behaviour that ensues seems fairly elaborate. It is as if the robot scurries away from the attack, hiding for a while under the tables that are also in the environment. When the coast appears clear, the robot is tempted out of its hiding spot by the light.

This raises one interesting theme that can be explored with the LEGO Tortoise: the differences between synthetic and analytic theories of its behaviour. From a synthetic approach, we have constructed the robot, and produced its program. The result is a fairly simple machine that uses a handful of loosely co-operating reflexes to navigate through its world. Any behaviour that we observe — simple or complex — is going to be explained by appealing to this knowledge. In contrast, the analytic approach only has available to it observations of robot behaviour, and must use these observations to infer internal processes. What kind of theory might this produce? Will it be more complicated than the synthetic one? How much will it have to change as more and more environments are explored? Let us examine some more robot behaviour to consider this issue in more detail.

7.14 THE TOTAL TORTOISE

7.14.1 Repeating History

In Chapter 6, we saw a number of photographs that are the records of the behaviour of Elsie and Elmer (Holland, 2003a). Can we reproduce functionally similar behaviour with the LEGO Tortoise?

7.14.2 Search for an Optimum

Video 7-2.mpg, available from the website that supports this book (http:// www.bcp.psych.ualberta.ca/~mike/BricksToBrains/). demonstrates the behaviour of the LEGO Tortoise in a number of situations that were inspired by Grey Walter's studies (Grey Walter, 1963). The video begins with the *search for an optimum* (see Figure 6-2 and the discussion in Section 6.4). In this situation, the robot's environment consists of a single light bulb on the floor, illuminating an otherwise dark room. When placed in this situation, the behaviour of the original Tortoise was described as follows: "Attracted at first by a distant bright light, the creature ... circles round it at a respectful distance, exhibiting a search for optima rather than maxima — the idea of moderation of the classical philosophers."

The LEGO Tortoise generates similar behaviour. In the dark, it oscillates around, seeking light. When the periscope mirror points toward the light source, the robot quickly moves toward it. However, when it comes too close to the light, it is dazzled. It then proceeds to circle the light at a safe distance. Knowing how this fairly simple robot has been constructed and programmed, how would you explain this behaviour? How do you think that this behaviour would be explained by someone who was not familiar with how the robot was constructed, and could only analyze what he or she observed?

7.14.3 Free Will

Now consider an environment in which two light sources are present on the floor. Grey Walter observed his Tortoise first circle one light, and then move to circle the other, demonstrating choice behaviour (see Figure 6-3 in Section 6.5), and rising above Buridan's ass. Similar choice behaviour is demonstrated by the LEGO Tortoise in the video's next segment. It begins by respectfully circling the light at the bottom of the video screen. When the circle is mostly complete, it quickly departs, and moves to the other light, which it also circles. This second "inspection" complete, it returns to the first light. "By scholastic definition the creature appears endowed with 'free will'. It approaches and investigates first one goal and then abandons this to investigate the other." Our synthetic methodology allows us to explain the robot behaviour without appealing to free will — but to what causes would an analytic approach appeal? Would they be the same as those appealed to in an analytic theory of Section 7.15.3?

7.14.4 Discernment

Grey Walter explored discernment by combining positive and negative tropisms. For example, he combined a single attracting light with a single repelling obstacle in one study (see Figure 6-4 in Section 6.6). His robot — and the LEGO Tortoise, as seen in the video — avoided the obstacle, maintaining a brief memory of its presence before proceeding to investigate the light. Again, consider the different accounts of this complex behaviour that would be produced by synthetic and analytic methodologies. Might analytic theories become more complex as behavioural complexity increases? Is this true of our synthetic theory?

7.14.5 Self-Recognition

Grey Walter's Tortoise's are perhaps most famous for performing the mirror dance (see Figure 6-6 in Section 6.8). When seeking light in front of a mirror, Elsie was attracted to the reflection of her pilot light. However, when the reflection was approached, her pilot light turned off (because the steering motor was off), causing new seeking behaviour to be produced. The result was a complex trajectory along a mirror that Grey Walter noted could be interpreted as evidence for self-awareness. In the video, the LEGO Tortoise produces a similar mirror dance, in the dark and in the light. The dance stops as soon as the mirror is removed. We possess a simple, synthetic account of this behaviour. How complicated would an analytic theory of it be?

7.15 TORTOISE IMPLICATIONS
7.15.1 Grey Walter's Legacy

Grey Walter's Tortoises have been described as the first biologically inspired robots, and as the first example of behaviour-based or "new" robotics (Holland, 2003b).

Grey Walter's robots serve as inspirations for embodied cognitive science at a number of different levels. His general purpose was to create machines "that would imitate a living creature in performance, as distinguished from appearance" (Grey Walter, 1963, p. 122). Such imitation was to be produced by exploiting a small number of simple principles. "Two ideas, goal-seeking and scanning, had combined as the essential mechanical conception of a working model that would behave like a very simple animal" (p. 125). Furthermore, these simple principles were capable of producing emergent behaviour because of feedback between the robot and its environment. "This again illustrates an important general principle in the study of animal behaviour—that any psychological or ecological situation in which such a reflexive mechanism [feedback] exists, may result in behaviour which will seem, at least, to suggest self-consciousness or social consciousness" (p. 130).

In short, Grey Walter's work illustrated that seemingly complex behaviour, often attributed to complex internal processes, might actually be the result of simple internal processes interacting with a complex environment. This is one of the fundamental ideas of embodied cognitive science.

7.15.2 The LEGO Tortoise

The LEGO Tortoise is important because it permits us to obtain hands-on experience with Grey Walter's influential ideas.

First, when the robot is activated, we are armed with a great deal of knowledge about its structure and mechanisms, because we have created this machine. As a result, whenever we observe complex or surprising behaviour, we are in a position to explain it by appealing to these known mechanisms. Our synthetic approach should produce simpler accounts of this complex behaviour than would be achieved by analyzing the behaviour without having built the robot (Braitenberg, 1984).

Second, with the robot we can easily demonstrate environmental contributions to behavioural complexity. That is, behavioural complexity should increase by leaving the machine alone, and by modifying its environment. This message was pioneered by early classical cognitive scientists (Simon, 1969), then largely ignored by classical cognitive science, but championed by embodied cognitive science (Clark, 1997, 2003; Dourish, 2001; Norman, 2002).

One example of exploring environmental complexity with the LEGO Tortoise was recently provided by one of my students. She created a square enclosure for the Tortoise, where each side was either opaque or mirrored. She manipulated environmental complexity by manipulating the number and location of mirrored sides. As more mirrors were added, the robot's environment became increasingly complicated because of the proliferation of reflections of its pilot lights. The result was a steady progression in the elaborateness of the mirror dance that the robot produced.

7.15.3 Degrees of Embodiment

The LEGO Tortoise also points us in the direction of the next issue to be explored in our study of embodied cognitive science using simple robots. It has been argued that robots can be differentiated in terms of their degrees of embodiment (Fong, Nourbakhsh, & Dautenhahn, 2003). A robot is not merely embodied by being constructed. Rather, its embodiment is reflected in the degree to which it can be perturbed by its environment, and can in turn affect is environment. Thus, the LEGO Tortoise is only moderately embodied: it can sense its environment, but only changes its environment accidentally (e.g., by bumping a light obstacle out of the way). In the next chapter, we will consider what might be gained by increasing embodiment, when we describe a robot that is no more complex than the Tortoise, but which is explicitly designed to modify its environment as it moves through it.

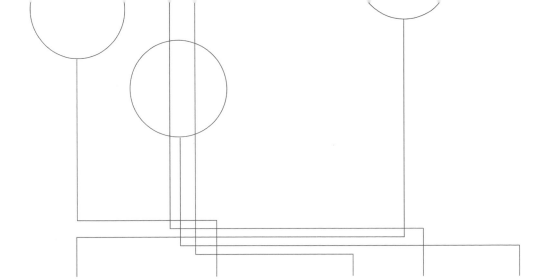

Chapter 8
Embodiment, Stigmergy, and Swarm Intelligence

8.0 CHAPTER OVERVIEW

The LEGO robots in previous chapters have exhibited various degrees of situatedness — from the thoughtless walkers, which (charitably) can "sense" elementary forces like gravity, to Braitenberg's Vehicle 2, which has sensors for measuring light, and ending with the LEGO Tortoise, which can sense both light and obstacles. While these robots map out a continuum of situatedness, at first glance it would seem that they are all equally embodied, because they are all constructed out of the same types of building blocks.

However, some would argue that embodiment means more than just being physically constructed; it has been claimed the degree of an agent's embodiment reflects the extent to which the agent can alter or manipulate its environment (Fong et al., 2003). From this perspective, a continuum of embodiment is also possible. However, the preceding robots do not map out this continuum particularly well, because they all react to — and fail to manipulate — the world in which they operate. The current chapter describes a new robot, the Lemming, which is designed not only to sense its environment, but also to change it. Like the Tortoise, the LEGO Lemming can sense and avoid obstacles. The Lemming also uses a light sensor. However, the function of this sensor assumes that the Lemming operates in a world in which coloured objects have been scattered on the floor. When an object is encountered by the robot, its colour is detected, and this controls the robot's behaviour. In particular, the sensed colour determines whether the object will be moved and deposited near a wall at the outskirts of the Lemming's domain, or will be placed near other bricks at the interior of the Lemming's world. This creates a higher degree of embodiment than was exhibited by any of the earlier LEGO robots. This is important, because the notion of cognitive scaffolding that was introduced in Chapter 3 requires that agents be able to manipulate their world. The "mind" of the Lemming has leaked into its world, because the coloured objects that it moves can be described as an external memory.

This chapter explores the Lemming's "leaky mind" in two contexts. The first involves a single Lemming that manipulates its external memory of coloured objects. The second involves a small colony of Lemmings that manipulate the collective memory of the colony. This leads us to consider in more detail both the notion of embodiment and the notion of stigmergy (which was introduced in Chapter 1), and to explore some of the ideas that are fundamental to collective intelligence.

8.1 TRAVELLING SALESMEN
8.1.1 The Travelling Salesman Problem

The travelling salesman problem, or TSP, is one of the most famous and important problems in the combinatorial optimization literature (Gutin & Punnen, 2002; Lawler, 1985). The problem itself is easy to express: Imagine a salesman who must visit a sequence of cities, stopping at each only once. In what order should the salesman visit the cities, so as to travel the shortest (and presumably least expensive) route?

The TSP has been studied for a very long time. While it was first named by Menger in 1932, its form was first defined by Voight in 1831 (Laporte & Osman, 1995). The extent of modern research on the problem is indicated by the existence of a bibliography of 500 references relevant to it (Laporte & Osman, 1995).

One reason for the long history of research on the TSP is because of its importance; the TSP is applicable to a wide variety of real-world problems (Punnen, 2002). These include scheduling tasks on a machine to minimize the cost of setting the machine up for each new job, and assigning a different frequency to each of a network of transmitters so that interference between transmitters is minimized. Punnen also notes that other areas to which the travelling salesman formulation is relevant include data analysis in psychology, X-ray crystallography, overhauling gas turbine engines, warehouse order-picking problems, and wall paper cutting.

A second reason for the long history of research on the TSP is its difficulty. The TSP is a famous example of an NP-complete problem (Kirkpatrick, Gelatt, & Vecchi, 1983). This means that as the number of cities involved in the salesman's tour increases linearly, the computational effort for finding the shortest route increases exponentially. For N cities, the number of possible routes to consider when doing an exhaustive search for the shortest route is ½ $(N-1)!$ This means that for a 4-city tour, one needs only consider 3 different routes to find the shortest. However, for an 8-city tour, the shortest route is but one of 2,520 possibilities; there are approximately 4.421e+30 routes to compare to find the shortest tour of 30 different cities!

8.1.2 Solving the TSP

Given the importance and the difficulty of the TSP, a number of different approaches to its solution have been explored. Many of these approaches are algorithms that have a long history in the numerical optimization literature (Bellmore & Nemhauser, 1968).

Some more recent solutions to such problems have been inspired by physical metaphors. Annealing is a physical process, describable using statistical mechanics, by which an optimal structure is obtained by bringing a substance to a high temperature, and then slowly cooling it. Optimality is discovered because at high temperatures the state of the substance (e.g., arrangements of atoms) can be moved out of local minima with high probability; the slow cooling can result in the state achieving its most stable configuration (i.e., a global minimum). Simulated annealing, where the state being optimized is the cost of the tour, has been successfully used to provide excellent solutions to the TSP (Kirkpatrick et al., 1983).

Other approaches to the TSP are biologically inspired. Neural networks have been used to discover TSP solutions (Hopfield & Tank, 1985; Siqueira, Steiner, & Scheer, 2007). Evolutionary programming techniques, such as genetic algorithms (Holland, 1992; Mitchell, 1996), have also been successfully applied (Braun, 1991; Fogel, 1988). Even molecular computers, which encode problem states using DNA molecules, have been explored (Lee, Shin, Park, & Zhang, 2004).

Approaches to the TSP have also been inspired by observing how insects deal with real-world situations (Tarasewich & McMullen, 2002). These approaches are of interest to us because they raise the possibility of using teams of simple robots to solve problems that might be beyond the capability of any individual member of the team. Let us now consider these solutions and their implications.

8.2 SWARM INTELLIGENCE
8.2.1 Economical Ants

One reason for the travelling salesman problem's importance is that being able to find the shortest route provides enormous advantages for a wide variety of human endeavors. However, the importance of this ability is not restricted to humankind. Any animal that must move regularly between two different locations, such as a nest and a food source, would benefit by identifying and using the most economical route (Goss, Aron, Deneubourg, & Pasteels, 1989). Is there any evidence that they do so?

For one example, consider the Argentine ant *Iridomyrmex humilis*. Goss et al. (1989) studied a laboratory colony of these ants by using a series of bridges that linked their nest to a food supply. In this bridge system there were two locations at which the ants had to choose between two different routes. At each decision point, one choice would lead to a route that was much longer than the one that would be followed if the other choice had been made. When the bridge system was first put in place, food was discovered in a matter of minutes. At this early stage, ants went in each direction at both decision points with equal probability. However, shortly afterward, a strong preference emerged: almost all of the ants chose the path that produced the shortest journey.

How do ants determine the shortest route between two locations? The answer to this question is rooted in local, computationally simple, ant behaviour. *Iridomyrmex humilis* leaves a trail of pheromones as it moves in either direction along a path between food and its nest. An ant that chooses the shortest path will return along it, and add to the pheromone trail at the decision points, sooner than an ant that has taken a longer route. This means that ants that arrive later at a decision point will find a stronger pheromone trail in the shorter direction, will be more likely to choose this direction, and will themselves add to the pheromone signal. "Each ant that passes the choice point modifies the following ant's probability of choosing left or right by adding to the pheromone on the chosen path. This positive feedback system, after initial fluctuation, rapidly leads to one branch being 'selected'" (Goss et al., 1989, p. 581).

The ability of ants to find shortest routes inspired a new approach to solving the travelling salesman problem (Dorigo & Gambardella, 1997). Dorigo and Gambardella programmed a colony of simulated ants to leave and follow pheromone trails, which also had a working memory that stored cities that had already been visited, so that the artificial ants would travel to a new city. They studied a number of different versions of the problem, and found that the simulated ant colony produced solutions that were as least as good as, and often better than, solutions produced by a variety of other algorithms, including neural networks and genetic algorithms.

8.2.2 Emergent Intelligence

The ability of ants—simulated or otherwise—to choose shortest routes does not, importantly, require a great deal of computational power within each individual. Individual ants do not determine optimal routes; it is the ant colony as a whole that solves the problem. "The selection of the shortest branch is not the result of individual ants comparing the different lengths of each branch, but is instead a collective and self-organizing process, resulting from the interactions between the ants marking in both directions" (Goss et al., 1989, p. 581).

8.3 COLLECTIVE CONTRIBUTIONS
8.3.1 Swarm Advantages

In Section 1.7.2, we saw that an organism could be defined as a coordinated system of activities that could obtain environmental resources, produce new activities, and adapt to environmental disturbances (Wheeler, 1911). This permitted entomologists like Wheeler to define the colonies of social insects as superorganisms, from which emerged more complex results (such as elaborate nests) than one would predict from examining the capabilities of individual colony members. Swarm intelligence is an interesting evolution of the idea of the superorganism. It offers advantages that may not be provided by other computational methods. "Nature-inspired intelligent swarm technology deals with complex problems that might be impossible to solve using traditional technologies and approaches" (Hinchey, Sterritt & Rouf, 2007, p. 113). What is provided by swarm intelligence that might be missing from traditional approaches?

Importantly, a swarm's components are only involved in local interactions with each other. This characteristic is the source of many of the advantages of swarm intelligence (Balch & Parker, 2002; Sharkey, 2006). For instance, a computing swarm is *scalable*—it can be comprised of varying numbers of agents, because the same control structure (i.e., local interactions) is used regardless of how many agents are in the swarm. For the same reason, a computing swarm is *flexible*—agents can be added or removed from the swarm without reorganizing the entire system. The scalability and flexibility of a swarm make it *robust*—it can continue to compute when some of its component agents no longer function properly. A second source of robustness comes from the nature of the swarm's agents themselves. For instance, if each agent is autonomous, and is capable of reacting or adapting to environmental changes, then these individual advantages will be inherited by the swarm as a whole.

8.3.2 Robot Collectives

When a swarm is composed of autonomous, embodied, and situated robots, it may be particularly well suited to solving some important real-world problems (Beni, 2005; Brooks & Flynn, 1989). One reason for this is that a robot collective would have all of the advantages of swarm intelligence that were mentioned in Section 8.3.1. A second reason for this is that robot collectives are capable of manipulating real-world objects and environments, and therefore can serve as real-world tools.

For example, NASA is interested in preparing landing sites on distant planets. A swarm of robots provides one possible solution to this problem (Parker, Zhang, & Kube, 2003). Parker et al. were inspired by a behaviour in some ants, called "blind bulldozing" (Franks, Wilby, Silverman, & Tofts, 1992), in which nests are constructed stigmergically by pushing material away from a nest site. Parker et al. designed a robot collective for blind bulldozing. An individual robot in the collective is usually in a *plowing state*, in which it moves straight in some heading, pushing debris as it moves. When the friction caused by an accumulation of debris exceeds a threshold, the robot switches into a *finishing state*, which causes it to turn a random amount before re-entering the plowing state. The robot could also switch into a *colliding state* when it randomly turns because of a collision with another robot in the collective. Parker et al. created variously sized robot collectives that created "nests" by pushing away gravel while following this algorithm. They found that a nest could be constructed by a single robot, but that the use of multiple robots decreased the time that was required to accomplish the task.

Robot collectives are not appropriate for all tasks, but are ideally suited for many (Balch, 2002). As we shall see, typical tasks for robot collectives include foraging, material transport, and sorting. It has also been argued that a collection of simple, mass-produced robots that do not require central control provide an ideal and inexpensive medium for conducting exploration of remote planets (Brooks & Flynn, 1989).

8.4 CRITICAL NUMBERS OF AGENTS
8.4.1 When Is a Swarm Intelligent?

In swarm intelligence, a problem's solution emerges from the activity of a collection of agents, suggesting that having a collection of agents is better than having a single agent working on the problem. However, swarm intelligence depends on more than mere numbers of agents. For a swarm to be considered intelligent, the whole must be greater than the sum of its parts. This idea has been used to identify the presence of swarm intelligence by relating the amount of work done by a collective to the number of agents in the collection (Beni & Wang, 1991).

Consider, for example, a collection of completely independent agents foraging for food. As the number of agents increased, one would expect that the collective would forage faster. However, if the agents worked completely independently of one another — if the whole were equal to the sum of its parts — then there would be a linear relationship between the amount of work accomplished and the number of agents, as shown in the dashed line in Figure 8-1. Beni and Wang (1991) would take this linear relationship to indicate the *absence* of swarm intelligence.

In contrast, consider agents that can interact with each other. A small

8-1

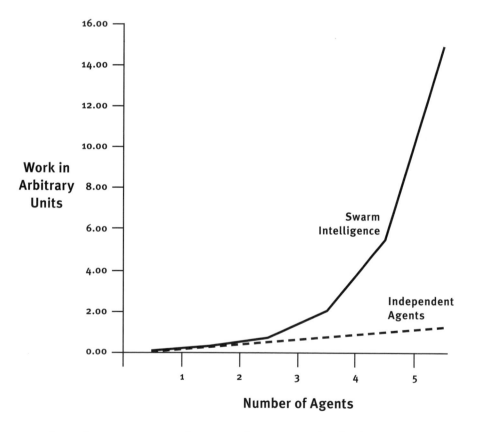

number of agents may not have much opportunity for interaction, and therefore may not perform better than the same number of independent agents. However, after some critical number of agents is reached, agent interaction becomes more likely, and makes the swarm more efficient than a non-interacting collective. This is shown in the solid non-linear function in Figure 8-1. With 2 or 3 agents, there is little difference between the solid and dashed functions. However, when there are more than 3 agents, the interactive collective is far more efficient than the non-interactive one. The non-linear relationship between the number of agents and the amount of work accomplished is taken by Beni and Wang (1991) to indicate the presence of swarm intelligence.

8.4.2 A Foraging Example

One study of robot foraging tested Beni and Wang's (1991) theory (Sugawara & Sano, 1997). In this study, obstacle-avoiding robots moved through an arena, collecting pucks, which they then brought back to a home location. In some conditions in this experiment the robots interacted: when a puck was encountered, the robot stopped for a set period of time and emitted a light that attracted other robots to that location.

Sugawara and Sano (1997) found a non-linear relationship between the number of robots and the number of pucks foraged over time, but only when robots interacted, supporting the theory of Beni and Wang (1991). When robots did not interact, only a linear relationship between these two variables was observed. Interestingly, though, robot interactions were not always helpful—they only improved efficiency if the pucks were unevenly distributed throughout the environment. Under such conditions, a robot emitting light would attract other robots to a high concentration of pucks. However, if the pucks were evenly distributed throughout the arena, then interactions actually caused a decrease in efficiency relative to a swarm of non-interacting robots. Thus, direct communication in a swarm does not always lead to improved performance. In the next section, we will see that an important issue in swarm intelligence is the degree to which communication is used to coordinate the activities of swarm members.

8.5 COORDINATION, COMMUNICATION, AND COST
8.5.1 Costly Coordination

Early research on robot teams studied small groups of homogenous robots (Gerkey & Mataric, 2004). Modern research examines much more sophisticated robot collectives that can consist of different types of robots that carry out diverse tasks at varying locations or times (Balch & Parker, 2002; Schultz & Parker, 2002). "It is no longer sufficient to show, for example, a pair of robots observing targets or a large group of robots flocking as examples of coordinated robot behavior. Today we reasonably expect to see increasingly larger robot teams engaged in concurrent and diverse tasks over extended periods of time" (Gerkey & Mataric, 2004, p. 939).

It is desirable to coordinate the actions of a team of diverse robots (Gerkey & Mataric, 2002, 2004; Mataric, 1998). One must determine the tasks carried out by individual robots in a team at any given time, in order to optimally achieve some global goal. This called the multi-robot task allocation problem.

With the hardware capabilities of modern robots, one general approach to solving the multi-robot task allocation problem is to employ *intentional co-operation* (Balch & Parker, 2002; Parker, 1998, 2001). Intentional co-operation is achieved by adopting some form of communication between robots so that task allocation can be negotiated, or so that one robot will be aware of what others are doing so that it does not unnecessarily duplicate efforts or work in opposition to the current efforts of other team members.

Intentional co-operation provides some particular advantages. If it is possible to have intentional co-operation amongst robots, then it should also be possible have a robot team co-operate with the needs of a human user (Gerkey & Mataric, 2004). As well, a robot team governed by intentional co-operation should be more efficient than one that is not (Sugawara & Sano, 1997). "By sharing information and leveraging each others' skills, a group of robots can truly be more than the sum of its parts" Gerkey & Mataric, 2002, p. 758).

A robot team governed by intentional co-operation may be extremely efficient at performing a task, and may also be able to use communication to structure robot activities so that a single collective can efficiently solve a diversity of problems. However, these advantages are not achieved without cost. First, the extent to which intentional co-operation imposes central control on the members of the robot team is the extent to which many of the advantages of robot collectives described in Section 8.3 are diminished. For instance, communication between robots is costly, and as more robots are added to a communicating team there is likely to be a "communications bottleneck" that makes the team less scalable (Kube & Zhang, 1994). Second, as communication makes the functions carried out by individual team members more specialized, the robustness of the robot collective might be jeopardized (Kube & Bonabeau, 2000).

In short, while the study of intentional co-operation for dealing with multi-robot task allocation is ongoing and important, other approaches are still worthy of consideration. In particular, is it possible for a robot collective to coordinate its component activities, and solve interesting problems, in the absence of direction communication?

8.5.2 A Stigmergic Solution

One answer to this question can be provided by studying the extent to which stigmergy can be used to control a team of robots (Kube & Bonabeau, 2000). As we saw in Chapters 1 and 3, stigmergy occurs when there is indirect communication between agents via the environment. An agent makes some change to the environment, which in turn signals another agent to perform a different behaviour than it would have in the absence of this signal. We have seen that stigmergy provides important control over the behaviours of social insects; these insect behaviours have inspired many studies of robot teams, which have shown that such collectives are capable solving interesting problems without the need of intentional co-operation. Let us now consider some prototypical examples of such research.

8.6 CO-OPERATIVE TRANSPORT
8.6.1 Robots that Push Boxes

Consider the New World army ant *Eciton burchelli* (Couzin & Franks, 2003). A foraging party of 200,000 ants from a colony will kill as many as 30,000 prey in a dawn-to-dusk swarming raid, and move this food back to the nest, often using *co-operative transport* (Franks, Sendova-Franks, & Anderson, 2001). Co-operative transport involves a group of agents working together to move a large object.

Roboticists have studied co-operative transport using the *box-pushing task*, in which a box, intentionally too heavy for a single robot to move, must be pushed to a goal location. Imagine a group of five small robots sitting in the corner of a laboratory room. In the middle of the room is a large box (Kube & Bonabeau, 2000). A spotlight hanging from the ceiling of the room is turned on, and the robots begin to move throughout the room. Then, a light in the middle of the box is turned on. Suddenly, the behaviours of the robots change, as shown in this video available from http://www.cs.ualberta.ca/~kube/ra97/multi-robot.mpeg.

When the box is lit, the robots move toward it and come into contact with it. Some of the robots remain in contact with the box, and push it toward the part of the room lit by the spotlight. Other robots are in the wrong place to accomplish this, and move around the box to take up a more effective position. Still other "robots leave the task, seemingly at random, and wander off only to return and join the group effort in transporting the box towards its goal" (Kube & Bonabeau, 2000, p. 99).

The robots move the box toward the goal, but not smoothly: the box may veer off in an incorrect direction, or sometimes stop moving. However, at each juncture the robots realign themselves, and eventually push the box to the intended goal. At this point, if the spotlight is turned off, and if a different spotlight is turned on, then the robots will reorganize themselves, and again — in a moderately erratic fashion — the box will be pushed toward the new goal location.

8.6.2 Stigmergic Co-operation

The robots' box-pushing behaviour mimics some of the co-operative transport behaviours observed in ants, and therefore might serve as a model of this insect ability. However, this co-operative behaviour emerges without direct communication between robots. "Rather a form of indirect communication (stigmergy) takes place through the environment by way of the object being manipulated" (Kube & Bonabeau, 2000, p. 100). This is important because other approaches to solving the box-pushing problem usually involve direct communications between robots, so (for example) one robot "knows" not to duplicate the efforts of another (Mataric, 1998; Parker, 1998, 2001).

Kube and Bonabeau (2000) achieved stigmergic control of box pushing by providing robots with behaviours that were elicited by simple stimuli. Robots used both touch and infrared sensors to detect (and avoid) other robots. Light sensors were used to locate the box as well as the goal spotlight. If a touch sensor was depressed, and the box-detecting light sensor was above threshold, then the robot had detected that it was in contact with the box. If it was in such contact, and could see the goal, then box-pushing behaviour was initiated. If it was in contact with the box, but could not see the goal, then other movements were triggered resulting in the robot finding contact with the box at a different position.

These behaviours caused the robots to seek the box, push it toward the goal, and do so co-operatively by avoiding other robots. Furthermore, as the robots acted on the environment, and changed the position of the box, this could change the situation sensed by other robots, and produce corresponding changes in behaviour. For instance, a robot pushing the box might lose sight of the goal because of box movement, and would therefore leave the box and use its other exploratory behaviours to come back to the box and push it from a different location. "Cooperation in some tasks is possible without direct communication" (Kube & Bonabeau, 2000, p. 100).

8.7 COLLECTIVE SORTING
8.7.1 Spatial Sorting by Ants

We have seen that ant colonies are able to find the shortest routes, to clear nests using blind bulldozing, and to move large objects by employing co-operative transport. Ants are also very adept at sorting objects — producing order from disorder — into useful and interesting spatial patterns. These patterns are frequently described as the products of collective intelligence (Franks & Sendova-Franks, 1992; Holland & Melhuish, 1999; Sendova-Franks, Scholes, Franks, & Melhuish, 2004).

One example of such behaviour is called *patch sorting*. In patch sorting, two or more classes of objects are placed in separate clusters, so that each cluster contains only one class of objects, and each cluster is spatially separated from the others (Holland & Melhuish, 1999). Ants exhibit patch sorting when they place eggs, larvae, and cocoons into separate piles (Deneubourg et al., 1991) or when they place corpses into different clusters when constructing a cemetery (Theraulaz et al., 2002).

Another example is the *annular sorting* of *Leptothorax* ant colonies (Sendova-Franks et al., 2004). A colony's brood is placed in a single cluster within the nest, but this cluster is highly structured: it is a set of concentric annuli, with each ring comprised of a different type of brood items. "Eggs and small larvae are in the middle whereas medium and large larvae are in concentric annuli increasingly further out towards the periphery" (Sendova-Franks et al., 2004, p. 1095). As a result, brood items that require more care are more easily accessed because they are in the outer rings of the sorted structure.

8.7.2 Stigmergic Sorting by Robots

Researchers have investigated whether interesting spatial patterns be produced by sorting procedures that are completely under stigmergic control (Holland & Melhuish, 1999). Robots used a "gripper" in front to capture a Frisbee lying on the floor. They could also sense Frisbee colour, and had proximity detectors that could register the presence of obstacles such as walls or other robots. The gripper also had a micro-switch that would not be triggered if the robot gripped a single Frisbee, but would be triggered if two or more Frisbees were pushed by the robot.

Holland and Melhuish (1999) explored a number of different algorithms that were used to control the behaviour of each of a collection of 10 robots. One, the *pullback algorithm*, consisted of three simple rules. First, if the gripper held a Frisbee, and an obstacle was encountered, then the robot made a random turn away from the obstacle. Second, if the gripper held a Frisbee, and another Frisbee was encountered, the robot would pull the Frisbee it held back a distance that depended on the Frisbee's colour, drop it, and turn away from it. Third, if no object was held, then the robot simply moved forward. This algorithm sorted Frisbees into a single cluster that was roughly annular in organization—one colour of Frisbee was in the centre of the cluster, which was surrounded by Frisbees of a different colour. Simpler sets of rules—for instance rules that were not sensitive to Frisbee colour—produced patch sorting.

Importantly, this sorting behaviour is produced by stigmergy. None of the robots directly communicate with one another. Instead, they indirectly communicate by moving Frisbees to different locations; these newly positioned Frisbees will in turn alter the behaviour of the robots, and spatial sorting emerges from this stigmergic system.

Variations of the pullback algorithm are capable of producing more striking annular sorting, and the sorting capabilities of robots and ants have been directly compared (Melhuish, Sendova-Franks, Scholes, Horsfield, & Welsby, 2006; Scholes, Wilson, Sendova-Franks, & Melhuish, 2004; Wilson, Melhuish, Sendova-Franks, & Scholes, 2004). This research demonstrates an interesting interplay between disciplines in which ant behaviour inspires robotics research, which in turn is being used to develop theories about ant behaviour. It also demonstrates that stigmergy is capable of producing spatially organized patterns.

8.8 STIGMERGY AND DEGREES OF EMBODIMENT
8.8.1 Extending the Mind into the World

In Chapter 3, we were introduced to an important idea in embodied cognitive science, the leaky or the extended mind (Clark, 1997, 1999, 2003, 2008; Wilson, 2004, 2005). According to the extended mind hypothesis, the mind and its information processing is not separated from the world by the skull. Instead, the mind interacts with the world in such a way that information processing is both part of the brain and part of the world—the boundary between the mind and the world is blurred, or disappeared. We saw in Chapter 3 that this can occur because of cognitive scaffolding. A simple example of cognitive scaffolding is extending memory by using external aids. However, full-blown information processing can be placed outside the traditional mind, into the world, by using appropriate artifacts. We saw an example of this in our discussion of the nomogram (Hutchins, 1995) in Section 3.10.1.

In order for cognitive scaffolding to occur—in order for the mind to extend itself into the world—cognitive agents must be able to interact with and alter the physical world. This was the reason that Chapter 3 argued for a central role of action in the series of cognition.

The extended mind hypothesis can be applied to single cognitive agents. However, when information processing leaks into the world then the extended mind hypothesis can also be applied to a group of agents that operate in a shared environment (Hutchins, 1995). The examples of stigmergy that have been described in the current chapter, as well as those that were introduced in Chapter 1, are also examples of scaffolded or extended group cognition. In particular, stigmergy places the control structure of the information-processing collective into the environment, removing the need for members of the collective to directly communicate with one another. Section 3.12 argued that such stigmergic control could also be considered to be a central characteristic of a modern production system.

Importantly, stigmergic control also requires that agents be able to manipulate the environment in which they are situated. Ants can only solve the travelling salesman problem by laying down a pheromone trail. Robots can only solve the box-pushing problem by contacting and moving the box in order to communicate to other robots that they need to move and push the box from a different location.

8.8.2 Degrees of Embodiment

The Lego robots from previous chapters of this book make it clear that there are different degrees of situation. The walking robots described in Chapter 5 are minimally situated, because they have no sensors and only passively react to physical forces. Vehicle 2, investigated in Chapter 4, is more situated because it uses light sensors to modify motor speeds. The Lego Tortoise, detailed in Chapters 6 and 7, is even more situated because it uses both light and touch sensors.

While these robots illustrate degrees of situation, at first glance it would seem that they are all equally embodied, in the sense that they are all physical artifacts. However, there are other definitions of embodiment that suggest that agents can be embodied to different degrees (Fong et al., 2003).

Fong et al. (2003, p. 149) argue that "embodiment is grounded in the relationship between a system and its environment. The more a robot can perturb an environment, and be perturbed by it, the more it is embodied." As a result, not all robots are equally embodied. A robot that is more strongly embodied than another is a robot that is more capable of affecting, and being affected by, the environment. Clearly an extended mind, or a stigmergic league–controlled collective, requires strong embodiment. The robots that we have built are all equally embodied, but that is because none of them is designed to affect the environment. We now turn to describing a more strongly embodied robot, which can therefore be used to explore ideas such as the extended mind and swarm intelligence.

8.9 THE LEMMING

8.9.1 Lemming Situation

Lemmings was a video game that was introduced in 1991 by Psygnosis for the Commodore Amiga. A player of this game had to assign abilities to a small number of lemmings so that they could alter the environment, and the behaviour of other lemmings, in such a way as to prevent mass migrations that would lead to disaster. In honour of this game, and the behaviours of the agents within it, we have named our more strongly embodied LEGO robot the Lemming. This is because we hoped that it could affect the behaviour of other LEGO Lemmings by manipulating its environment—in particular, by moving light and dark LEGO bricks to different locations in an enclosed arena.

In order to do this, the Lemming (shown below in Figure 8-2) is situated in its environment using three different sensors. An ultrasonic sensor mounted on the top of the Lemming is used to detect and avoid walls and other obstacles, such as other Lemmings. A second ultrasonic sensor mounted near the base of the Lemming is used to detect the presence of to-be-moved bricks. Finally, a light sensor mounted inside the "brick catcher" at the front of the robot is used to analyze a captured brick. In particular, the light sensor detects the colour of the brick, which is then used to elicit an appropriate colour-dependent behaviour from the Lemming.

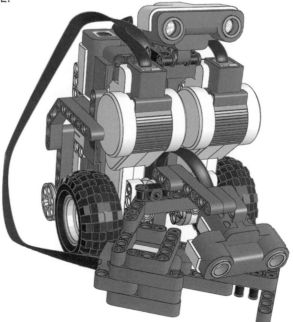

8.9.2 Lemming Embodiment

The embodiment of the Lemming is also critical. In particular, the "brick catcher" at the front of the robot has a very definite shape. First, its shape is such that when an object is contacted by the brick catcher, it is moved against the light sensor so that its colour can be detected. Second, the shape of the brick catcher is asymmetrical. As a result, if the Lemming turns to its left, the object will remain trapped inside and can be pushed to a new location. However, if the Lemming turns to its right, the object is released from the catcher, and can therefore be left behind in some new position where it can affect the behaviour of a Lemming that encounters it later.

8.10 FORAGING FOR ROBOT PARTS AND WORLD PARTS
8.10.1 Robot Parts

The Lemming is a fairly simple robot to build, and requires the parts that are illustrated below in Figure 8-3. The pages that follow describe how to construct this robot. If the reader would prefer to use wordless, LEGO-style instructions, they are available as a pdf file from the website that supports this book (http://www.bcp.psych.ualberta.ca/~mike/BricksToBrains/).

8-3

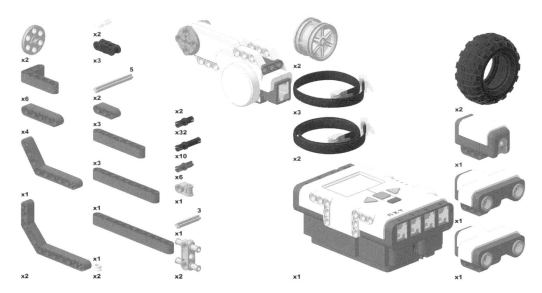

8.10.2 Bricks to Move

In addition to the robot itself, it is necessary to build objects that are placed in the world for the Lemming to manipulate. These objects are two LEGO bricks high, 4 studs long, and 4 studs wide, as illustrated in Figure 8-4. We built each of these objects using four 2 × 4 LEGO bricks.

Importantly, half of these objects are all black, and the other half are all white, because the colour of the object affects robot behaviour. We constructed 28 different black objects and 28 different white objects, requiring a total of 112 black and another 112 white 2 × 4 LEGO bricks.

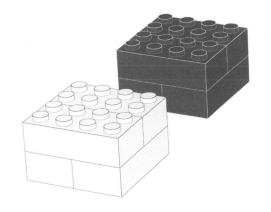

8.11 CHASSIS AND REAR WHEELS
8.11.1 NXT Brick as Chassis

Other NXT robots described in this book (Vehicle 2 in Chapter 4, the Tortoise in Chapter 6, and antiSLAM in Chapter 9) are constructed by building a central spine of beams and liftarms that serves as a chassis to which other components, such as the NXT brick, are attached. A different approach is illustrated in the Lemming, which uses the NXT brick itself as the primary chassis to which other components are attached. As well, in this robot the NXT brick is positioned vertically instead of horizontally, as shown in Figure 8-5. The first step in building the robot is illustrated in this figure. Liftarms are attached to the NXT brick in order to support two small rear wheels that will help keep the robot stable.

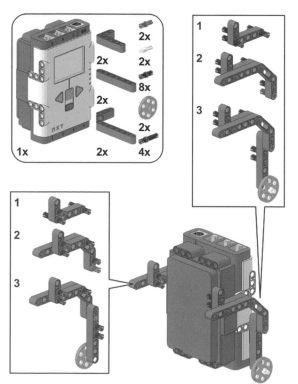

8.12 MOUNTING MOTORS

8.12.1 Motors and Cables

The liftarms that were used to create mounts for the rear wheels are also used to attach two motors to the front of the robot, as shown in Figure 8-6. Note that additional parts are attached to each motor; these parts are used later to mount other Lemming components. As well, some of the cables that are later plugged into sensors are best run between the motors and the NXT brick, and they should be attached in this second step for this reason. The ports into which each of the four cables used in this step are inserted are indicated in Figure 8-6.

8-6

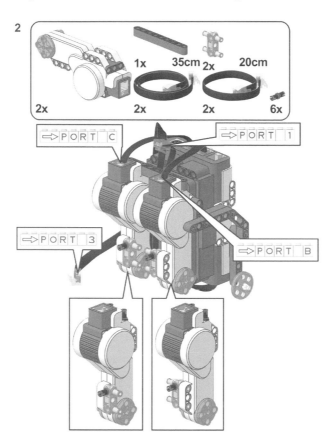

8.13 UPPER ULTRASONIC SENSOR AND FRONT WHEELS

8.13.1 The Upper Ultrasonic

The next step in constructing the Lemming is to mount an ultrasonic sensor to the top of the robot, following the instructions that are provided at the top of Figure 8-7. This sensor will be pointing slightly upward, and is mounted high on the robot, so that it *won't* detect objects on the floor in front of the robot. (If the sensor refuses to stay pointed upward when mounted, a paper clip can be used to keep it pointing in the desired direction.) The purpose of this ultrasonic sensor is to detect larger obstacles, such as the walls that define the arena in which the robot operates.

8.13.2 Front Wheel Drive

The Lemming is a front wheel drive machine. The wheels are LEGO 56 × 26 tires mounted onto hubs that are the standard wheels for NXT robots. They are attached to axles that are inserted directly into each motor, and held in place with a half bush, as illustrated at the bottom of Figure 8-7.

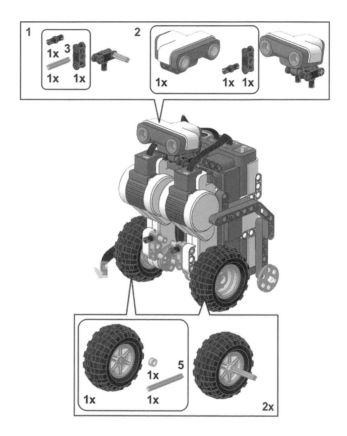

8-7

8.14 MOUNTING THE LOWER ULTRASONIC SENSOR
8.14.1 Angled Ultrasonics

One of the goals of the Lemming's design was to have it situated in its world in such a way that it could detect, and move toward, objects on the ground in front of it. This was accomplished by mounting a second ultrasonic sensor to the robot; this one is mounted on the front of the robot near its bottom, as illustrated in Figure 8-8. Note that this sensor is attached in such a way that it points downward toward the floor. The angle at which this sensor is tilted is important, because it affects how far away the objects are that can be detected. When this sensor is angled as shown in Figure 8-8, then it should be able to detect objects as far away as 50 cm. The tension of the cable used to connect this sensor to the NXT brick should be sufficient to hold it at the desired angle.

8-8

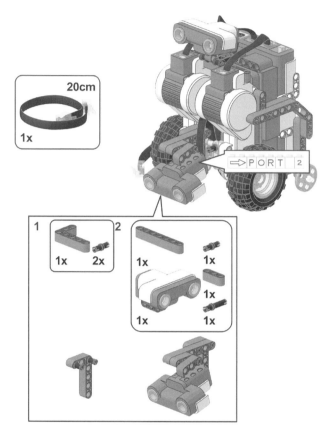

8.15 DESIGNING THE BRICK CATCHER
8.15.1 Important Embodiment

The main function of the Lemming is to trap, and to move around, coloured square objects that are in its environment. This is accomplished by making a "brick catcher" that is attached to the front of the robot, and is pushed along the floor by the Lemming. The embodiment of this Lemming component is important in several ways. First, it must be large enough to be able to trap a brick (Figure 8-4) that is encountered, regardless of the orientation of the brick with respect to the brick catcher. Second, it must not be too large—only one brick will be trapped at a time. Third, when a brick is in place, the shape of the brick completes the shape of the brick catcher, converting it from a catcher into a plow. Fourth, once trapped, the brick must slide into a position that permits its colour to be examined by the robot. Fifth, the shape of the brick catcher is critical for later robot behaviour. The brick catcher is shaped in such a way that if the robot turns to the right, the brick remains trapped, and can be moved by the robot. However, it is also shaped in such a way that if the robot turns to the left, the brick will escape from the brick catcher, and can be left behind by the robot.

The brick catcher begins by attaching various beams and liftarms together in the fashion that is illustrated in Figure 8-9.

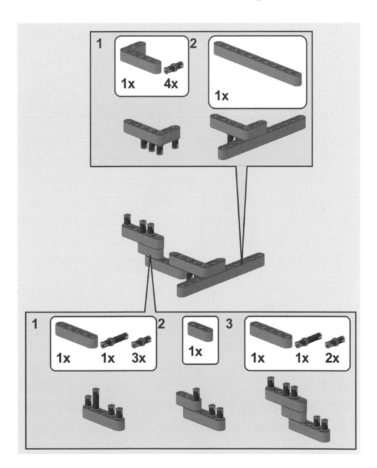

8.16 BRICK CATCHER, BRICK PROCESSOR
8.16.1 Embodiment and Situation

Figure 8-10 illustrates how to complete the remainder of the brick catcher. The upper part of this completed structure permits it to be attached to the Lemming's chassis. The lower part of this completed structure defines the shape that permits bricks to be captured, analyzed, moved, and released as was noted in Section 8.15. Note that one critical aspect of this shape is a light sensor that forms one of the "walls" used to push a captured brick. The idea behind this design is that the movement of the robot will force the brick directly against this light sensor. Once there, the light sensor can measure whether the captured object is white or black; the sensed colour will determine the subsequent behaviours of the Lemming.

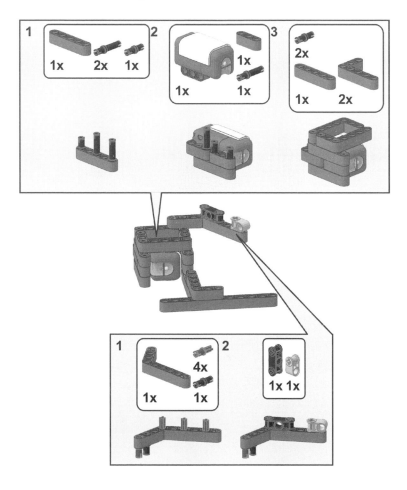

8.17 COMPLETING THE LEMMING
8.17.1 Final Construction

The Lemming is completed by attaching the brick catcher to the front of the robot as illustrated in Figure 8-11, and by making sure that all sensors and motors have the appropriate cables connected to the ports that were earlier indicated in Figures 8-6 and 8-8. With this construction completed, we can now turn to developing a subsumption architecture for the Lemming that will permit it to drive forward, avoid walls, and detect, capture, analyze, move, and release coloured objects.

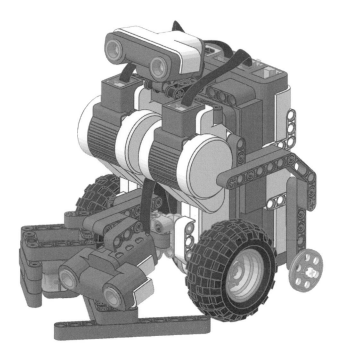

8.18 LEVEL 0: DRIVE AND CALIBRATE
8.18.1 Driving

The most basic level in the subsumption architecture for the Lemming enables it to move forward. This is accomplished by turning both motors on in the forward direction, as indicated in the short NXC program for this level that is listed below. Note that this program is separate from the calibration program that is listed below it after the two double lines of ==.

The Level 0 program uses the OnFwdSync command, which includes a Sync variable to control both motors. The OnFwdSync command is a useful NXC command that permits two motors to be coordinated in a single line of code. However, the default value for Sync in Level 0 does not cause the robot to move straight ahead. Instead, it causes the motors to run at slightly different speeds so that the robot gently turns to the right. The value for Sync is set in Level -1 (described in Section 8.22) which is used to set motor speeds by pooling conditions detected by higher levels in the architecture. The value that is used requires that each Lemming be calibrated.

8.18.2 Calibration

In some instances, the Lemming will detect objects and move straight toward them. However, no two NXT motors are exactly the same, so an individual Lemming must be calibrated to find, in essence, the precise value for Sync that will produce straight movement. The program that is listed below the Level 0 code is a separate program for calibrating a Lemming. The program is run, and the behaviour of the Lemming is watched. If it does not go straight ahead, the value for straight_calibration is changed as described. When the Lemming moves in a straight line, the value for straight_calibration is recorded and used in the main task that is described later.

```
/*===== Level 0: Drive the Lemming Forwards =========== */
task Drive(){
        while(true){
                OnFwdSync(BothMotors, DriveSpeed, Sync);
        }
} /* task Drive ends here! */
/* ============================================*/
/* ============================================*/
/* CALIBRATION PROGRAM BELOW
calibrate so the Lemming can drive straight
============================================ */
#define LeftMotor OUT_B
#define RightMotor OUT_C
#define BothMotors OUT_BC
/*======== Drive===============================*/
int speed = 65; //set all the lemmings at this speed
int straight_calibration = 0;
/*use this value in your code. Adjust it so that the lemming will
drive in a straight line when you run the code. Try
zero first. If not zero it will probably be between -5 and 5 */
task main(){
        while(true){
                OnFwdSync(BothMotors, speed, straight_calibration);
        }
}
```

8.19 LEVEL 1: DODGE OBSTACLES
8.19.1 The Lemming's *Umwelt*

A Lemming moves around white and black bricks that it detects on the floor of a room. The room's walls are obstacles that can damage the robot if it collides with them. Other Lemmings might also be part of the environment; they too need to be avoided.

8.19.2 Avoiding Obstacles

Level 1 in the Lemming's subsumption architecture uses the top ultrasonic sensor to detect to-be-avoided obstacles. If an obstacle is detected, the robot spins around away from it. The direction in which it spins depends upon whether it is carrying a brick, and upon the colour of that brick. There are two subtleties in the code below.

First, because the Lemming can exist in an environment that includes other Lemmings that are also emitting ultrasonic signals, it checks its ultrasonic signal twice, with a short delay between checks, to ensure that it is not responding to the signal sent by another Lemming. The constant movement of the robots makes it unlikely that such a "rogue ultrasonic" signal will be detected twice!

Second, in rare instances a Lemming might be pointed at such an angle so that it will encounter a wall, but cannot receive an echo from it. The CheckStuck routine is used to solve this problem. It too works by checking the ultrasonic signal twice. Because Lemmings constantly move, it is unlikely that they will receive identical ultrasonic signals a second or so apart. If this does occur, the CheckStuck routine treats this as being stuck, and initiates avoidance behaviour by setting the collision variable.

```
/*===== Level 1: Dodge ==============================
If the wall is within the wall sensor's range, spin for a little while. */
int wallDistance, lastReading, SpinDirection;
const int KEEP_DIRECTION = 1, DUMP_DIRECTION = -1, HALF_SPIN = -50, COLLISION = 27;
int spinTime = 750, stuckTime = 1500, NearWall = COLLISION;
bool collision;
task Dodge(){
    collision = false; SpinDirection = DUMP_DIRECTION;
    while(true){
            until (collision || (SensorUS(WallSensor) < NearWall));
            Wait(50);
            if (collision || (SensorUS(WallSensor) < NearWall)){ //double check to make sure it is not a US
                error!
                    collision = true; Wander = 0;
```

```
                              DriveSpeed *= SpinDirection; Spin = SpinDirection * HALF_SPIN;
                              Wait(spinTime);
                              Wander = 1; DriveSpeed = BaseSpeed;
                              Spin = STRAIGHT; SpinDirection = DUMP_DIRECTION;
                              NearWall = COLLISION;
                              collision = false;
                      }
              }
      }
      task CheckStuck()
      {
              while(true){
                      lastReading = SensorUS(WallSensor);
                      Wait(stuckTime);
                      if(lastReading == SensorUS(WallSensor)) collision = true;
              }
      }
```

8.20 LEVEL 2: SEEK BRICKS

8.20.1 Brick Attraction

The main purpose of the Lemming is to move large bricks around in
its environment. In order to do this efficiently, it would be convenient
if the Lemming could sense the presence of bricks in front of it, and
move itself toward these detected objects. In short, it would be nice if
bricks could attract the Lemming.

8.20.2 Using the Lower Ultrasonic

Level 2 uses the lower ultrasonic sensor to accomplish "brick attraction."
The lower ultrasonic sensor is mounted in such a way that it can detect
the presence of one or more of the large bricks (Figure 8-4) when they
are 50 cm or more away from the Lemming. Brick attraction is accomplished by having the Lemming move straight (i.e., stop its gentle turning) when the lower ultrasonic detects an object. It moves straight for
a set period of time that is long enough for the Lemming to physically
encounter the object that was sensed by the ultrasonic sensor.

The NXC code for Level 2 is provided below. Note that it essentially
works by setting the variable Wander to 0 for a set period of time if a
brick has been sensed; Wander is set to 1 when no brick is sensed and
when the straightTime has elapsed. Wander is a value that is passed to the
to-be-described Level -1, which uses this information, as well as information from other levels, to determine motor speeds at any given time.

```
/*===== Level 2: Seek Bricks ===========================
If the brick sensor detects a brick, straighten out. The sensor is forward-
mounted, so this will result in a brick attraction. */
int brickthreshold = 120, brickDistance, straightTime = 1000;
task Seek(){
      while(true){
              if(SensorUS(BrickSensor) < brickthreshold){
                      Wander = 0; Wait(straightTime);
              }
              else Wander = 1;
      }
}
```

8.21 LEVEL 3: PROCESS BRICK COLOURS
8.21.1 Bricks and Behaviour

While the general purpose of the Lemming is to move bricks around, its behaviour is more specific because where it pushes a brick to, and where it leaves the brick, depends on the colour of the brick. That is, once a brick is trapped by the brick catcher, the brick is pushed against the light sensor mounted on the bottom of the Lemming. The light sensor is used to classify the brick as being either light (e.g., white) or dark (e.g., black). If the brick is light, the Lemming acts as a blind bulldozer. It pushes the light brick forward until it encounters a wall (via Level 1). It then spins from the wall in such a way that the light brick is left at the wall. However, if the brick is dark, the behaviour of the Lemming is quite different. It might push the brick to a wall, but when the wall is detected it will turn in the opposite direction, so that the dark brick is not left by the wall. Instead, the Lemming will only spin in a direction that deposits the dark brick when it detects another brick using the lower ultrasonic sensor (Level 2). In short, the Lemming attempts to cluster dark bricks together in the interior of its environment.

These behaviours are accomplished by setting the direction of the spin on the basis of detected brick colour, as shown in the NXC code below. The variables set in the code below affect the motor behaviours that are controlled by Level -1, which is described in the next Section.

```
/*===== Level 3: seeBricks ===========================
If carrying a dark brick, drop it at other bricks but keep it at walls.
If carrying a light brick, drop it at walls but keep it at other bricks. */
// to cope with inaccuracy of sensor at close range.. collision detected from further away..
// drive straight, then, dump
const int FAR_COLLISION = 65;
```

```
int noBrick = dark_threshold, lightBrick = light_threshold, brickCollision = 30;
int currentBrick, wallFar = FAR_COLLISION, untilDump = 500;
bool foundDark, foundLight;
task seeBrick(){
     while(true){
             until(ColourSensor > noBrick);
             if (ColourSensor > lightBrick){
                     SpinDirection = DUMP_DIRECTION; foundLight = true;
                     NearWall = COLLISION; // With light brick, come close to wall
             }
             else if((!collision) && (SensorUS(BrickSensor) <= brickCollision) &&
(SensorUS(WallSensor) >
                     wallFar)){
                             PlayTone(587,200);
                             Wander = 0; Wait(untilDump);
                             if (ColourSensor < lightBrick){ //double check to make sure
it is not a light brick
                                     SpinDirection = DUMP_DIRECTION; collision = true;
foundDark = true;
                             }
                             until((!collision) || (ColourSensor > lightBrick));
             }
             else {
                     SpinDirection = KEEP_DIRECTION;
                     NearWall = FAR_COLLISION; // With dark brick, keep far from wall
             }
     }
}
```

8.22 LEVEL -1: INTEGRATE LEVELS TO CONTROL MOTORS
8.22.1 Multiple Motor Influences

Levels 0, 1, 2, and 3 of the Lemming's subsumption architecture are all very simple, but they lead to different effects on the motors. Level 0 moves the robot forward, and requires it to turn slightly as it moves. If Level 1 detects an obstacle, then the motors must spin the robot to avoid it. If Level 2 detects a brick, then the Lemming is steered straight toward it. If Level 3 has detected a coloured brick, then this will affect the direction that it spins when it next encounters a wall or another brick.

The purpose of Level -1 is to integrate these different motor control signals so that the behaviour of the Lemming reflects the demands of the other levels of the architecture. The NXC code for this level is provided below. Interestingly, this is done by computing a value for Sync that reflects the signals that are coming from the other levels. Sync is the constant that controls the relative speeds of the two motors, and is used to control motor speeds in a single line in Level 0 (Section 8.18). The Speed task sets the Sync variable by considering several different variables that will affect the relationship between motors: Should the robot be moving straight, or should it be wandering with a slight curve? Should it be avoiding an obstacle by spinning? This level computes Sync with a simple equation that combines variables that are affected by higher levels, and which produces the desired motor behaviour of the robot.

```
/*===== îLevel -1î: Integration ============================================
Sets the two motor speeds by combining the upper layersí controls. */
const int SET_MINOR_TURN = 10;
/* the Lemming turns slightly to the right when not detecting a brick, SET_MINOR_TURN
determines how much it turns. default = 10 */
const int STRAIGHT = straight_calibration;
const int FULL_SPIN = 100, MINOR_TURN = STRAIGHT + (-1 * SET_MINOR_TURN);
int DriveSpeed, BaseSpeed, Wander, Spin, Sync, BaseSync = MINOR_TURN;
task Speed(){
      DriveSpeed = BaseSpeed;
      BaseSync = MINOR_TURN;
      while(true){
              Sync = (BaseSync * Wander) + Spin; //Compute Sync by combining variables
      }
}
```

8.23 PUTTING ALL THE LEVELS TOGETHER
8.23.1 The Main Task

The complete program for the Lemming is available from the website that supports this book (http://www.bcp.psych.ualberta.ca/~mike/BricksToBrains/). The main task, whose NXC code is listed below, is used to define the constants that are used in the various levels, to assign values for parameters that calibrate individual Lemmings, and to start all the previously described levels running. Note that one of these tasks is keepAwake, whose code is also provided below. The behaviour of a Lemming is usually observed by running it in its test environment for a considerable length of time (30–40 minutes). The NXC brick will automatically turn itself off after a few minutes in order to conserve battery power. The keepAwake task is run to reset the brick's timer to prevent this from happening.

Note too that the straight_calibration value, obtained from playing with the calibration routine that was described in Section 8.18, is assigned in this final section of code. For the Lemming that was to be tested using this subsumption architecture, the value of this variable was found to be equal to 3. The light_threshold and dark_threshold values are used to identify light bricks and dark bricks, and could be modified to permit the Lemming to classify bricks whose colours were other than black or white.

```
#defne WallSensor S1
#defne BrickSensor S2
#defne ColourSensor SENSOR_3
#defne LeftMotor OUT_B
#defne RightMotor OUT_C
#defne BothMotors OUT_BC
/*===== Calibration: Set the lemming specifc constants here. */
const int straight_calibration = 3, light_threshold = 540, dark_threshold = 340;
/*===== KeepAwake ========================================================
======
Reset the sleep timer so the lemming does not shut off automatically! */
const int TEN_MINUTES =36000000;
task keepAwake()
{
    while(true)
    {
        Wait(TEN_MINUTES);
        ResetSleepTimer();
    }
}
/*===== Main Task =====================================================*/
task main(){
    SetSensorType(WallSensor, SENSOR_TYPE_LOWSPEED);
    SetSensorType(BrickSensor, SENSOR_TYPE_LOWSPEED);
    SetSensorType(S3, SENSOR_TYPE_LIGHT_ACTIVE);
    SetSensorMode(S3, SENSOR_MODE_RAW);
    Wander = 1;
    Spin = 0;
    BaseSpeed = 65;
    start Speed;
    start Drive;
    start Dodge;
    start CheckStuck;
    start Seek;
    start seeBrick;
    start keepAwake;
}
```

8.24 THE LONELY LEMMING

8.24.1 Lemming Behaviour

Now that the Lemming has been constructed and programmed, we are able to observe its behaviour. Furthermore, we can construct copies of the Lemming, give each of them the same program, and explore whether a collection of Lemmings behaves any differently than does a single machine.

To start our behavioural explorations, let us create a checkerboard pattern of 28 black and 28 white squares (each square constructed as shown in Figure 8-4) and place this pattern in the middle of a small testing room that was 2.4 metres long and 1.65 metres wide. Figure 8-12 is a photograph of the floor of the room, taken from above, before the robot is released.

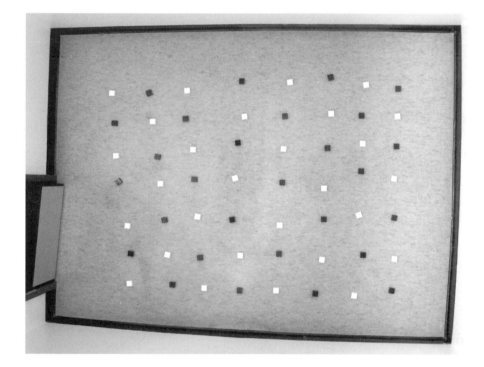

Now, a single Lemming is released into the room. We expect that it will rearrange bricks on the floor. Given the program that we have created, we might also expect that it will create a roughly annular rearrangement of these objects, by pushing the white squares out toward the walls, and by keeping the black squares in the middle of the room. The behaviour of a single Lemming in this environment is provided as part of Video 8-1, which is available from the website.

An examination of the behaviour of the Lemming in the early part of this video supports our expectations. The Lemming moves about the environment, pushing the objects aside as it wanders. On occasion, a square will be captured in the brick catcher. The behaviour that immediately follows this event depends on the colour of the captured object. If it is white, then the robot moves across to the wall that it is facing, and deposits the brick near the wall. If it is black, then the behaviour is less regular. The robot seems to spin a bit, seeking a direction, moving in this direction, and then suddenly choosing a different path. Usually after behaving like this for a moment, the black square is deposited in the middle of the room, near other bricks.

However, if the single Lemming is left to run long enough (70 minutes in the example below), the resulting sorting of the bricks is not quite as expected, as is illustrated in Figure 8-13. Most of the black bricks remain in a loose cluster in the middle of the room, which is expected. However, the majority of the white bricks are not just near a wall: instead, these "junk" bricks have been bulldozed into one of the corners of the room! The Lemming was certainly not intentionally designed to do this: none of its sensors or programmed tasks were designed to detect corners. How is it able to push most of the white objects, and few of the black ones, into the corners of its environment?

8-13

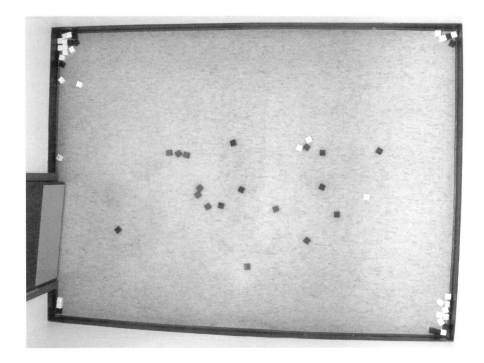

8.25 COLLECTIVE COLLECTING
8.25.1 Two Lemmings

The bricks on the floor of the testing room provide the potential for stigmergic control of Lemming behaviour. That is, the behaviour of one Lemming is determined in part by the colour of an object that it has captured; when it moves this object to a different location in the testing room, this has the potential to affect the future of this behaviour, or the behaviour of some other Lemming that might be sharing the environment.

To begin to explore the collective behaviour of Lemmings, the objects in the testing room were set up in the pattern that is illustrated in Figure 8-12, and two Lemmings were released into the room. The behaviour of a pair of Lemmings in this situation is also demonstrated in Video 8-1.

After 30 minutes, the objects are arranged in the room as shown in Figure 8-14. This final arrangement is very similar to the one produced by a single Lemming (Figure 8-13): almost all of the white bricks are in one of the four corners of the room, and almost all of the black bricks are in a loose cluster in the middle. The only noticeable difference between this end state and the one provided in Figure 8-14 is that in the latter all of the white objects have been removed from the middle of the room. That is, the white bricks that are not in corners are very close to walls, and appear to be quite far from black bricks.

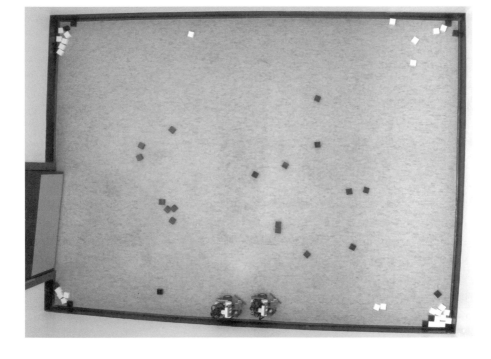

8-14

8.25.2 Three Lemmings

Further explorations were conducted by letting three Lemmings work in a room that began as shown in Figure 8-12. When done, they too had arranged the bricks by placing most of the white ones in the corners, and a loose cloud of the black ones in the middle of the room. Their work is shown in Figure 8-15. There is very little to distinguish this result from the one illustrated in Figure 8-14, except the time of the run: almost all of the white bricks had been moved into corners after Lemming activity that only lasted about 11 minutes. In the next section, we use comparisons of Lemming sorting speeds to consider whether colonies of this robot seem to exhibit collective intelligence.

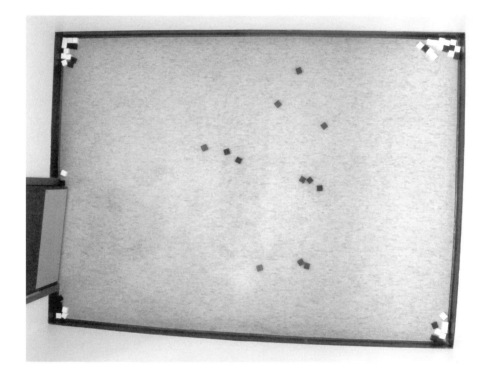

8.26 EXPLAINING SORTING INTO CORNERS
8.26.1 Corner Analysis

One of the surprising behaviours of the Lemming is its strong tendency to push white bricks into the corners of the testing room. Consider this behaviour from an analytic perspective: Imagine giving a designer some photographs that depicted the starting condition of the testing room (Figure 8-12), as well as the goal condition of the room (any of Figures 8-13, 8-14, or 8-15). One could also provide the designer with a pre-constructed Lemming. The designer's task would be to program the robot to convert the starting state of the room into the goal state.

In this scenario, the designer is likely to adopt an analytic approach. He or she would examine the photographs in an attempt to determine the difference between the start and goal states, and use these to sketch a general algorithm. A plausible algorithm might be: *Find a brick. If it is black, leave it in the middle of the room. If it is white, leave it in a corner. Repeat.*

This algorithm raises some challenges. How does one program a Lemming to find a corner of the room? From the analytic perspective, developing the desired program would be difficult.

8.26.2 Corners for Free

From our synthetic knowledge of the Lemming, it should be clear that it does not use the analytic algorithm described above. We know that its program does not identify corners and move white objects to them. If this behaviour is not programmed directly into the robot, then how does it occur?

Consider a completely different system that is also constrained by "walls": an artificial neural network that is called the *brainstate-in-a-box* (Anderson, Silverstein, Ritz, & Jones, 1977). This network consists of a set of processors that send signals to one another. Each processor produces internal activity values, but they cannot be lower than -1, or higher than +1.

In the brainstate-in-a-box, the set of neurons can be represented as a vector in a space, where the vector coordinates are the activity values of each processor in the network. When the processors signal one another, this vector grows in length. Soon, one or more of the neurons reaches a value of -1 or +1. This is equivalent to the vector (the brainstate) hitting a "wall" of a box that surrounds the vector. This wall pushes the vector along it. Eventually, the brainstate hits other walls, and finally is pushed into a corner of the surrounding box. In this corner, the brainstate is forced to stop growing, and its activities represent a memory that has been recalled from the network.

The testing room in which the Lemming roams restricts its movement in much the same way. When the robot wanders, it has a tendency to gently turn to its right. However, when it encounters a wall of the room, its movement is restricted. The result is that the Lemming moves along the wall that it encounters. Many examples of this behaviour—which emerges from an interaction between the Lemming and its environment—can be seen in Video 8-1. (The difference between the Lemming and the brainstate is that the Lemming has the ability to turn away from corners!)

How does this cause white bricks to be pushed into corners? The answer to this question comes from realizing that these bricks are not usually pushed directly into corners. Instead, they are first pushed near one of the walls of the testing room. Later, when a Lemming is moving along this wall (because of the interaction between its wandering behaviour and the structure of the testing room) it will catch this brick. It will then push this brick along the wall until it encounters another wall. Where will this wall be? It will be at a corner of the testing room. Now the white brick will be deposited at its final location, the corner, from which it will not be moved because the Lemmings tend not to venture into corners.

In short, sorting white bricks into corners emerges from the interaction of two Lemming behaviours: pushing white bricks to walls, and moving along walls when wandering.

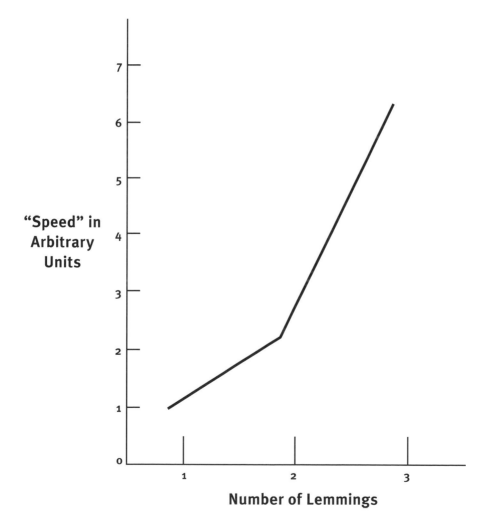

8-16

8.27 DO LEMMINGS HAVE COLLECTIVE INTELLIGENCE?
8.27.1 "Speed" of Work

As was noted earlier, Beni and Wang (1991) argued that one of the characteristics of collective intelligence was that as more agents were added to the collective, the amount of work done rose exponentially (instead of linearly, see Figure 8-1). Is this the case for a collective of Lemmings?

Typically, a single Lemming is finished sorting the blocks (i.e., has moved about 85% of the white bricks into corners) after approximately 70 minutes. In contrast, two Lemmings achieve the same result after about 30 minutes, and three Lemmings take only 11 minutes.

We converted these times to "speed" of work by taking the single Lemming's time as the standard, and dividing each of the times noted above into this standard value 70. If there is no collective Lemming intelligence, then this value would be equal to the number of Lemmings working together. However, the results that are plotted in Figure 8-16 show a much different pattern. Note that the graph in Figure 8-16 is very similar to the graph in Figure 8-1 that represents Beni and Wang's (1991) definition for collective intelligence. Thus it would appear that the phrase "collective intelligence" can be applied to a colony of Lemmings.

8.28 EXPLAINING COLLECTIVE INTELLIGENCE
8.28.1 Brick Dynamics

The results presented in Section 8.27 indicate that a small colony of Lemmings demonstrates collective intelligence. That is, three Lemmings sort the room considerably more than three times faster than does a single Lemming. Such collective intelligence is not directly programmed into a colony of Lemmings. The program that was described earlier controls the behaviour of a single robot. It does not detail any changes in this behaviour if other robots or detected, nor does it provide any explicit communication between robots. How, then, does collective intelligence emerge in a colony of such simple machines?

Clues to answering this question are provided by examining the parts of Video 8-1 that compare the behaviour of 1, 2, or 3 Lemmings working together. The first clue comes from ignoring the movement of the robots, and just paying attention to very general changes in brick positions over time.

How do the positions of the bricks in the room change from start to finish? The dynamics of the bricks are similar regardless of the number of robots involved (with the exception of the time course). First, the bricks start out in a regular grid. Second, the grid is destroyed, both by catching bricks and plowing bricks, so that white and black bricks are intermingled in a more compressed and random arrangement. Third, bricks migrate away from the middle of the room, as both black and white bricks are pushed to varying distances from the four walls. Fourth, white bricks migrate from walls to corners (see Section 8.26), while black bricks migrate back toward the middle of the room.

The second clue to accounting for collective intelligence comes from considering Lemming movement. The brick dynamics discussed above indicate that a general tendency in brick sorting is for bricks to be pushed away from the middle of the room, and then for some bricks (i.e., the black bricks) to be pushed back into the middle. Both of these results require that Lemmings move through the middle of the room. However, the likelihood of this happening is one of the striking differences between the behaviour of a single Lemming and a small colony of Lemmings.

When a single Lemming begins in the testing room, it has a marked tendency to remain at the room's outskirts, and rarely moves through the middle of the room. In contrast, when a group of Lemmings starts in the room, one or more of the robots moves through the room's middle almost immediately, and robots are frequently seen in this area. As a result, the progression of brick dynamics that is associated with sorting occurs much more rapidly. It seems that an increased tendency to explore the middle of the room is the primary cause of accelerated sorting. But why is it that a group of Lemmings shows this tendency, while a single Lemming does not?

8.28.2 Interaction and the Middle

There are two types of interactions that cause groups of Lemmings to venture into the testing room's middle more frequently than does a single machine.

First, all Lemmings turn away from obstacles, but the only obstacle that a single Lemming will encounter is a wall of the room. In contrast, in groups of robots, each robot serves as an additional, moving obstacle, increasing the frequency avoidance behaviour, changing the location of this behaviour, and causing Lemmings to venture into the middle of the room very early.

Second, once Lemmings have moved through the middle of the room, the bricks in the starting grid are pushed into tighter groups, and these groups are more likely to be detected by the lower ultrasonic sensor of the robot. That is, clustered groups of bricks are more likely to attract a robot to them than are bricks in the original grid. The early destruction of the starting grid of bricks by a colony produces clusters of bricks that attract robots and accelerate the dynamics of brick sorting. As a single robot destroys the starting grid much later — because of its avoidance of the room's middle — brick dynamics proceed at a much slower pace.

8.29 IMPLICATIONS AND FUTURE DIRECTIONS
8.29.1 Implications

One idea emerging from embodied cognitive science is the extended mind: the idea that when external resources are used to scaffold cognition, the mind has extended beyond the confines of the skull, and has leaked into the world (Clark, 1997, 2008; Clark & Chalmers, 1998; Wilson, 2004). This idea was discussed in more detail earlier in Chapter 3.

A cognitive agent with an extended mind must be able to manipulate its environment. It must exhibit a higher degree of embodiment, in the sense of Fong et al. (2003), than do the LEGO robots described in preceding chapters. The Lemming is an example of a simple LEGO robot that is more embodied than these previous machines, because it is capable of moving objects from one location to another. This embodiment was demonstrated when the Lemming "blind bulldozed" white bricks to the outer edges of its environment, while it moved black bricks to a more central location. This task was inspired by studies of sorting in ants (Deneubourg et al., 1991; Sendova-Franks et al., 2004; Theraulaz et al., 2002) and robots (Holland & Melhuish, 1999; Melhuish et al., 2006; Scholes et al., 2004; Wilson et al., 2004).

The sorting behaviours of collectives are cited as examples of stigmergy (Holland & Melhuish, 1999), because sorting can be accomplished by groups of agents that do not communicate with one another directly, but do so indirectly by manipulating the environment (i.e., by changing the locations of to-be-sorted objects in the shared world). The Lemming's sorting behaviour illustrates stigmergy in this sense. Clearly, changing the location of a brick in the testing room had the potential of altering the future behaviour of a Lemming that might encounter this brick in its new location. An unexpected result of the stigmergic interactions of the robots was the dramatic acceleration of sorting that resulted when more than one Lemming was turned loose in the testing room. The fairly simple notion of stigmergy demonstrated by the Lemming can produce an interesting example of collective intelligence, according to one definition of this phenomenon (Beni & Wang, 1991).

8.29.2 Future Directions

Nevertheless, the kind of stigmergy illustrated by the Lemming is very simple. Fortunately, the Lemming is a platform that could be easily extended to explore the notions of stigmergy and embodiment in more complicated ways.

Some of these extensions would entail altering the subsumption architecture of the robot, and possibly "tweaking" its sensory abilities. For instance, what would be the effect of altering the behaviour of a machine as it wanders, permitting it to turn to the right or to the left? Would it be possible to change the brick release behaviour of a Lemming so that it more closely resembled some of the algorithms for annular sorting, such as the pullback algorithm (Holland & Melhuish, 1999), and as a result produce more compact clusters of bricks at the end of sorting? The current version of the Lemming works by distinguishing light from dark bricks. Might it be possible to program the Lemming to distinguish more than two brick types — and generate more than two types of behaviour — by elaborating its light sensing routines?

Other interesting extensions would entail recognizing that the extended mind can involve both a scaffolding world and internal representations. Imagine a Lemming that keeps a memory of the last one or two bricks that it has captured, and whose behaviour depends not only on the colour of the current brick, but also these memories. Would such a device be capable of carrying out simple computations, using the floor of the testing room as an external memory or scratchpad?

That the Lemming could be used to explore such questions suggests that it is much more than a LEGO toy. The conception of the LEGO robots as toys, totems, or tools is the topic of the next and final chapter of this book.

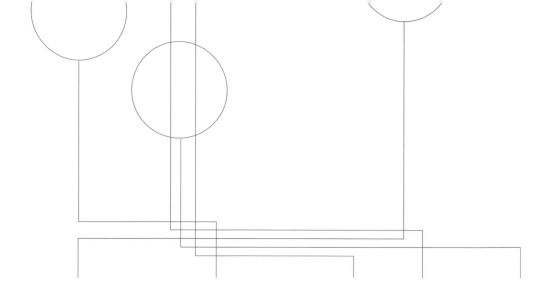

Chapter 9
Totems, Toys — Or Tools?

9.0 CHAPTER OVERVIEW

When Grey Walter reflected upon his robots, one question that he asked was whether they were totems, toys, or tools (Grey Walter, 1963). The robots that we have described in this book also require that this issue be explored. Some of these machines, such as Vehicle 2, the passive dynamic walker, the Strandbeest, and the Tortoise, are modern variants of historically important machines. All of our robots are constructed from toy parts. Are our machines more than just replicas of other devices, or more than mechanical toys? We answer these questions by making a point similar to that made by Grey Walter: our robots are more than totems, and more than toys — they are tools that can be used to explore many of the fundamental ideas of embodied cognitive science. Furthermore, they are tools that can be used to contribute new insight into current research issues. We demonstrate this with a final robot, called antiSLAM. This robot was initially designed to freely explore its environment, following walls and avoiding obstacles. However, we realized that it could generate some of the important regularities in a popular paradigm, called the reorientation task, used to study spatial cognition. Traditional theories of this task are strongly representational. AntiSLAM is of interest because it can produce many key reorientation task behaviours, but does so without any spatial representations or cognitive maps. This chapter provides some background on the reorientation task, and details the construction and programming of antiSLAM. It then describes antiSLAM's behaviour in several different versions of the reorientation paradigm. The implications of these results clearly indicate that our LEGO robots provide both "hard fun" and "hard science."

9.1 ARE OUR ROBOTS MORE THAN TOTEMS?
9.1.1 Uncanny Machines

The robot Tortoises provided "mimicry of life" (Grey Walter, 1963, p. 114). Grey Walter's worry was that they were merely totems, and that his investigations of them were without meaning. "We are daily reminded how readily living and even divine properties are projected into inanimate things by hopeful but bewildered men and women; and the scientist cannot escape the suspicion that his projections may be psychologically the substitutes and manifestations of his own hope and bewilderment" (p. 115). Such a concern arises whenever one simulates the real (Baudrillard, 1994). While a symbol or simulation begins as a reflection of reality, it evolves into having "no relation to any reality whatsoever: it is its own pure simulacrum" (Baudrillard, 1994, p. 6). Grey Walter's concern was that Tortoise behaviour might be meaningless because it would not refer to the behaviour of the real.

This issue is rooted in the seventeenth-century comparison of man and machine (Grenville, 2001; Wood, 2002). The view that man was a machine governed by universal, mechanistic principles was a central tenet of Cartesian philosophy (Descartes, 1637/1960). Eighteenth-century applications of this philosophy appeared in the form of elaborate, life-mimicking, clockwork automata, such as the androids constructed by Pierre and Henri-Louis Jaquet-Droz that wrote, sketched, or played the harpsichord (Wood, 2002). More famous automata of Jacques de Vaucanson, — including a flute player and a food-digesting duck — were in circulation for a century.

In their day, clockwork automata raised serious tensions between science and religion. In 1727, androids of Vaucanson's that served dinner and cleared tables were deemed profane, and his workshop was ordered destroyed (Wood, 2002). Pierre Jaquet-Droz was imprisoned by the Spanish Inquisition — along with his writing automaton! Such tensions were salved by Descartes' dualism: animals were machines; men were machines that also had souls, machines that also thought (see also Chapter 1).

Modern machines that mimic life still raise serious questions about what it is to be human. The emotions that they can provoke have long been exploited in literature and film. Freud examined how the feeling of the uncanny was used as a literary device (Freud, 1919/1976). He noted that the uncanny requires that the familiar be presented in unfamiliar form. The source of the uncanny that Freud identified mirrors Baudrillard's analysis of symbols. "The cyborg is uncanny not because it is unfamiliar or alien, but rather because it is all too familiar. It is the body doubled — doubled by the machine that is so common, so familiar, so ubiquitous, and so essential that it threatens to consume us, to destroy our links to nature and history" (Grenville, 2001, pp. 20–21).

Modern robotics shifts the uncanny from fiction to fact, as did eighteenth-century automata. Current humanoid robots produce a phenomenon called the *uncanny valley* (MacDorman & Ishiguro, 2006; Mori, 1970): our acceptance of androids suddenly plummets when their appearance grows to be very lifelike, but can still be differentiated from biological humans. The uncanny valley concerns "what troubles us when we are faced with certain versions of ourselves—bionic men, speaking robots, intelligent machines, or even just a doll that moves" (Wood, 2002, p. xxvii). Wood notes that all automata are presumptions "that life can be simulated by art or science or magic. And embodied in each invention is a riddle, a fundamental challenge to our perception of what makes us human." Usually such troubling riddles are addressed by discovering what differentiates us from machines.

This is the *opposite* of the problem that faced Grey Walter. For his robots to be accepted scientifically, they must at some level be *equivalent* to living organisms. Only if this were true could their mimicry advance the understanding of living beings as Grey Walter intended. Why should we believe that his robots—and the LEGO devices that have been described in preceding chapters—are not merely totems? Why should we believe that their mimicry can tell us something new about adaptive, biological agents?

9.2 ARE OUR ROBOTS MORE THAN TOYS?
9.2.1 The Tortoise as Toy

When one sees images of historical automata (Grenville, 2001; Standage, 2002; Wood, 2002), it is hard to imagine that they led their audience to the uncanny valley. While they often took humanoid form, it would be impossible to confuse these machines with living organisms on the basis of appearance. They were marvelous, not for their resemblance to life, but rather because their intricate actions seemed beyond the ability of ordinary machines. As living dolls, eighteenth-century automata defined "the golden age of the philosophical toy" (Wood, 2002, p. 17).

Toys are central to other devices that we have discussed in this book. For example, the passive dynamic walkers that were introduced in Chapter 5 were inspired by walking toys (McGeer, 1990a; Wisse & Linde, 2007), one of which was patented in 1888 by Faris, the other—the "Wilson Walkie"—in 1938 by Wilson.

Grey Walter argued that the scientific merit of his Tortoises required that they be more than totems. However, their intended usefulness also required that they be more than toys. While he made this argument too (Grey Walter, 1963), his robots' appearances were such that it was hard not to consider them to be children's toys.

Grey Walter's efforts to publicize his robotic research (Holland, 2003b) certainly did little to dispel this notion. Newspaper accounts of his work used the researcher's own words to minimize its scientific intent: "Toys which feed themselves, sleep, think, walk, and do tricks like a domestic animal may go into Tommy's Christmas stocking in 1950, said brain specialist Dr. Grey Walter in Bristol last night" (Holland, 2003b, p. 352). The Grey Walter Online (http://www.ias.uwe.ac.uk/Robots/gwonline/gwonline.html) website reports that one of Grey Walter's robots was sent to an American company in 1953 with the express purpose of creating a new kind of toy. The 1972 documentary *Future Shock*, narrated by Orson Welles, includes footage of Grey Walter saying of one of his Tortoises, "This looks rather as though it was a child's toy, and I suppose it might be." While he goes on to say, "but in fact it is rather a serious model of my ideas about behavior," the damage is done.

9.2.2 LEGO Is a Toy!

When the British Association of Toy Retailers chose its "Toy of the Century," what lucky toy beat out the stiff competition provided by the teddy bear, Action man, and Barbie? It was LEGO, of course. One component of the LEGO world, the NXT Mindstorms robotics system, has itself been deemed an award-winning toy, receiving the Canadian Toy Testing Council Best Bet 2007 and the Oppenheim Toy Portfolio Platinum Award 2007.

To distinguish Tortoises from toys, Grey Walter could at least note that his earliest devices were constructed from war surplus parts. We are in a less enviable position with our robots in this book, because they are literally constructed from children's playthings. The seeds that became LEGO Mindstorms were first sown as part of educational research at MIT in the 1960s (Martin, Mikhak, Resnick, Silverman, & Berg, 2000), and the programmable brick that emerged from this work is an educational toy (Resnick, Martin, Sargent, & Silverman, 1996). LEGO robots were designed for children. "Designing tools that allow children to add computation to traditional construction—and recognizing the learning opportunities afforded by this activity—has been the focus of our work over the last number of years" (Martin et al., 2000, p.10).

Martin et al. (2000, p. 10) ask, "when does something stop being a machine and start being a creature?" It is exactly this question that Grey Walter pondered when he argued that the Tortoises were more than totems; perhaps this question is even more telling when the machine in question is literally a children's toy, such as the LEGO robots that we have been investigating. What arguments did Grey Walter provide that permit us to separate the Tortoises — and our LEGO creations — from totems and toys?

9.3 FROM TOTEMS AND TOYS TO TOOLS

9.3.1 Tortoise as Tool

Why did Grey Walter (1963) believe that his machines were more than totems? He argued that totems became potent symbols because they resembled that which they represented. "Until the scientific era, what seemed most alive to people was what most *looked* like a living being. The vitality accorded to an object was a function primarily of its form" (Grey Walter, 1963, p. 115). In contrast, his robots were not concerned with reproducing appearances, but instead with imitating behaviour and performance.

What classes of behaviour would be the target of scientific imitation? Grey Walter (1963, p. 120) provided an intimidating list: "exploration, curiosity, free-will in the sense of unpredictability, goal-seeking, self-regulation, avoidance of dilemmas, foresight, memory, learning, forgetting, association of ideas, form recognition, and the elements of social accommodation. Such is life."

Grey Walter's (1963) Tortoises demonstrated many of these characteristics (see also Chapters 6 and 7). They were more than toys, because (in his view) toys could not produce any of these behaviours. "The technical genius of the Swiss watchmakers was really wasted on their delicate clockwork automata; they arouse only a passing interest because they are neither sacred nor, like life, unpredictable, their performance being limited to a planned series of motions, be it a boy actually writing a letter or a girl playing a real keyboard instrument" (p. 115).

By situating his embodied machines, Grey Walter (1963) moved them beyond totems and toys. They produced behaviours that were creative and unpredictable because they were governed by the relationships between their internal mechanisms and the surrounding, dynamic world. "The important feature of the effect is the establishment of a feedback loop in which the environment is a component" (p. 130).

9.3.2 Pedagogical and Scientific Tools

We believe that Grey Walter's arguments that his machines were tools apply equally well to our LEGO creations because of their varying degrees of situatedness and embodiment. Their ease of construction, and the popularity and wide availability of their components, mean that they can also serve admirably as another kind of tool: a pedagogical tool that permits students to explore some of the key ideas emerging in embodied cognitive science.

The issues raised by embodied cognitive science are not easily dealt with by more traditional approaches (Clark, 1997, 1999, 2003, 2008). As a result, Clark (1997, p. 103) asks, "What kind of tools are required to make sense of real-time, embodied, embedded cognition?"

Programmable bricks are devices that serve as "things to think with." (Resnick et al., 1996, p. 450). Mindstorms bricks present computation in a new light—not as the programming of a disembodied, stationary desktop computer, but as the development of sense–act relations in creatures that move freely in the world (Resnick et al., 1996). LEGO robots, then, appear to have been designed to provide a rich medium for exploring embodied cognitive science. Students can use them to "realize that sophisticated behaviours can emerge from interactions of rules with a complex world, but at the same time, are still captivated by the wonder of a machine acting like a pet" (Martin et al., 2000, p. 10).

However, can LEGO robots be more than pedagogical tools? Grey Walter (1963, p. 132) noted of his machines that "as tools they are trustworthy instruments of exploration and frequent unexpected enlightenment." When used to illustrate historically important robots, LEGO robots are instruments of exploration. But retracing the steps of others limits their possibility for enlightenment. We need to use our LEGO devices to explore new ideas in embodied cognitive science. The remainder of this chapter illustrates this possibility, by showing how LEGO robots can contribute to current debates arising in the study of human and animal navigation.

9.4 ANIMAL NAVIGATION AND REPRESENTATION
9.4.1 Navigational Organisms

"Navigation is the process of determining and maintaining a course or trajectory from one place to another" (Gallistel, 1990, p. 35). There are many examples of extraordinary navigational feats. One is the small blackpoll warbler's 1,575-km nonstop flight over the western Atlantic as it migrates from its New England staging grounds to South America or the West Indies (Baird, 1999; Drury & Keith, 1962). Another example is provided by the long-distance navigators of Micronesia who use seagoing canoes to routinely complete voyages of 150 miles or more without sight of land, and without the use of charts or instruments (Finney, 1976; Gladwin, 1970; Hutchins, 1995). However, many more mundane tasks depend on navigation, and are critical to the survival of most organisms. As a result, there has been an intensive study of navigation that has revealed a tremendous amount of information about the spatial information that animals exploit to move about in their environment.

For instance, consider one basic element of navigation, determining a heading, which is a direction of movement relative to some external coordinate system (Gallistel, 1990). There is an abundance of evidence that many animals are sensitive to direction. The tail-wagging dance of bees communicates the location of food sources using directional information (Frisch, 1966, 1967). In particular, the dance specifies the angle between the food source and the azimuth of the sun, with the beehive at the origin of the angle. Birds like Clark's nutcrackers and pigeons can localize food caches by encoding directional relationships amongst multiple landmarks (Jones & Kamil, 2001; Kamil & Cheng, 2001; Kamil & Jones, 1997, 2000; Spetch, Rust, Kamil, & Jones, 2003). The rat's hippocampus contains head direction neurons that respond strongly when the rat's head points in the cell's preferred direction (Redish, 1999; Sharp, Blair, & Cho, 2001; Taube & Muller, 1998).

9.4.2 Sense–Think–Navigate

At the heart of most research on animal navigation is the notion that it is representational, in the strong sense proposed by classical cognitive science (see Chapters 2 and 3). That is, navigation is typically viewed as a sense–think–act process, where the thinking involves the use of various representations of space. As a result, it is not surprising that one particular interest of scientists is determining the properties of spatial representations in animals and humans (Healy, 1998).

For instance, many theories of bird navigation — concerning both local homing and long-distance migration — assume a map-and-compass model, in which birds use some form of navigational map to determine their current location relative to a goal, and then use a celestial or magnetic compass to set their heading toward that goal (Mouritsen, 2001; Wiltschko & Wiltschko, 2003). Similarly, there is a great deal of evidence that a cognitive map of the world (Tolman, 1948), specifying both locations and directions, is encoded in the rat hippocampus (O'Keefe & Nadel, 1978), although there is considerable debate about its specific properties (Burgess, Jeffery, & O'Keefe, 1999; Dawson et al., 2000; McNaughton et al., 1996; Redish, 1999; Redish & Touretzky, 1999; Touretzky & Redish, 1995; Touretzky, Wan, & Redish, 1994). Micronesian navigators do so by superimposing a number of mental images that are anchored by the rising and setting points of various stars (Hutchins,

1995). Gallistel (1990, p. 121) notes that "orienting towards points in the environment by virtue of the position the point occupies in the larger environmental framework is the rule rather than the exception and, thus, cognitive maps are ubiquitous."

The dominance of sense–think–act theories of navigation has two implications for the current chapter. First, biologically inspired robotic models of navigation are frequently representational, as the next section shows. Second, given the themes developed in this book, we can wonder whether sense–act theories of navigation are possible, and whether such theories could be explored using synthetic methodologies to construct different kinds of robots.

9.5 REPRESENTATION AND ROBOT NAVIGATION
9.5.1 Animals to Animats

The intensive study of human and animal navigation has led to theories that are predominantly representational. For the most part, these theories are tested and refined using traditional methodologies from experimental psychology and neuroscience. However, in a growing number of cases these theories are also tested by using them to develop autonomous, navigating robots. These biologically inspired robots can be thought of as artificial creatures, often called animats, that have been used to test the strengths and weaknesses of various representational theories of navigation.

For example, the existence of place and head-direction cells in the rat's hippocampus has inspired a number of different navigational robots (Arleo & Gerstner, 2000; Burgess, Donnett, Jeffery, & O'Keefe, 1997; Milford, 2008). The Burgess et al. (1997) robot employs a control system that includes "sensory cells" that encode a robot's distance from walls via infrared sensing, and "place cells" that are used to encode the robot's location when it is reinforced. Simple learning routines are used to modify connection weights between the various components of the control system. Tests of the robot demonstrated that it could use this modelled cognitive map to localize its position and direction when it moved around a rectangular environment. It remembered locations where it was reinforced at a particular location, and could return to them even when started from different locations.

Arleo and Gerstner (2000) have developed an animat that is similar to the Burgess et al. (1997) machine, but includes senses of self-motion. The robot associates locations in its "hippocampal map" with rewards, and will navigate to them. If it is not rewarded after it arrives at an intended location, the unrewarded location will be forgotten from its map.

9.5.2 SLAM and AntiSLAM

Because most theories of animal navigation are representational, it is not surprising that when they are transferred to animats, as in the preceding examples, the resulting machines are sense–think–act devices. However, robot navigation is often construed as *necessarily* being representational. "Low level robots may function quite adequately in their environment using simple reactive behaviours and random exploration, but more advanced capabilities require some type of mapping and navigation system" (Milford, 2008, p. 10).

Because of this assumption, one of the central problems being explored by roboticists is *simultaneous localization and mapping* (Jefferies & Yeap, 2008), or SLAM. Assume that robots find their place in the world by relating their current sensed location to some place on an internal map. However, if they are placed in a novel environment then no such map exists, and self-localization is impossible. Methods must be developed for the agent to simultaneously build a new map of the novel environment and locate itself using this map. This is a difficult problem, and robotics researchers are turning to studies of biological navigation to help solve it (Jefferies & Yeap, 2008). For example, Milford (2008) suggests that simultaneous localization and mapping can be accomplished by using a hippocampus-inspired model that uses both place cells and head-direction cells.

The SLAM problem is predicated upon the assumption that navigation involves representation. Some researchers who study animal navigation have begun to question aspects of this assumption (Alerstam, 2006). To what extent might a completely reactive, sense–act robot be capable of demonstrating interesting navigational behaviour? We now turn to exploring this question. First we will introduce a simple task that has been used to study navigation in local environments, and has inspired representational theories. Second, we will explore some recent concerns about such theories. Third, we will investigate non-representational theories of this task, which include some synthetic studies that use our own reactive LEGO robot, which we — for obvious reasons — call antiSLAM.

9.6 SPATIAL BEHAVIOUR AND THE REORIENTATION TASK
9.6.1 Navigational Cues

How do organisms find their place in the world? One approach to answering this question is to set up small, manageable indoor environments. These environments can be customized to provide a variety of different cues to animals that learn to navigate within them. For instance, like some of the robots described in the preceding section, an animal might be reinforced for visiting a particular location. What information does an animal use to return to this location in the hope of receiving more rewards?

Studies of navigation in indoor environments have found that humans and animals exploit various *geometric cues* as well as *feature cues* (see Cheng & Newcombe [2005] for a recent review). Geometric cues are relational, while feature cues are not: "A geometric property of a surface, line, or point is a property it possesses by virtue of its position relative to other surfaces, lines, and points within the same space. A non-geometric property is any property that cannot be described by relative position alone" (Gallistel, 1990, p. 212). One question of considerable interest is the relative contributions of these different cues for navigation.

9.6.2 The Reorientation Task

One approach to answering this question is the *reorientation task*. In this paradigm, an agent is placed within an "arena" that is usually rectangular. Metric properties (wall lengths, angles between walls) combined with an agent's distinction between left and right (e.g., the long wall is to the left of the short wall) provide geometric cues.

Other arena properties can provide feature cues. For example, Figure 9-1 illustrates an arena that has one blue wall, while all the other walls are black; the distinctive colour is a feature cue. Or, one could place unique objects at different locations in the arena. This is shown in Figure 9-2, where each letter in the figure stands for a unique object (e.g., a coloured panel) that distinguishes each corner from the others. These objects also provide feature cues.

9-1

In the reorientation task, an agent learns that a particular place — usually a corner of a rectangular arena — is a goal location. Imagine that when placed in either arena illustrated in Figure 9-1 or Figure 9-2, the agent is rewarded when it visits the corner labelled 4, but is not rewarded when it visits any other corner. The agent learns that corner 4 is the goal location.

The agent is then removed from the arena, disoriented, and returned to an arena, with the task of using the available cues to relocate the goal. Of particular interest are experimental conditions in which the arena has been altered from the one in which the agent was originally trained.

For example, in the new arena the feature cues might have been moved to different locations than was the case when the subject originally learned the goal location. This places feature cues in conflict with geometric cues. Will the agent move to a location defined by geometric information, or will it move to a different location indicated by feature information? Extensive use of the reorientation task has revealed some striking regularities.

9.7 BASIC FINDINGS WITH THE REORIENTATION TASK
9.7.1 Rotational Error

First, consider the case in which agents are trained that corner 4 is a goal location using an arena like Figure 9-1 or 9-2. Then, the agent must reorient itself in a new arena that only provides geometric cues. Such an arena has no local features that can be used to distinguish one location from another, as illustrated in Figure 9-3.

Geometric cues do not uniquely specify a target location in such an arena. For instance, the geometric cues available at Location 4 of Figure

9-3 are identical to those available at Location 2 of the same figure: 90° angle, longer wall to the left and shorter wall to the right. However, these geometric cues can distinguish Location 4 from either Location 1 or Location 3.

Under such conditions, one of the basic findings is *rotational error* (Cheng, 1986, 2005). When rotational error occurs, the trained animal goes to the goal location (e.g., Location 4 in Figure 9-3), as well as the corner that is geometrically identical to it (Location 2), which is located at a 180° rotation through the centre of the arena, at above chance levels. Rotational error is usually taken as evidence that the agent is relying upon the geometric properties of the environment.

9.7.2 Mandatory Geometry

The second main regularity that governs the reorientation task occurs when feature cues, such as the distinct objects illustrated in Figure 9-2, are available during training. Such feature cues uniquely identify a goal location — that is, it is possible for an agent to learn where the goal location is by only using these cues, and by ignoring geometric cues. However, the evidence suggests that agents still learn about the geometric properties during this training, even though these cues are irrelevant or unnecessary in this version of the task. That is, geometric cues still influence behaviour even when such cues are not required to solve the task. It as if the processing of geometry is mandatory.

This regularity is supported by several pieces of evidence. First, in some cases subjects continue to make some rotational errors even when a feature disambiguates the correct corner (Cheng, 1986; Hermer & Spelke, 1994).

Second, when features are removed following training, subjects typically revert to choosing both of the geometrically correct locations (Kelly, Spetch, & Heth, 1998; Sovrano, Bisazza, & Vallortigara, 2003).

Third, consider the case when features are moved after training—for instance, after being trained in the arena illustrated in Figure 9-2, the animal must reorient in the arena illustrated in Figure 9-4, where all of the objects have been moved in a clockwise direction. This produces a conflict between geometric and feature cues; control by both types of cues is often observed in such conditions (e.g., Brown, Spetch, & Hurd, 2007; Kelly, Spetch, & Heth, 1998). That is, there will be an increased tendency to visit Corner 1 than was the case during training, because it is now marked by the correct feature. However, Corner 4 will still be visited (because it still has the correct geometric cues), as will the geometrically equivalent Corner 2.

9-4

9.8 REPRESENTATIONAL THEORIES OF REORIENTATION
9.8.1 The Geometric Module

The reorientation task has inspired a number of different theories related to reorientation and navigation. For instance, Gallistel (1990) viewed the solution of the reorientation task as a two-stage process. The first stage occurs when an agent is first placed in an arena: it encodes the shape of the arena by attending to metric cues, such as wall lengths and angles between walls, as well as to sense cues (i.e., the distinction between left and right). The purpose of encoding the arena's shape is that this information is then used by the agent to determine its heading: that is, the arena's shape provides the reference frame for the agent's ability to orient itself.

The second stage occurs when an agent is disoriented, and then placed in an arena once again. In this stage, the agent uses a representation of the shape of the previously encountered arena as a mental map. The agent "gets its heading and position on its map by finding the rotation and translation required to produce a congruence (shape match) between the currently perceived shape of the environment and a corresponding region of its map" (Gallistel, 1990, p. 220). If the only sources of information used to create such maps are sense and geometric cues, one consequence of this theory is rotational error in rectangular arenas.

A key assumption of the Gallistel (1990) model is that the processing of environmental shape is *modular* (Fodor, 1983). According to Fodor, a module is a neural substrate that is specialized for solving a particular information-processing problem. Modules operate in a fast, mandatory fashion; they exhibit characteristic breakdown patterns when they fail because of their specialized neural circuitry; and they operate independently of the influence of the contents of higher-order beliefs — that is, they are cognitively impenetrable (Pylyshyn, 1984). It has been argued (Cheng, 1986; Gallistel, 1990) that the geometric computations in Gallistel's model are modular because they are mandatory and because they are not influenced by "information about surfaces other than their relative positions" (Gallistel, 1990, p. 208).

Why would there be a module for processing geometric cues? Gallistel (1990) proposes two reasons. First, reorientation can be accomplished by using a fairly simple algorithm for bringing the shape of the new arena, and the shape of the remembered arena, into register. Such an algorithm is not subject to combinatorial explosion when the shape of the arena changes (e.g., in size or complexity). These computational advantages are substantial, and therefore it may be important to 'reify' them as modular properties.

Second, from an evolutionary point of view, geometric modularity might take advantage of the fact that overall shape of an animal's typical environment is not likely to change dramatically, even though many visual features within the environment might change from day to day. "In relying on overall shape alone, the nervous system finesses the problem of finding the optimal weights for mediating the trade-offs between changes occurring along incommensurable sensory dimensions" (Gallistel, 1990, p. 212). This is an example of modularity being used to solve the frame problem that can be frequently encountered by representational systems (Dawson, 1998; Pylyshyn, 1987).

9.8.2 Geometry and Representation

The geometric module is an influential theory designed to account for the regularities in the reorientation task. For the purpose of the current chapter, it is important to stress that it is a representational theory: "Rats have a representation of the shape of the environment that includes the uniquely metric relations and sense" (Gallistel, 1990, p. 219). Recently, though, questions have been raised about the nature — and even possible existence — of the geometric module. We shall see that such questions can inspire research that explores possible non-representational accounts of the reorientation task.

9.9 WHITHER THE GEOMETRIC MODULE?
9.9.1 Modifying Modularity

Recently, some researchers have begun to question the geometric module. One reason for this is that the most compelling evidence for claims of modularity comes from neuroscience (Dawson, 1998; Fodor, 1983), but such evidence about the modularity of geometry in the reorientation task is admittedly sparse (Cheng & Newcombe, 2005). As a result, most arguments about modularity in this context are based on behavioural data. However, the data obtained from the reorientation task is consistent with many different notions of modularity (e.g., Cheng & Newcombe, 2005, Figure 3).

For this reason, some researchers have proposed alternative notions of modularity when explaining reorientation task regularities (e.g., Cheng, 2005; Cheng & Newcombe, 2005). Cheng (2005, p. 17), suggests that "geometric and feature information are encoded together in one record for localization. This process is non-modular." Cheng then attempts to preserve modularity by arguing that different types of information might be stored in the same location, but when certain devices access this common store, they only access particular types of information, and are thus modular in nature. In short, Cheng conjoins "a modular process and a non-modular representational structure." This approach is very similar to that exemplified in the production system architectures discussed in Chapter 3 (Anderson, 1983; Anderson et al., 2004; Newell, 1973, 1990) if one views working memory as a "non-modular representation structure" and individual productions as special purpose "devices."

9.9.2 Non-modular Reorientation

While some theories (e.g., Cheng, 2005) are attempts to preserve geometric modularity by redefining it, others reflect more radical approaches

(Cheng, 2008b): several new theories of the reorientation task completely reject the existence of a geometric module.

For instance, one model of the reorientation task uses a general theory of associative learning in which geometric and feature cues are not treated differentially (Miller & Shettleworth, 2007, 2008). Organisms learn what cues are present at a reinforced location, and then are more likely to approach locations with similar cues at a later time. This model does not require the reorientation task to be solved by applying geometric operations to global representations of arena shape. Similarly, simple neural networks can generate reorientation task regularities by modifying associations involving locally available cues only—such as the length and colour of a particular wall—while at the same time having absolutely no representation of global arena shape (Dawson, Kelly, Spetch, & Dupuis, 2008).

Another theory assumes that agents reorient by maximizing the visual similarity (e.g., unprocessed pixilated images of locations) of locations in the new arena to the image of the goal location in the original arena (Cheung, Stuerzl, Zeil, & Cheng, 2008; Stuerzl, Cheung, Cheng, & Zeil, 2008). In this theory, the metric of visual similarity does not make explicit the geometric properties (i.e., arena shape) that were central to earlier theories of the task.

While these newer theories reject the geometric module, they still share Gallistel's (1990) assumption that the reorientation task is solved by representational mechanisms. The Cheung et al. (2008) theory is obviously representational, because it involves matching visual input to remembered visual information. Associative models like Miller and Shettleworth's (2007, 2008) or Dawson et al.'s (2008) are also representational because associative strengths or neural network connection weights are representations of previous experience (Dawson, 2004).

However, if representations of locally visible cues are sufficient to deal with the reorientation task, then it is a small step to ask whether non-representational, sense–act processes are also sufficient. Some recent robotic studies of the reorientation task have explored this very question (Nolfi, 2002; Nolfi & Floreano, 2000). Let us now consider some examples of this reactive research.

9.10 REACTIVE ROBOTS AND THEIR EVOLUTION
9.10.1 New Wave Robotics

The pioneering autonomous, self-navigating robots were sense–think–act devices, planning their future movements using some sort of internal map (Nilsson, 1984). As we have seen, this tradition persists in much of modern research on robot navigation (Filliat & Meyer, 2003; Milford, 2008; Trullier, Wiener, Berthoz, & Meyer, 1997). However, there is an alternative trend in robotics, a movement that has been called "new wave" (Sharkey, 1997). New wave robotics strives to replace representation with reaction (Brooks, 1999); robots are created with direct links between sensors and actuators, so that they are better described as sense–act systems than as sense–think–act devices. All the LEGO NXT robots that we have described earlier in this book are simple examples of new wave, or reactive, robots.

While reactive roboticists do not necessarily deny the existence or importance of representations, they recognize that "embodied and situated systems can solve rather complicated tasks without requiring internal states or internal representations" (Nolfi & Floreano, 2000, p. 93). Of particular interest to us is the use of reactive robots to investigate behaviour in the reorientation task (Lund & Miglino, 1998). Lund and Miglino conducted a study in which their robots were evolved to accomplish the reorientation task, and to mimic animal behaviour when governed by the geometric cues of a rectangular arena. Before discussing their robot, let us briefly discuss the evolutionary approach that they adopted.

9.10.2 Evolving Robots

While we have been exploring LEGO Mindstorms robots, robotics researchers have conducted their work using a variety of different platforms. One of these is the Khepera miniature robot (Mondada & Floreano, 1995; Mondada, Franzi, & Ienne, 1994). This small robot has two motor-driven wheels, and 8 infrared proximity detectors arranged about its puck-shaped chassis. When used as a reactive robot, the speeds of its two motors are determined by the signals from the various proximity detectors. To set speed, the signals are weighted; the roboticist's task is to find a set of signal weights that cause the robot to perform some task of interest.

Evolutionary robotics is one approach to discovering an appropriate set of signal weights to control robot behaviour (Nolfi & Floreano, 2000). Evolutionary robotics is inspired by a more general form of evolutionary computation, genetic algorithms (Holland, 1992; Mitchell, 1996).

In general, the evolution of a Khepera robot design begins by defining a fitness function that measures how well a robot is performing a task of interest (Nolfi & Floreano, 2000). An initial population of different control systems (e.g., different sets of sensor-to-motor weights) is then produced. Each of these control systems is evaluated using the fitness function, and the control systems that produce higher fitness values are maintained. They are also used as templates for new control systems in the next generation of control systems, which are created by "mutating" the surviving control systems in prescribed ways. The whole process is then repeated with the next generation of controllers. It is expected that average fitness will improve with each new generation. The evolutionary process can be stopped when, for instance, fitness stops changing appreciably over a series of generations.

Of course, evolutionary methods are not restricted to the Khepera robot, although this is the platform used by Lund and Miglino (1998). Robot evolution has played an important role in the development of other robots that have been discussed in earlier chapters. Jansen used genetic algorithms to discover his holy numbers that were introduced in Chapter 5 (Jansen, 2007), and later evolved his Strandbeest by racing different walking robots on the beach, disassembling the losers, and using their parts to make (possibly mutated) copies of the winners. Evolutionary techniques were also central to Braitenberg's proposal for the development of his more advanced vehicles (Braitenberg, 1984).

9.11 REACTIVE ROBOTS AND ROTATIONAL ERRORS
9.11.1 Reactive Reorientation

Lund and Miglino (1998) evolved controllers for Khepera robots to perform the reorientation task in a rectangular arena without feature cues. A fitness function for this task is easily created; a fitter robot will find itself closer to the goal than will a less fit robot. The robots were started from eight different locations in the arena. A satisfactory controller for the robot was achieved after 30 generations of evolution, navigating the machine (from any of the starting locations) to the goal on 41% of trials. This controller also produced rotational error—it navigated the robot to the corner 180° from the goal on another 41% of the test trials. This data is very similar to results obtained from rats (e.g., Gallistel, 1990).

Lund and Miglino (1998) noted it was impossible for their Khepera robots to represent arena shape. "The geometrical properties of the environment can be assimilated in the sensory-motor schema of the robot behavior without any explicit representation. In general, our work, in contrast with traditional cognitive models, shows how environmental knowledge can be reached without any form of direct representation" (p. 198).

How does a completely reactive robot achieve this performance? When the robot is far enough from the arena walls that none of the sensors are detecting an obstacle (about 4 cm), the controller weights are such that it moves in a gentle curve to the left. As a result, when the robot leaves from any of the eight starting locations, it never encounters a short wall! When a long wall is encountered, the robot turns left, and follows the wall until it stops in a corner. A robot that always turns left, and never encounters a short arena wall, must necessarily produce rotational errors!

9.11.2 Representative Reaction

One problem with Lund and Miglino's (1998) robot is that it, like the thoughtless walker discussed in Chapter 5, depends critically upon its environment. If it is placed in an arena that is even slightly different in size, its emulation of rat reorientation behaviour diminishes (Nolfi, 2002). The robot would be more representative of animals if it were not limited to a particular arena.

Nolfi (2002) addressed this problem by evolving robots in arenas of varying sizes, and using a much wider variety of starting positions and orientations. Nolfi required 500 generations of robot evolution to achieve satisfactory performance in this more difficult task. However, at this point his robots produced rotational errors, and could do so in different-sized arenas.

How did Nolfi's (2002) robot deliver such performance in the absence of a cognitive map? First, the robot tends to move forward, avoiding walls, eventually encountering a corner. In this situation, signals from the walls cause it to adjust itself until it is at a 45° angle from one wall, and then it will turn either clockwise or counterclockwise (depending upon whether the sensed wall is to the robot's left or the right).

The final turn away from the corner means that the robot will be pointing in such a way that it will follow a long wall. This is because sensing a wall at 45° provides an indirect measurement of whether the wall is short or long! "If the robot finds a wall at about 45° on its left side and it previously left a corner, it means that the actual wall is one of the two longer walls. Conversely, if it encounters a wall at 45° on its right side, the actual wall is necessarily one of the two shorter walls. What is interesting is that the robot 'measures' the relative length of the walls through action (i.e., by exploiting sensory–motor coordination) and it does not need any internal state to do so" (p. 141). In other words, the sensory states of the robot permit it to indirectly measure the relative lengths of walls without directly comparing or representing length. It will use this sensed information to follow the long wall, which will necessarily lead the robot to either the goal corner or the corner that results in a rotational error, regardless of the actual dimensions of the rectangular arena.

9.12 REORIENTING LEGO ROBOTS
9.12.1 Motivating AntiSLAM

We have seen that there is tremendous interest in studying navigation, and that one paradigm used to conduct this study is the reorientation task. We have also seen that much of the general study of navigation is consistent with sense–think–act models of cognition (Healy, 1998; Milford, 2008); this is also true of the reorientation task (Gallistel, 1990; Miller & Shettleworth, 2007, 2008). However, more recent work shows that reorientation task regularities can be produced by reactive robots that are incapable of building and using spatial representations (Lund & Miglino, 1998; Nolfi, 2002).

One theme being explored in this chapter is the use of LEGO robots as scientific tools. To illustrate this possibility, we now describe a LEGO machine developed to navigate through an environment, with the hope that it too can provide insight into the reorientation task. One of the important problems faced by roboticists who develop autonomous, navigating robots is SLAM: simultaneous localization and mapping. The robot that we are about to describe is a very simple, reactive device that is not capable of creating or exploiting internal representations of the world. For this reason, we call it antiSLAM.

The antiSLAM robot began its development as a machine designed to follow walls, explore its environment, and avoid obstacles (see Section 9.31). However, we discovered that it could also serve as an alternative reactive robot for the reorientation task (Lund & Miglino, 1998; Nolfi, 2002). It differs from these robots in several respects. First, it uses a far simpler (and cheaper) architecture: it is a LEGO robot that uses far fewer sensors than the Khepera robots that have been described. Second, rather than using evolutionary techniques to develop controllers for this machine, we instead developed a subsumption architecture for navigation. Third, the most advanced version of antiSLAM uses both ultrasonic and light sensors that permit it to react to both local feature cues as well as the overall geometry of its environment.

9.12.2 Ultrasonic Sensors

The Khepera robots for the reorientation task dealt with rectangular arenas by using eight infrared sensors that would detect obstacles when the robot was fairly close to them. A single LEGO NXT brick does not permit this many sensors to be used at one time. Our alternative strategy was to start with a robot that had only two sensors, and to use sensors that had a longer range than the ones described in preceding pages (Lund & Miglino, 1998; Nolfi, 2002).

Our initial robot used two LEGO ultrasonic sensors. These sensors essentially act as sonar devices, sending out ultrasonic signals and listening for an echo (Astolfo et al., 2007). The timing of the echo is used as a measure of the distance between the sensor and the obstacle that reflected the signal; the maximum range of these sensors is around 100 inches (Boogaarts, 2007). As is the case with other sensors, the ultrasonic sensor can be set to return values in different modes — for instance, in inches or in centimetres. We decided to use the sensor in raw mode, where it returns a value of 255 when a reflecting obstacle is at the maximum range of the sensor, and decreases to a minimum of 0 as the sensor moves closer and closer to the obstacle.

One issue with ultrasonic sensors is that LEGO builders do not recommend using more than one at a time. "It cannot be used in an area in which another ultrasonic sensor is already at work because the signals sent out by the two sensors will interfere with each other and cause misreading" (Boogaarts, 2007, p. 39). However, because we needed our robot to be simultaneously sensitive to walls at different positions (e.g., when facing a corner of a rectangular arena), we needed to use more than one ultrasonic sensor. Thus, our first version of antiSLAM explored whether two ultrasonic sensors could be successfully used in Level 0 of a subsumption architecture for navigation.

9.13 ANTISLAM OVERVIEW

9.13.1 Modifying Vehicle 2

The antiSLAM robot (Figure 9-5) can be thought of as a descendant of a Braitenberg Vehicle 2 (whose construction was described in detail in Chapter 4). Indeed, the initial construction of antiSLAM is identical to the first several steps used to build Vehicle 2. As well, both robots include two light sensors that can be used to independently control the speed of two rear motors, steering the robot around an environment. However, antiSLAM differs from Vehicle 2 in two important ways. First, its light sensors point outward from the robot. Second, antiSLAM includes two additional ultrasonic sensors that also point outward. Compare Figure 9-5 to Figure 4-1 in Section 4.2 to see the similarities and differences between the two robots.

AntiSLAM, in its complete programmed form, uses its ultrasonic sensors to follow walls in a rectangular arena, slowing to a halt when these sensors detect a corner. It then initiates a turning routine to exit the corner and continue exploring. Its light sensors can be used to process local features — for instance, it can have a preference to approach an illuminated location. It uses these capabilities to reliably find a target location that is associated with particular geometric and local features. When local features are removed, it navigates the arena using geometric cues only, and produces rotational errors. In short, it produces some of the key features of the reorientation task — however, it does so without creating a cognitive map, and even without representing a goal.

AntiSLAM's subsumption architecture is also interesting. We will see that as specific layers are added to this architecture, antiSLAM transforms from a Vehicle 2 to a Vehicle 3 and upward through Braitenberg's evolutionary progression of machines (Braitenberg, 1984).

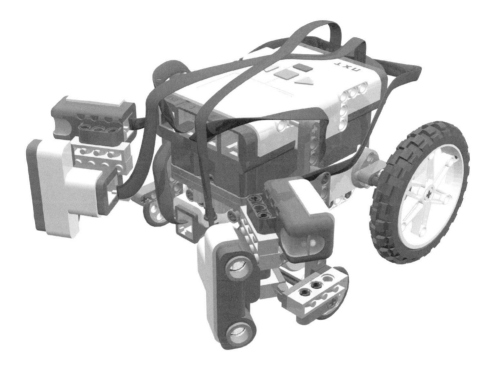

9.14 FROM VEHICLE 2 ONWARD
9.14.1 Foraging for Parts

As Braitenberg explored different vehicles in his thought experiments, he did so by following an evolutionary path, where one machine could be described as being the previous device with an additional specialization or capability. For example, Vehicle 3 is a Vehicle 2 that includes more than one kind of sensor, with a myriad of connections (inhibitory and excitatory, ipsilateral and contralateral) between sensors and motors (Braitenberg, 1984). Vehicle 4 is a Vehicle 3 that incorporates non-linear relationships between sensor readings and motor behaviours.

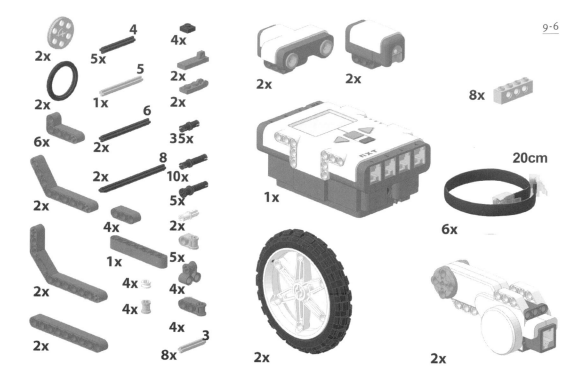

When it is constructed and programmed, it will be apparent that antiSLAM is an example of a Braitenberg Vehicle 4. The parts that are required to construct an antiSLAM robot are provided in Figure 9-6. The pages that follow provide words and images that describe how to construct this machine. If the reader would prefer to use wordless, LEGO-style instructions, they are available as a pdf file from the website that supports this book (http://www.bcp.psych.ualberta.ca/~mike/BricksToBrains/).

9.15 A SPINE FOR antiSLAM

9.15.1 Creating a Chassis

As antiSLAM can be viewed as an evolutionary descendant of Vehicle 2, it should not be surprising to find that antiSLAM's chassis — its internal spine — is identical to the chassis that was constructed for Vehicle 2 and described in Sections 4.4 and 4.5. The images below indicate how to construct the chassis.

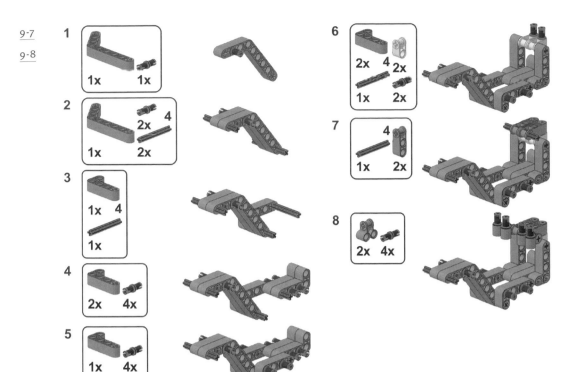

9.16 STRUCTURE FROM MOTORS
9.16.1 Motors and Axles

The next step in building antiSLAM is to attach two motors to the chassis, and then to attach two axle assemblies to each of the motors, as shown in Figure 9-9. At this time additional pins are added to the robot; soon these pins will be used to permit the NXT brick to be attached to the chassis. As was the case for Vehicle 2, the physical structure of each motor is incorporated into the chassis design in order to reinforce the chassis.

9.17 SENSOR SUPPORTS AND FRONT WHEELS
9.17.1 Creating Sensor Supports

AntiSLAM requires that four different sensors be mounted near the front of the robot. They are supported by two double-bent liftarms that are pinned to the NXT brick — which is also added at this stage — as is illustrated in Figure 9-10. The additional axles and pins that are inserted into the double-bent liftarms will be used to support wheels, ultrasonic sensors, and light sensors on either side of the robot.

9

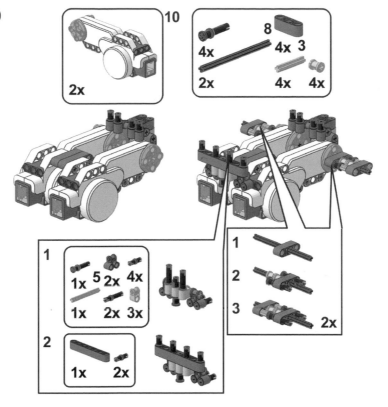

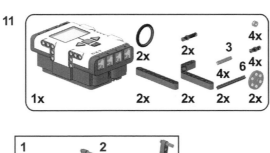

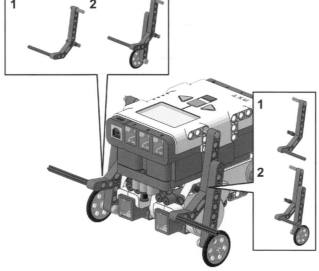

9.17.2 Front Wheels

While Vehicle 2 used a "front slider" to support the front weight of the robot, antiSLAM works best if two small wheels are used to keep the front stable as it scurries about the environment. The wheels are wedge belt wheels, with tires, that are mounted onto beams that in turn are attached to the double-bent liftarms, as shown in Figure 9-11.

9.18 SENSOR ARRAYS

9.18.1 Mounting Sensors

The sensor mounts that were added in Section 9.17 are used to support two ultrasonic sensors. It is these sensors that permit antiSLAM to respond to "geometric features" when reorienting itself. Each sensor is mounted on a LEGO hinge so that it can be pointed outward from the front of the robot, as shown in Figure 9-11. A light sensor, that also points outward, is also mounted just above and behind each ultrasonic sensor, as is also illustrated in Figure 9-11.

9-11

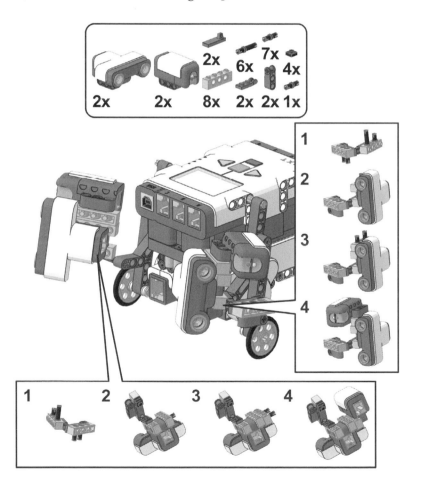

9.19 antiSLAM's REAR WHEELS AND CABLES

9.19.1 Rear Wheels

As was the case for Vehicle 2, 81.6 × 15 LEGO motorcycle wheels, with tires, are used as the rear wheels for antiSLAM, as shown in Figure 9-12. Each of these wheels is independently powered by its own motor, so the robot can be steered by manipulating robot speed. These are very large wheels, and will be able to move the robot along at a fairly high speed.

9.19.2 Connecting Cables

The final step in building antiSLAM is to use cables to connect the various sensors and motors to the NXT brick's ports. The light sensor on the robot's right is connected to Input Port 4, and the ultrasonic sensor on the robot's right is connected to Input Port 3. The light sensor on the robot's left is connected to Input Port 1, and the ultrasonic sensor on the robot's left is connected to Input Port 2. The robot's right motor is connected to Output Port B, while the robot's left motor is connected to Output Port C.

When Vehicle 2 was cabled in Chapter 4, it was noted that there was a choice between an ipsilateral and contralateral relationship between sensors and motors (Section 4.14). Furthermore, these relationships were put in place by changing the ports that the motors were plugged into. Because antiSLAM uses a larger number of sensors, the cables will not be manipulated after they are connected to the ports described in the previous paragraph. It is still the case that relationships between sensor types can be ipsilateral or contralateral, but the nature of the sensor-to-motor mapping will be handled by software in this robot.

9-12

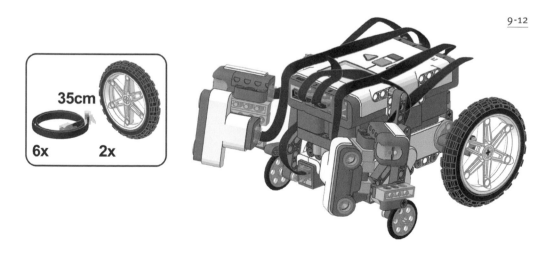

35cm

6x   2x

9.20 antiSLAM level 0: drive
9.20.1 Subsumption Architecture

With Section 9.17, the construction of antiSLAM is complete. However, we need to program this robot in order to bring it to life. As was the case with the LEGO Tortoise (Chapter 7) and the LEGO Lemming (Chapter 8), we have adopted a subsumption architecture to do so.

The code below provides Level 0 of this subsumption architecture; this level provides the basic capability "drive." That is, Level 0 converts a signal from an ultrasonic sensor into a motor speed, and drives the motor forward at that speed. The right motor is controlled by the task DriveRight(), while the left motor is controlled by DriveLeft().

The Level 0 code that is provided does not explicitly state how motor speeds are calculated. This is actually accomplished by a lower-level task that is affected by several higher levels in the architecture; this "Level -1" task will be described momentarily. For the time being, recognize that as an obstacle gets closer to an ultrasonic sensor, the value output by the sensor decreases. This behaviour is mapped into motor speed, so that as a sensor gets closer and closer to a wall in a rectangular arena, the motor that the sensor controls will slow, eventually coming to a virtual halt.

The relationship between ultrasonic sensors and motors in antiSLAM is contralateral. Thus, Level 0 essentially causes antiSLAM to become a version of Vehicle 2 that uses ultrasonic sensors instead of light sensors. The robot will steer away from walls—if started in a corridor, it will move quickly down the hallway, keeping as far away from the walls on both sides as best it can. However, eventually the corridor will end. As the robot approaches the end, it will begin to slow, and it will also begin to turn. Eventually the relationship between its ultrasonic sensors and its motors will cause it to turn into a corner of the corridor, where it will come to a halt.

```
/*=====Level 0: Drive===================================
Feed the distance from each ultrasonic sensor to the motor.
The robot is wired contralaterally, and thus avoids all walls equally. As a result, when it
reaches a corner, it slows down and ends up
stopping, getting corner detection ïfor freeî.
*/
task DriveRight(){
     while(true){
            OnFwd(RightMotor, RightSpeed);
     }
}
```

```
task DriveLeft(){
    while(true){
        OnFwd(LeftMotor, LeftSpeed);
    }
}
```

9.21 LEVEL 1: ESCAPE
9.21.1 Importance of Escaping

Our earlier discussion of reactive robots that were capable of performing the reorientation task in a similar fashion to animals (Nolfi, 2002; Nolfi & Floreano, 2000) noted that rotational errors were produced by the methods used by the robot to leave a corner. Level 1, whose code is provided below, is a simpler version of this approach — it causes the robot to spin out of a corner, but the actual spin is accomplished without using sensors to control precise movements.

Level 1 begins by taking advantage of the rotational sensors that are built into NXT motors. The task Retreat() examines the output of these sensors, and determines if the wheels are rotating less than a threshold amount. If so, the task assumes that an obstacle is impeding the robot from all frontward directions, and so the motors are controlled in such a way that the robot spins in place. When the spin is completed, the robot will be pointing approximately 45° away from the obstacle. Just prior to spinning, the routine causes the robot to emit a sound. This sound can be used by the experimenter as an objective measure of when this routine is called, which will be important when the behaviour of the robot is measured.

The robot's spin is accomplished by manipulating variables used in the still-to-be-described Level -1 to alter motor speed. When Sensitivity is set to 0, motor speed will not be affected by other sensors, such as the ultrasonic detectors. When Reverse is set to 35, this will cause the motors to move in opposite directions, producing the spin, which is simply conducted for a set period of time using the Wait function.

Note that Level 1 fits into the subsumption architecture by sending signals that directly modifies the behaviour of one lower level (Level -1), which in turn affects the behaviour of another lower level (Level 0).

Note too that Level 1 converts antiSLAM from a Vehicle 2 into a Braitenberg Vehicle 3 (Braitenberg, 1984), because now motor speeds are affected by two different types of sensors (ultrasonic sensors and rotation sensors), which can have different, and even opposing, sensor-to-motor relationships.

```
/*=====Level 1: Escape ===============================
If the motors move less than a stated threshold over a delay period, the robotís sensors are
temporarily overridden as it spins around. It ends up pointing approx. 45 degrees from the
corner when normal operation resumes. */
int Threshold, Delay;
void Spin(){
    ResetRotationCount(LeftMotor); ResetRotationCount(RightMotor);
    Sensitivity = 0; Reverse = 35; //Disable sensors, enable spin term
    Wait(4000); //Time to spin in milliseconds
    Reverse = 0; Sensitivity = 1; //Return to default settings
    ResetRotationCount(LeftMotor); ResetRotationCount(RightMotor);
}
task Retreat(){
    long RotCount;//Tracks motor rotation.
    while(true){
            RotCount = MotorRotationCount(LeftMotor) + MotorRotationCount(RightMotor);
            Wait(Delay);
            if(((MotorRotationCount(LeftMotor)+MotorRotationCount(RightMotor)-RotCount)
                    < Threshold) && Reverse==0){//If motors slow down while not spinning
                        PlayTone(440, 500); //Beep to indicate spinning and data point.
                        Spin();
                        Wait(500);
            }
    }
}
```

9.22 LEVEL 2: FOLLOWING WALLS

9.22.1 Biasing Lower-level Behaviour

As was previously described, the Level 0 behaviour makes all walls equally aversive—antiSLAM attempts equally far away from walls on the left and walls on the right when its driving behaviour is primarily controlled by Level 0. Level 2 introduces a bias in the robot's behaviour that causes it to have a wall-following preference. For instance, one can set this bias so that the robot keeps closer to the wall on its right as it moves through the arena. A change in the bias—a change in the robot's "handedness"—can result in the robot keeping closer to the wall on its left.

It is the combination of robot "handedness" with its corner-escaping behaviour that should produce rotational errors in the reorientation task. That is, given its position when spinning out of a corner, and its preference to follow a wall on one side, the robot should move from that corner to the corner that is diagonally opposite in a reorientation task arena.

Note that Level 2 operates by manipulating variables (LeftBias, Right-Bias) that are used by the to-be-described Level -1 to determine motor speed. This is because one can give the robot a preference to be nearer to a wall on one side by providing a tendency or bias to turn in that direction, which can be accomplished by "tweaking" motor speeds. That is, by making the motor on the robot's left turn faster than the motor on the robot's right, the robot will have a tendency to turn to the right. When the robot gets too close to a wall on its right because of this turning, it will be straightened out because of the ultrasonic signals that detect and avoid the wall (Level 0). In short, the turning bias defined here in Level 2, combined with the obstacle avoidance achieved in Level 0, will interact to make the robot keep a wall closer on one side than the other as it moves.

```
/*=====Level 2: Follow ================================
Introduce a bias to the robotís motors, causing it to ípreferî one motor (a sort of
íhandcdnessî). The difference in motor power induces a turn, pushing the robot closer to one
wall. Level 0 will straighten it out after it gets close enough, resulting in a robot that follows
walls on one side. */
int Nearest(bool hand){ /*This returns the value of the sensor nearest the wall.
                              Note: ínearestî is defned by the robotís ípreferredî side*/
    if (hand) return SensorUS(RightEar);
    else return SensorUS(LeftEar);
}
bool preferred; //Determines which side the robot prefers. True = right turns.
int bias; //The strength of the bias term. See the main task.
task Seek(){
    //Set the bias to the appropriate side.
    while(true){
            if (preferred) {RightBias = bias; LeftBias = 0;}
            else {LeftBias = bias; RightBias = 0;}
    }
}
```

9.23 LEVEL 3: USING LIGHT AS A LOCAL FEATURE
9.23.1 Local Feature Sensitivity

Nolfi's robots (Nolfi, 2002; Nolfi & Floreano, 2000) used an array of sensors to detect proximity to walls, but did not use any sensors to process local features, which is a key ingredient of the reorientation task. One of our goals for antiSLAM was to use its light sensors to detect a particular local feature (wall "colour"). This ability is provided by Level 3, whose code is given below. Our expectation is that when Level 3 is operating, if a goal corner is marked by local features (i.e., if it was illuminated by lights, and therefore was brighter than other locations in the arena), then this information can be used by the robot to prevent rotational error from occurring.

How is such feature sensitivity to be accomplished? The two light sensors mounted on antiSLAM can be used to influence motors in a fashion similar to that used in Vehicle 2 (Chapter 4). In antiSLAM the light sensors are connected in such a way that they influence the motor on the contralateral side of the robot. Thus, when light is detected, the robot has a tendency to turn toward it and approach it, accelerating as the light gets brighter. The only difference between antiSLAM and Vehicle 2 is that in antiSLAM motor speeds are affected by ultrasonic sensors as well. As a result, rather than affect motor speed directly, Level 3 computes values that are passed on to Level -1, which combines readings from all sensors to determine motor speed. Level -1 is described on the next page.

Note that in the code below the sensor values are modified by the term Vision. This provides a weight to light sensor information (relative to a different weight applied to ultrasonic sensor information). In the experiments reported later in this chapter, the light sensors were given a 60% weight, and the ultrasonic sensors were given a 40% weight.

One important characteristic of antiSLAM is that its embodiment affects what Level 3 detects when the robot gets very close to a brightly lit corner. Because the light sensors are spread apart, and because they are angled outward, the robot has a "blind spot" to nearby light directly in front of the machine. So, when it moves into a brightly lit corner, when the light enters the bright spot the light sensors detect little light, and the robot slows down.

```
/*=====Level 3: Feature ==============================
Enables and reads the light sensors (eyes) as a percentage based on ìVisionî
(a sensitivity term), such that more light = more speed. Since the connection is contralateral,
this results in the robot turning toward sources of light.
However, level -1 weighs this visual sense with the earlier ultrasonic sense,
allowing both terms to infuence the robotís fnal behavior. */
int Vision; //The strength of the light sensors in percent.
task See(){
    //Sets the strength of the robotís visual response to a scaled percentage.
    while(true){
            LVis = Sensor(LeftEye)*Vision/100;
            RVis = Sensor(RightEye)*Vision/100;
    }
}
```

9.24 LEVEL -1: DETERMINING MOTOR SPEEDS
9.24.1 Finally, Level -1

The behaviour of antiSLAM is completely determined by the speeds of its two motors. The speeds of these motors depend on the signals being sent by ultrasonic, rotation, and light sensors. The speeds are directly affected by combinations of these signals, as well as by other factors, such as the robot's bias to follow walls on one side or the other.

The actual calculation of motor speed requires that these various factors be combined, and this is accomplished by Level -1, whose code is provided below. All of the variables that are used in the calculations given below have values that depend either directly on sensor readings or on signals that are sent down to this level by higher levels in the subsumption architecture, as has been noted in the preceding pages of this chapter. Note that the ultrasonic sensors are weighted by the Hearing variable in this code. This variable provides a weight to the ultrasonic signal; a similar weight was applied to the light sensor signal in the code that was described for Level 3.

```
/*=====îLevel -1î: Integration ==========================================
Each of the terms (Sensitivity, Reverse, LeftBias, RightBias) is part of a later levelís
connection
to the motors. See the main task to see their defaults.
On its own, this task does nothing. However, it will function at every level
without modifcation. */
int Sensitivity,Reverse, LeftSpeed,RightSpeed, LeftBias,RightBias, LVis,RVis;
int Hearing;
task Drive(){
    while(true){
            //îHearing/255î converts from responsive raw ultrasonic to % motor speed.
            RightSpeed = ((SensorUS(RightEar)*(Hearing-LeftBias)/255)+RVis)
                    * Sensitivity+Reverse;
            LeftSpeed = ((SensorUS(LeftEar)*(Hearing-RightBias)/255)+LVis)
                    * Sensitivity-Reverse;
    }
}
```

9.25 THE MAIN TASK
9.25.1 Putting It Together

In order to get all of the levels in the subsumption architecture working, the main task is used to initialize some important variables and to call the various tasks that define each level. Note that by modifying the main task — for instance, by commenting out a call to a task — one can selectively remove layers from the subsumption architecture in order to investigate changes in behaviour. Note too that the variables that are used to weight the light sensor and ultrasonic sensor signals (Vision and Hearing) are set in the main task. Note that the complete program for this robot is available from the website that supports this book (http://www.bcp.psych.ualberta.ca/~mike/BricksToBrains/).

```
//Anti-SLAM v8
//Convention: Left and Right refer to the SENSOR/MOTOR, NOT THE WIRING.
//See the wiring diagrams in Chapter 9.
#defne LeftMotor OUT_B
#defne RightMotor OUT_C
#defne LeftEar S2
#defne RightEar S3
#defne LeftEye S1
#defne RightEye S4
//=== Main Task ============================================================
task main(){
//Set up ultrasonic sensors and speed calculation weights.
    SetSensorLowspeed(LeftEar);
    SetSensorLowspeed(RightEar);
    SetSensorMode(LeftEar, SENSOR_MODE_RAW);
    SetSensorMode(RightEar, SENSOR_MODE_RAW);
    Sensitivity = 1; //Level 0 connection: Ultrasonic sensitivity. Default 1.
    Reverse = 0; //Level 1 connection: Lets robot spin and escape. Default 0.
    LeftBias = 0; //Level 2 connection: Causes robot to prefer left turns. Def. 0.
    RightBias = 0;//Level 2 connection: As above, but prefers right turns. Def. 0.
    LVis = 0; //Impact of the left eye on movement. Zero at this level.
    RVis = 0; //Impact of the right eye on movement. Zero at this level.
    Hearing = 100; //Strength of ultrasonic sense. (Overridden at level 3.)
    start Drive; //Starts mapping the motor speeds to the collective input.
    //Level 0.
    start DriveRight; //Turn the right motor on.
    start DriveLeft; //Turn the left motor on.
    //Level 1. (Delete below this line for a level 0 robot.)
    Threshold = 360; //Combined motor movement to be considered ëoní. Default 360.
    Delay = 5000; //How long the robot needs to have been stopped. Default 5000ms.
    start Retreat; //Allow the robot to escape corners.
    //Level 2. (Delete below this line for a level 1 robot.)
    preferred = true; //True for left-handed (right-following), false otherwise.
    bias = 40; //Fixed value for handedness bias. Default 40.
    start Seek; //Follow the wall on your preferred side.
```

```
//Level 3. (Delete below this line for a level 2 robot.)
//Set up eyes.
SetSensorType(LeftEye, SENSOR_TYPE_LIGHT_INACTIVE);
SetSensorMode(LeftEye, SENSOR_MODE_PERCENT);
SetSensorType(RightEye, SENSOR_TYPE_LIGHT_INACTIVE);
SetSensorMode(RightEye, SENSOR_MODE_PERCENT);
Hearing = 40;// % of ultrasonic sense that feeds to the motors. Default 40.
Vision = 60;// % of light sense that feeds to the motors. Default 60.
start See;
}
```

9.26 PRIMITIVE BEHAVIOURS
9.26.1 Levels -1 + 0

With the robot constructed and programmed, we are now in a position to observe its behaviour. As was the practice in Chapter 7, let us consider how antiSLAM's behaviour changes as more and more layers of its subsumption architecture are added. Examples of the behaviour described in the following pages are provided in Video 9-1; this video is provided by the website for this book (http://www.bcp.psych.ualberta.ca/~mike/BricksToBrains/).

The most primitive behaviour of antiSLAM is produced when only Level -1 (which computes motor speed) and Level 0 (which uses the two ultrasonic sensors to influence motor speed) of the subsumption architecture are operational. When placed in a long hallway, antiSLAM turns itself so that it is pointing down the length of the hallway, accelerating as it points away from nearby walls. It then quickly propels itself down the hallway, keeping to its centre. During this journey, the walls of the hallway may not be perfectly uniform, because of doorways or small alcoves. When these variations are encountered, they influence the ultrasonic sensors, and the robot veers slightly toward the open space.

If the hallway is not completely clear, more interesting behaviour is observed. The legs of chairs, tables, and people all serve as obstacles in the hallway, and are detected by antiSLAM's ultrasonic sensors. The robot nimbly steers around these obstacles, and resumes its dash down the middle of the hallway when the obstacles are behind it.

The hallway, however, is not infinitely long. Eventually antiSLAM encounters a set of doors that block its way and that cannot be avoided. As the robot nears the end of the hallway, it begins to decelerate. It slows and turns, and eventually is trapped in a corner from which it cannot escape. It approaches the corner, its wheels barely turning, and its journey has ended.

Thus, the most primitive antiSLAM is very capable of moving through its world, turning away from obstacles whenever possible, and keeping as great a distance between it and walls as possible. Eventually, though, it finds itself in a corner. Note that it does all of this without having a cognitive map, and without any explicit knowledge about hallways, obstacles, or corners. All of this behaviour is in essence the product of a cousin of Vehicle 2 whose two motors slow down when obstacles are near, and speed up when the path is clear. That is, this behaviour is the result of some basic rules, chance, and the structure of the environment. It is not the result of spatial representations.

9.26.2 Levels -1 + 0 + 1

The most primitive version of antiSLAM is capable of finding corners, but when it succeeds at this task, it is trapped. This problem is solved by adding Level 1 to the mix. Recall that this level uses the rotation sensors in the two motors to detect when antiSLAM has slowed to the point that it has essentially stopped. In this situation, Level 1 manipulates the motor speed equation of Level -1 in such a way that the robot reverses itself and turns around. The robot emits a brief tone to inform its observers that Level 1 has detected the circumstances that cause it to change the robot's behaviour. After the evasive manoeuver has been performed, the robot reverts to its normal exploratory behaviour.

When this more advanced robot is run, its behaviour is very similar to the robot described in Section 9.27.1. It too moves down the middle of hallways, avoids obstacles, and stops at discovered corners. However, once a corner has been found, after remaining stationary for a bit the robot beeps, turns away from the corner, and points at an angle down the hallway. Then it accelerates, and continues its journey in the opposite direction. Again, this behaviour is produced without the need of explicit spatial representations or memories.

9.27 BIAS AND REORIENTATION
9.27.1 Levels -1 + 0 + 1 + 2

The next stage in the evolution of antiSLAM is to add Level 2, which adds a bias that causes the robot to be closer to the wall on its right than it is to the wall on its left. This is accomplished by having the robot always turn slightly to its right; when it gets too near the wall, the activity of Level 0 causes it to straighten out.

When this version of the robot is observed, its behaviour is very similar to its ancestor described in Section 9.27.2, with one exception: the robot no longer keeps to the middle of the hallway as it explores.

Instead, it has a marked tendency to be closer to the wall on its right than on its left. Again, it avoids obstacles, and finds (and then escapes) corners. However, after watching the robot perform its exploration for a while, it becomes evident that of the four corners of the hallway that it could discover, it has a strong tendency to only find two, and these two corners are geometrically equivalent. Is this simple robot — that only reacts to obstacles and has a turning bias, and does not have a cognitive map or spatial representation — capable of producing a key finding in the reorientation task, that of rotational error?

9.27.2 Rotational Error and AntiSLAM

In order to answer this question, we conducted an experiment that was similar to the one used to examine the spatial reorientation of a more sophisticated reactive robot (Nolfi, 2002; Nolfi & Floreano, 2000). Our robot was placed in a small, empty testing room that provided rectangular arena that was 2.4 metres long and 1.65 metres wide. Location 4 in Figure 9-15 was considered to be the goal corner. Adopting Nolfi's methodology, antiSLAM was placed in one of eight different starting locations. The eight locations are illustrated in Figure 9-13. In this figure, each location is represented by an arrow. The base of each arrow indicates where the back of antiSLAM was positioned, and the arrowhead indicates the direction in which antiSLAM was pointed. The robot was started, and then permitted to explore the arena for 5 minutes. Each time that the robot initiated its Level 1 "escape corner" routine, the location of the robot in the arena was recorded. The entire experiment involved recording the robot's behaviour after it was started once at each of the eight starting locations.

9-13

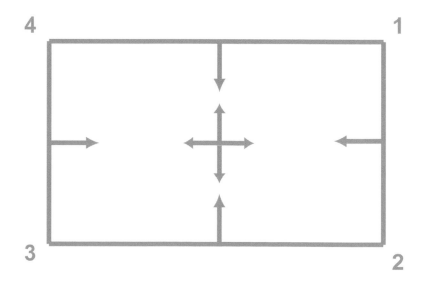

The results of the experiment are provided in Figure 9-14, where the number provided in each corner indicates the percentage of times that the "escape corner" routine was executed there. Note that while the robot would occasionally discover corners 1 or 3, most of the time it would stop at either the goal location (corner 4) or its geometric equivalent (corner 2). Furthermore, it visited these two locations roughly equally. As well, it would typically follow a path that took it directly between corner 4 and corner 2, and back again, once it had discovered one of these two corners, as indicated by the arrows in the figure. This pattern of results demonstrates that this version of antiSLAM produces rotational error!

9-14

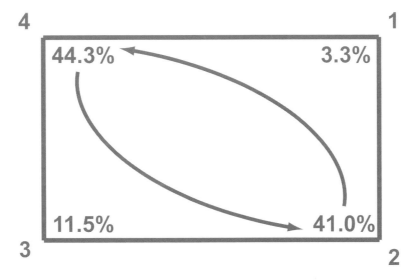

9.28 BEYOND ROTATIONAL ERROR
9.28.1 Nolfi and Beyond

It could be argued that antiSLAM is limited by its bias to the right—while this enables rotational errors between corners 4 and 2 (Figure 9-16), it prevents the robot from generating similar behaviour if corner 1 or corner 3 were the goal. However Level 2 can be easily used to convert antiSLAM's bias to the right into a bias to the left. One could imagine a higher level in its architecture (Level 4) that allowed the robot to explore all of an arena by occasionally changing its bias from right to left, and later from left to right. This could be associated with a learning routine: if the robot was reinforced when it stopped (e.g., by receiving a Bluetooth signal), it would be more likely to preserve its current "handedness."

9.28.2 Feature Sensitivity

AntiSLAM can be extended beyond Nolfi's robots by adding sensitivity to local features. In antiSLAM, this is accomplished by Level 3. Now, in addition to being sensitive to obstacles, antiSLAM will be attracted to lights. We can define corner 4 as the goal location by illuminating it with lamps.

We activated Level 3 in the robot, and defined Level -1 so that light sensitivity was given a weight of 60 and ultrasonic sensitivity was given a weight of 40. We hung two small lights over corner 4 and used them to project this local feature to this location. We then repeated the experimental methodology that was described in Section 9.28. The results are shown in Figure 9-15.

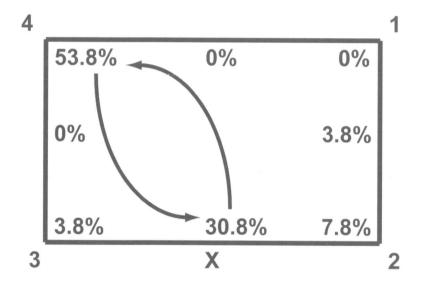

The addition of light sensitivity, and the lighting of corner 4, has changed the robot's behaviour markedly. Now, from any location, it quickly sees the light and stops in corner 4. When this corner is escaped, the robot moves away. However, now it rarely goes to corner 2, or to any other corner for that matter. Instead, it usually slowly veers and heads toward location X midway between the long wall between corners 3 and 2, as shown in Figure 9-15. Here it stops — in spite of this location not being a corner! — executes its escape routine, and heads back to corner 4. This is very similar to the behaviour of ants in the reorientation task, who — when featural information informs them that they are not going to the goal location — execute a U-turn and head back to a corner that is marked by the correct local feature (Wystrach & Beugnon, 2009).

The data reported in Figure 9-15 is atypical because it includes locations other than the four corners that are typically examined in the reorientation task. If we followed this convention, we would only be reporting the robot's visits to the corners, which account for 65.4% of the data in the figure. Focusing only on these trials, antiSLAM visited the goal corner 82.3% of the time (i.e., 53.8/65.4), visited corner 3 on 5.8% of these trials, visited corner 2 on 11.9% of this subset of trials, and never visited corner 1. When reported in this way, the data reveals a strong effect of the local feature; reported as in Figure 9-17, though, the data makes the same case and provides some more interesting sense on how the path taken by the robot was affected.

It has been suggested that information about paths might provide information about strategies for solving the task (Cheng, 2008a); of course, antiSLAM never changes its strategy — changes in behaviour reflect changes in the environment in which it is situated, changes to which it reacts.

9.29 MOVING THE LOCAL FEATURE
9.29.1 Moving the Light

It was previously noted that researchers were interested in the reorientation task because they could reposition local cues before an agent was reintroduced to the arena, placing local features and geometric features in conflict. How does this type of manipulation affect antiSLAM when all of its software levels are running?

To answer this question, we conducted the same experiment that was described in Section 9.29, but in this version of the experiment corner 3 was illuminated. Now the local feature was present in a corner that was not preferred by the robot because of its bias to keep walls on its right. The results shown in Figure 9-16 indicate that there was a clear conflict between the local feature and antiSLAM's geometric preferences.

The top portion of Figure 9-16 indicates the percentage of trials that antiSLAM executed its "escape" routine at various locations in the arena. Note that in addition to the four corners, the midpoints of all four walls now become locations of interest. If we restrict our attention to the corner locations, we note that the results indicate that antiSLAM's behaviour is guided by both local and geometric features. It is most likely to stop at corner 3, which is illuminated as the goal location, but is not preferred by antiSLAM's ultrasonic mechanisms (recall Figure 9-16). It still has a moderately strong preference to visit corner 4, which is nearly equal to its preference to visit the geometrically equivalent corner 2. Corner 1, which lacks the preferred geometry and the local feature, is rarely visited. Such results are typical of studies of humans and animals in this task (Cheng & Newcombe, 2005).

The lower part of Figure 9-16 illustrates an example journey of anti-SLAM in this condition. The arrows indicate the path taken by the robot; the letters at the base of each arrow indicate their temporal order. In this example, the robot starts at location W, and moves to location X before being attracted to the light at corner 3. Note that the complicated journey depicted in this figure is the result of the competing influences of the walls, the turning bias, the brightly illuminated corner, and light reflecting from this corner to other walls. It is not the result of planning, or strategies, or the use of a cognitive map.

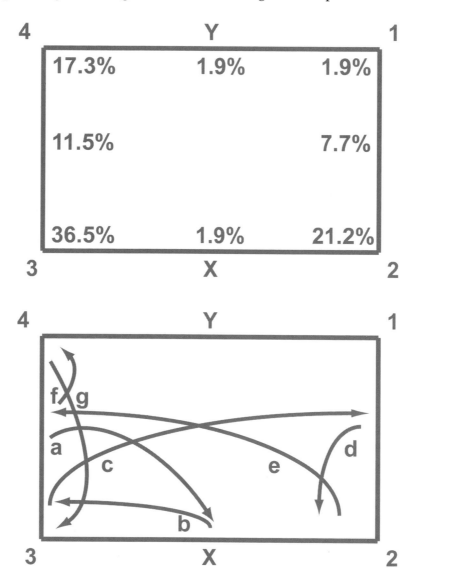

9.30 ALL LEVELS WITH NO LOCAL FEATURE
9.30.1 Turning Lights Off

In Section 9.28, we observed rotational error in antiSLAM when only ultrasonic signals were driving the motors. How does this robot behave when all of its levels are running, but when there are no lights turned on to serve as local features? We answered this question by observing the fully functional antiSLAM in a final version of the experiment where the conditions were identical to those explored with the less advanced robot in Section 9.28. The results are presented in Figure 9-17.

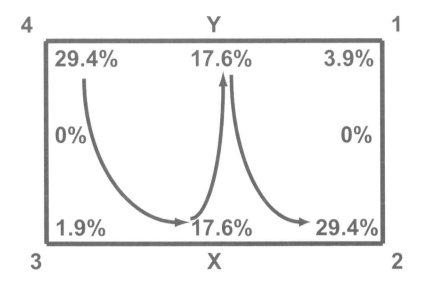

If we restrict ourselves to examining visitations to corner locations, which are almost always the locations of interest to reorientation task researchers, we see clear evidence of rotational error once again. Anti-SLAM has a strong preference to visit corner 4, an equally strong preference to visit corner 2, and rarely executes its "escape corner" routine in either of the other corners.

However, antiSLAM executes its "escape corner" routine at other locations too; specifically locations X and Y in Figure 9-17. The arrows in Figure 9-17 illustrate a complicated course that antiSLAM takes to produce rotational error results. For instance, it frequently starts at corner 4, and heads to location X. It then turns from location X and stops at location Y. From location Y it will turn and stop at corner 2. A journey in the opposite direction, with stops at Y and then X, is frequently undertaken to move antiSLAM from corner 2 to corner 4. This journey is not always taken. For instance, sometimes it will circle between corner 4 and location X, or between corner 2 and location Y.

Had we adopted the more typical practice of only reporting visitations to one of the four corner locations, our results would be very similar to those reported earlier in Figure 9-14. Clearly, though, the path that was taken by the robot in that figure is markedly different from the path that is illustrated in Figure 9-17. Unfortunately, except for rare instances (Wystrach & Beugnon, 2009), researchers do not report the paths taken by their biological subjects in reorientation arenas, so we cannot evaluate either Figure 9-14 or Figure 9-17 in terms of their fit to animal behaviour.

Why does the full-fledged robot produce this more complicated journey, compared to the more typical journey between corners that produced the data in Figure 9-14? The answer is that this robot is still functioning with the ultrasonic sensors weighted at 40, and the light sensors weighted at 60. There is a small amount of ambient light in the room, but it does not differentiate corners. However, the lower weighting of the ultrasonic signal means that its object-detecting abilities are diminished. As a result, these weaker abilities can be overwhelmed by its bias to turn toward the right. Much of the "clover leaf" trajectory that it takes between corners 4 and 2 is dictated by this bias overriding the other senses of the robot. In other words, the difference in paths between Figures 9-17 and 9-14 does not reflect a difference in strategy, but instead reflects different kinds of interactions between sense–act mechanisms.

9.31 REORIENTING REORIENTATION
9.31.1 Building a Better Mouse

One of my undergraduate students, Mike M., created a LEGO "mouse" as a robotics project. The robot had two rotation sensors mounted in front, to which were attached long "whiskers." When the robot bumped into an object, both whiskers were pushed backward. This caused the robot to back away from obstacles. However, when only one of the rotation sensors was active, this was usually because the robot was near a wall. In this case, the robot followed the wall by keeping its whisker in contact with it.

AntiSLAM did not begin as an agent for the reorientation task. Instead, it was our attempt to create an NXT version of Mike's mouse, one that followed walls using ultrasonic sensors. However, at the same time we were also working on another project involving artificial neural networks and the reorientation task (Dawson et al., 2008). It was only when we watched antiSLAM that we saw a connection between the two projects, and that antiSLAM was capable of generating rotational error. This led to our subsequent discovery of Nolfi's robots (Nolfi, 2002; Nolfi & Floreano, 2000), and the addition of Level 3's sensitivity to local features as we "tweaked" antiSLAM's later design with the reorientation task in mind.

AntiSLAM's development illustrates one of the key advantages of the synthetic methodology: getting surprising results "for free." AntiSLAM began as a set of simple capabilities that produced movement, wall following, and obstacle avoidance. It was never intended to produce rotational error, or provide insight into theories of spatial cognition — but we were fortunate enough to discover that it could accomplish these things.

9.31.2 Different Views of Reorientation

Another lesson from the history of antiSLAM's development impacts theoretical considerations about the reorientation task itself. Of course, both antiSLAM and Nolfi's robots indicate that some of the known regularities that govern this task can be produced by devices that simply react to their environment and do not employ spatial representations. Whether the same can be said of biological agents that perform this task is an open question (Cheng, 2008a). The reactive robots do raise important questions: clearly, benchmark results like rotational error do not necessarily imply the use of internal spatial representations. If biological agents are more than reactive, then additional results will be required to support this representational position.

Furthermore, theories that are specifically designed to explain the reorientation task are generally sense–think–act in nature, tacitly assuming that an agent has the primary goal of finding a previously reinforced location. So, researchers view the task in terms of possible goals — usually the four corner locations in a rectangular arena — and the features available at each location.

In contrast, because antiSLAM was designed to explore, it was not burdened by typical assumptions about the reorientation task. No goals were represented; no features present at corners were remembered. Reorienting behaviour emerged out of the interactions between simple sense–act reflexes and the environment. It is intriguing to consider how antiSLAM might behave in other traditional tasks used to study spatial behaviour, such as radial arm mazes.

Finally, the traditional view of the reorientation task usually places restrictions on what is relevant data. "The paths taken by animals have not appeared in print. Nolfi's work suggests that such paths can be relevant to interpreting what strategy an animal is using" (Cheng, 2008a, p. 155). AntiSLAM also shows that such non-traditional data is critical. Of particular interest were the paths that it took in the reorientation arena, and the fact that some locations — like midpoints between corners — were important determinants of its behaviour. Importantly, these trajectories say nothing about "strategy." Instead, they show that complex routes emerge in the reorientation arena when simple sense–act couplings react to the information provided, in the absence of plans, strategies, or spatial representations.

9.32 HARD FUN AND HARD SCIENCE
9.32.1 Hard Fun

Synthetic psychologists often emphasize the simple nature of their machines by using toy-like descriptions (see Section 9.2.1). For instance, Braitenberg (1984, p. 20) notes that "it is pleasurable and easy to create little machines that do certain tricks," making his synthetic psychology sound like fun.

There are, though, different kinds of fun. Consider the following anecdote (Negroponte, 1995). In 1989, elementary school children demonstrated their work with the MIT Media Lab's LEGO and Logo projects to the media. "A zealous anchorwoman from one of the national TV networks, camera lights ablazing, cornered one child and asked him if this was not just all fun and games. ... After her third repetition of the question and after considerable heat from the lights, this sweaty-faced, exasperated child plaintively looked into the camera and said, 'Yes, this is fun, but it's hard fun'" (p. 196).

"Hard fun" is an idea that has emerged from the study of how to use technology to enhance education (Papert, 1980, 1993). It is the idea that learners do not mind engaging in (and learning from) activities that are challenging provided that these activities are also fun, in the sense that they connect with learners' interests (Picard et al., 2004).

Learning about embodied cognitive science by building LEGO robots is another example of hard fun. The fun part, of course, is engaging in *bricolage* in order to produce a working, behaving, and lifelike machine. What is it, though, about this kind of fun that makes it hard?

Building these robots is hard in the sense that it requires releasing traditional sense–think–act approaches to cognition and behaviour. Traditional robot development is the top-down task of "making the robot do what I want" (Petre & Price, 2004). This is accomplished by thinking of goal behaviours, by hypothesizing a set of internal mechanisms to produce these behaviours, and by inserting this analysis into the robot's program. However, in order to make a LEGO robot "do what I want," this analytic approach must be abandoned, because it won't succeed given the simplicity and memory limitations of these machines. Instead, one must permit the environment to make its important contribution to the performance of the machine, by working bottom-up to build simple sense–act mechanisms into the machine, and by letting emergent behaviours produced in early stages of development guide later modifications.

Overcoming this natural analytic tendency is a key issue in embodied cognitive science. For instance, many of the chapters in a recent handbook of situated cognition (Robbins & Aydede, 2009) reject some of the more interesting ideas in embodied cognitive science, such as the extended mind, and attempt to morph the notions of embodiment and situatedness into very traditional, and representational, cognitive theories (Dawson, 2009). One reason for this is a strongly entrenched goal of explaining behaviour by appealing to internal mechanisms, such as neural processing (MacIver, 2008).

The hard fun of using LEGO robots to explore embodied cognitive science is an attempt to deal with this issue at two different levels. First, the successful development of these machines requires students to focus on a simple robot's immediate connection to its environment. Second, the hands-on experience of working with the robots and their environment provides a particular kind of scaffolding to support thinking about embodied and situated cognition. Simple environments scaffold the robots' abilities; the robots themselves scaffold our understanding of embodied cognitive science.

9.32.2 Hard Science

This is not to say that the LEGO robots are merely pedagogical tools. Hopefully, the discussion of antiSLAM in the current chapter has shown that even a very simple LEGO robot can be used to contribute new insights into current issues in cognitive science. We hope that the reader will be inspired by this book to develop new machines that will continue this tradition.

References

Agre, P. (1997). *Computation and human experience*. New York, NY: Cambridge University Press.

Agullo, M., Carlson, D., Clague, K., Ferrari, G., Ferrari, M., Yabuki, H., et al. (2003). *LEGO mindstorms masterpieces*. Rockland, MA: Syngress Publishing.

Albright, T. D. (1984). Direction and orientation selectivity of neurons in visual area MT of the macaque. *Journal of Neurophysiology, 52*, 1106–1130.

Alerstam, T. (2006). Conflicting evidence about long-distance animal navigation. *Science, 313*(5788), 791–794.

Alexander, R. M. (2005). Walking made simple. *Science, 308*(5718), 58–59.

Altendorfer, R., Moore, N., Komsuolu, H., Buehler, M., Brown, H. B., McMordie, D., et al. (2001). RHex: A biologically inspired hexapod runner. *Autonomous Robots, 11*(3), 207–213.

Altendorfer, R., Saranli, U., Komsuoglu, H., Koditschek, D., Brown, H. B., Buehler, M., et al. (2001). Evidence for spring loaded inverted pendulum running in a hexapod robot. *Experimental Robotics VII, 271*, 291–302.

Anderson, J. A., Silverstein, J. W., Ritz, S. A., & Jones, R. S. (1977). Distinctive features, categorical perception and probability learning: Some applications of a neural model. *Psychological Review, 84*, 413–451.

Anderson, J. R. (1983). *The architecture of cognition*. Cambridge, MA: Harvard University Press.

Anderson, J. R. (1985). *Cognitive psychology and its implications* (2nd ed.). New York, NY: W. H. Freeman.

Anderson, J. R., Bothell, D., Byrne, M. D., Douglass, S., Lebiere, C., & Qin, Y. L. (2004). An integrated theory of the mind. *Psychological Review, 111*(4), 1036–1060.

Anderson, J. R., & Matessa, M. (1997). A production system theory of serial memory. *Psychological Review, 104*(4), 728–748.

Andriacchi, T. P., & Alexander, E. J. (2000). Studies of human locomotion: Past, present and future. *Journal of Biomechanics, 33*(10), 1217–1224.

Aristotle. (1953/350 BC). *On the heavens*. Cambridge, MA; London, UK: Harvard University Press.

Arleo, A., & Gerstner, W. (2000). Spatial cognition and neuro-mimetic navigation: A model of hippocampal place cell activity. *Biological Cybernetics, 83*, 287–299.

Ashby, W. R. (1956). *An introduction to cybernetics*. London: Chapman & Hall.

Ashby, W. R. (1960). *Design for a brain* (2nd ed.). New York, NY: John Wiley & Sons.

Asimov, I. (2004). *I, robot* (Bantam hardcover ed.). New York, NY: Bantam Books.

Assayag, G., Feichtinger, H. G., Rodrigues, J.-F., & European Mathematical Society. (2002). *Mathematics and music: A Diderot mathematical forum*. Berlin, Germany; New York, NY: Springer-Verlag.

Astolfo, D., Ferrari, M., & Ferrari, G. (2007). *Building robots with LEGO Mindstorms NXT* (Updated ed.). Rockland, MA: Syngress.

Baerends, G. P. (1959). Ethological studies of insect behavior. *Annual Review of Entomology, 4*, 207–234.

Bailey, D. (1992). *Improvisation: Its nature and practice in music*. New York, NY: Da Capo Press.

Baird, J. (1999). Returning to the tropics: The epic autumn flight of the blackpoll warbler. In K. P. Able (Ed.), *Gatherings of angels: Migrating birds and their ecology* (pp. 63–77). Ithaca, NY: Cornell University Press.

Balch, T. (2002). Taxonomies of multirobot task and reward. In T. Balch & L. E. Parker (Eds.), *Robot teams* (pp. 23–35). Natick, MA: A. K. Peters.

Balch, T., & Parker, L. E. (2002). *Robot teams*. Natick, MA: A. K. Peters.

Ballard, D. (1986). Cortical structures and parallel processing: Structure and function. *The Behavioral And Brain Sciences, 9*, 67–120.

Bar-Cohen, Y. (2006). Biomimetics: Using nature to inspire human innovation. *Bioinspiration & Biomimetics, 1*, P1–P12.

Barnes, D. M., & Mallik, A. U. (1997). Habitat factors influencing beaver dam establishment in a northern Ontario watershed. *Journal of Wildlife Management, 61*(4), 1371–1377.

Bartsch, M. S., Federle, W., Full, R. J., & Kenny, T. W. (2007). A multiaxis force sensor for the study of insect biomechanics. *Journal of Microelectromechanical Systems, 16*(3), 709–718.

Baudrillard, J. (1994). *Simulacra and simulation*. Ann Arbor: University of Michigan Press.

Bateson, G. (1972). *Steps to an ecology of mind*. New York, NY: Ballantine Books.

Bayraktaroglu, Z. Y. (2009). Snake-like locomotion: Experimentations with a biologically inspired wheel-less snake robot. *Mechanism and Machine Theory, 44*(3), 591–602.

Bechtel, W., & Abrahamsen, A. A. (2002). *Connectionism and the mind: Parallel processing, dynamics, and evolution in networks* (2nd ed.). Malden, MA: Blackwell.

Bechtel, W., Graham, G., & Balota, D. A. (1998). *A companion to cognitive science*. Malden, MA: Blackwell.

Bellmore, M., & Nemhauser, G. L. (1968). The traveling salesman problem: A survey. *Operations Research, 16*(3), 538–558.

Beni, G. (2005). From swarm intelligence to swarm robotics. *Swarm Robotics, 3342*, 1–9.

Beni, G., & Wang, J. (1991, April 9–11). *Theoretical problems for the realization of distributed robotic systems*. Paper presented at the IEEE International Conference on Robotics and Automation, Sacramento, CA.

Benson, B. (2003). *The improvisation of musical dialogue: A phenomenology of music*. Cambridge, UK; New York, NY: Cambridge University Press.

Benson, D. J. (2007). *Music: A mathematical offering*. Cambridge, UK; New York, NY: Cambridge University Press.

Bertalanffy, L. v. (1967). *Robots, men, and minds*. New York, NY: G. Braziller.

Best, J. B. (1995). *Cognitive psychology*. St. Paul: West Publishing.

Blackwell, T. (2003). *Swarm music: Improvised music with multiswarms*. Paper presented at the AISB Symposium on Artificial Intelligence and Creativity in Arts and Science, Aberystwith, Wales.

Blackwell, T., & Young, M. (2004a). Self-organised music. *Organised Sound, 9*, 123–136.

Blackwell, T., & Young, M. (2004b). Swarm granulator. *Applications of Evolutionary Computing, 3005*, 399–408.

Bladin, P. F. (2006). W. Grey Walter, pioneer in the electroencephalogram, robotics, cybernetics, artificial intelligence. *Journal of Clinical Neuroscience, 13*, 170–177.

Blickhan, R., & Full, R. J. (1993). Similarity in multilegged locomotion: Bouncing like a monopode. *Journal of Comparative Physiology A, 173*, 509–517.

Boden, M. A. (2006). *Mind as machine: A history of cognitive science*. New York, NY: Clarendon Press.

Bonabeau, E., & Meyer, C. (2001). Swarm intelligence: A whole new way to think about business. *Harvard Business Review, 79*(5), 106–114.

Bonabeau, E., Theraulaz, G., Deneubourg, J. L., Franks, N. R., Rafelsberger, O., Joly, J. L., et al. (1998). A model for the emergence of pillars, walls and royal chambers in termite nests. *Philosophical Transactions of the Royal Society of London Series B-Biological Sciences, 353*(1375), 1561–1576.

Bonds, M. E. (2006). *Music as thought: Listening to the symphony in the age of Beethoven*. Princeton, NJ: Princeton University Press.

Boogaarts, M. (2007). *The LEGO Mindstorms NXT idea book: Design, invent, and build*. San Francisco, CA: No Starch Press.

Boole, G. (1854). *An investigation of the laws of thought, on which are founded the mathematical theories of logic and probabilities*. London, UK: Walton & Maberley.

Botez, M. I. (1975). Two visual systems in clinical neurology: Readaptive role of the primitive system in visual agnosis patients. *European Neurology, 13*, 101–122.

Braitenberg, V. (1984). *Vehicles: Explorations in synthetic psychology*. Cambridge, MA: MIT Press.

Braun, H. (1991). On solving traveling salesman problems by genetic algorithms. *Lecture Notes in Computer Science, 496*, 129–133.

Bregler, C., Malik, J., & Pullen, K. (2004). Twist based acquisition and tracking of animal and human kinematics. *International Journal of Computer Vision, 56*(3), 179–194.

Brentano, F. C. (1874/1995). *Psychology from an empirical standpoint* (Paperback ed.). London, UK; New York, NY: Routledge.

Bronowski, J. (1973). *The ascent of man*. London, UK: British Broadcasting Corporation.

Brooks, R., & Flynn, A. M. (1989). Fast, cheap and out of control: A robot invasion of the solar system. *Journal of The British Interplanetary Society, 42*, 478–485.

Brooks, R. A. (1989). A robot that walks: Emergent behaviours from a carefully evolved network. *Neural Computation, 1*, 253–262.

Brooks, R. A. (1999). *Cambrian intelligence: The early history of the new AI*. Cambridge, MA: MIT Press.

Brooks, R. A. (2002). *Flesh and machines: How robots will change us*. New York, NY: Pantheon Books.

Bruner, J. S. (1957). On perceptual readiness. *Psychological Review, 64*, 123–152.

Bruner, J. S., Postman, L., & Rodrigues, J. (1951). Expectation and the perception of color. *American Journal of Psychology, 64*(2), 216–227.

Burgess, N., Donnett, J. G., Jeffery, K. J., & O'Keefe, J. (1997). Robotic and neuronal simulation of the hippocampus and rat navigation. *Philosophical Transactions Of the Royal Society Of London, B, 352*, 1535–1543.

Cage, J. (1961). *Silence: Lectures and writings* (1st ed.). Middletown, CT: Wesleyan University Press.

Canetti, E. (1962). *Crowds and power*. London, UK: Gollancz.

Caudill, M. (1992). *In our own image: Building an artificial person*. New York, NY: Oxford University Press.

Cheng, K. (1986). A purely geometric module in the rat's spatial representation. *Cognition, 23*, 149–178.

Cheng, K. (2005). Reflections on geometry and navigation. *Connection Science, 17*(1–2), 5–21.

Cheng, K. (2008a). Geometry and navigation. In M. E. Jefferies & W. K. Yeap (Eds.), *Robotics and cognitive approaches to spatial mapping* (pp. 145–161). Berlin, Germany: Springer-Verlag.

Cheng, K. (2008b). Whither geometry? Troubles of the geometric module. *Trends in Cognitive Sciences, 12*(9), 355–361.

Cheung, A., Stuerzl, W., Zeil, J., & Cheng, K. (2008). The information content of panoramic images II: View-based navigation in nonrectangular experimental arenas. *Journal of Experimental Psychology-Animal Behavior Processes, 34*(1), 15–30.

Chirikjian, G. S., & Burdick, J. W. (1995). The kinematics of hyper-redundant robot locomotion. *IEEE Transactions on Robotics and Automation, 11*(6), 781–793.

Chomsky, N. (1980). *Rules and representations.* New York, NY: Columbia University Press.

Clague, K., Agullo, M., & Hassing, L. C. (2002). *LEGO Software Power Tools.* Rockland, MA: Syngress Publishing.

Clancey, W. J. (1997). *Situated cognition.* Cambridge, UK: Cambridge University Press.

Clark, A. (1989). *Microcognition.* Cambridge, MA: MIT Press.

Clark, A. (1993). *Associative engines.* Cambridge, MA: MIT Press.

Clark, A. (1997). *Being there: Putting brain, body, and world together again.* Cambridge, MA: MIT Press.

Clark, A. (1999). An embodied cognitive science? *Trends in Cognitive Sciences, 3*(9), 345–351.

Clark, A. (2003). *Natural-born cyborgs.* Oxford, UK; New York, NY: Oxford University Press.

Clark, A. (2008). *Supersizing the mind: Embodiment, action, and cognitive extension.* Oxford, UK; New York, NY: Oxford University Press.

Clark, A., & Chalmers, D. (1998). The extended mind (Active externalism). *Analysis, 58*(1), 7–19.

Clegg, B. (2007). *The man who stopped time.* Washington, D.C.: Joseph Henry Press.

Collins, S. H., Ruina, A., Tedrake, R., & Wisse, M. (2005). Efficient bipedal robots based on passive-dynamic walkers. *Science, 307*(5712), 1082–1085.

Collins, S. H., Wisse, M., & Ruina, A. (2001). A three-dimensional passive-dynamic walking robot with two legs and knees. *International Journal of Robotics Research, 20*(7), 607–615.

Conrad, R. (1964). Information, acoustic confusion, and memory span. *British Journal of Psychology, 55*, 429–432.

Conway, F., & Siegelman, J. (2005). *Dark hero of the information age: In search of Norbert Wiener, the father of cybernetics.* New York, NY: Basic Books.

Cooper, R. (1977). Obituary, W. Grey Walter. *Nature, 268*, 383–384.

Copland, A. (1939). *What to listen for in music.* New York, NY; London, UK: Whittlesey House, McGraw-Hill Book Company.

Cottingham, J. (1978). 'A brute to the brutes?': Descartes' treatment of animals. *Philosophy, 53*(206), 551–559.

Couzin, I. D., & Franks, N. R. (2003). Self-organized lane formation and optimized traffic flow in army ants. *Proceedings of the Royal Society of London Series B-Biological Sciences, 270*(1511), 139–146.

Craik, K. J. M. (1943). *The nature of explanation.* Cambridge, UK: Cambridge University Press.

Cummins, R. (1975). Functional analysis. *Journal of philosophy, 72*, 741–760.

Cummins, R. (1983). *The nature of psychological explanation.* Cambridge, MA: MIT Press.

Curtis, P. D., & Jensen, P. G. (2004). Habitat features affecting beaver occupancy along roadsides in New York state. *Journal of Wildlife Management, 68*(2), 278–287.

Cybenko, G. (1989). Approximation by superpositions of a sigmoidal function. *Mathematics of Control, Signals, and Systems 2*, 303–314.

Danckert, J., & Rossetti, Y. (2005). Blindsight in action: What can the different sub-types of blindsight tell us about the control of visually guided actions? *Neuroscience and Biobehavioral Reviews, 29*(7), 1035–1046.

Dawkins, M. S. (1993). *Through our eyes only? The search for animal consciousness.* Oxford, UK: W. H. Freeman.

Dawson, M. R. W. (1998). *Understanding cognitive science.* Oxford, UK: Blackwell.

Dawson, M. R. W. (2004). *Minds and machines: Connectionism and psychological modeling.* Malden, MA: Blackwell Pub.

Dawson, M. R. W. (2009). Review of Philip Robbins and Murat Aydede, The Cambridge handbook of situated cognition. *Canadian Psychology, in press.*

Dawson, M. R. W., Boechler, P. M., & Valsangkar-Smyth, M. (2000). Representing space in a PDP network: Coarse allocentric coding can mediate metric and nonmetric spatial judgements. *Spatial Cognition and Computation, 2*, 181–218.

Dawson, M. R. W., Kelly, D. M., Spetch, M. L., & Dupuis, B. (2008). Learning about environmental geometry: A flaw in Miller and Shettleworth's (2007) operant model. *Journal of Experimental Psychology-Animal Behavior Processes, 34*(3), 415–418.

Dawson, M. R. W., & Zimmerman, C. (2003). Interpreting the internal structure of a connectionist model of the balance scale task. *Brain & Mind, 4*, 129–149.

de Latil, P. (1956). *Thinking by machine: A study of cybernetics.* London, UK: Sidgwick & Jackson.

Delcomyn, F. (2004). Insect walking and robotics. *Annual Review of Entomology, 49*, 51–70.

Deneubourg, J. L., & Goss, S. (1989). Collective patterns and decision-making. *Ethology Ecology & Evolution, 1*(4), 295–311.

Deneubourg, J. L., Goss, S., Franks, N., Sendova-Franks, A., Detrain, C., & Chretien, L. (1991). *The dynamics of collective sorting robot-like ants and ant-like robots. Proceedings of the first international conference on simulation of adaptive behavior (From animals to animats)* (pp. 356–363). Paris, France: MIT Press.

Dennett, D. C. (1987). *The Intentional Stance.* Cambridge, MA: MIT Press.

Dennett, D. C. (1991). *Consciousness explained.* Boston, MA: Little, Brown.

Dennett, D. C. (2005). *Sweet dreams: Philosophical obstacles to a science of consciousness.* Cambridge, MA: MIT Press.

Descartes, R. (1637/1960). *Discourse on method and meditations.* Indianapolis, IN: Bobbs-Merrill.

Detrain, C., & Deneubourg, J. L. (2006). Self-organized structures in a superorganism: Do ants "behave" like molecules? *Physics of Life Reviews, 3*(3), 162–187.

Dickinson, M. H., Farley, C. T., Full, R. J., Koehl, M. A. R., Kram, R., & Lehman, S. (2000). How animals move: An integrative view. *Science, 288*, 100–106.

Devlin, K. (1996). Good-bye Descartes? *Mathematics Magazine, 69*, 344–349.

Dorigo, M., & Gambardella, L. M. (1997). Ant colonies for the travelling salesman problem. *Biosystems, 43*(2), 73–81.

Dourish, P. (2001). *Where the action is: The foundations of embodied interaction.* Cambridge, MA: MIT Press.

Downing, H. A., & Jeanne, R. L. (1986). Intraspecific and interspecific variation in nest architecture in the paper wasp *Polistes* (Hymenoptera, Vespidae). *Insectes Sociaux, 33*(4), 422–443.

Downing, H. A., & Jeanne, R. L. (1988). Nest construction by the paper wasp, *Polistes*: A test of stigmergy theory. *Animal Behaviour, 36*, 1729–1739.

Drury, W. H., & Keith, J. A. (1962). Radar studies of songbird migration in eastern New England. *The Ibis, 104*, 449–489.

Dubner, R., & Zeki, S. M. (1971). Response properties and receptive fields of cells in an anatomically defined region of the superior temporal sulcus in the monkey. *Brain Research, 35*, 528–532.

Dudek, D. M., & Full, R. J. (2006). Passive mechanical properties of legs from running insects. *Journal of Experimental Biology, 209*(8), 1502–1515.

Endo, G., Togawa, K., & Hirose, S. (1999). *Study on self-contained and terrain adaptive active chord mechanism.* Paper presented at the IEEE/RSJ International Conference on Intelligent Robots and Systems.

Evans, H. E. (1966). Behavior patterns of solitary wasps. *Annual Review of Entomology, 11*, 123–&.

Evans, H. E., & West-Eberhard, M. J. (1970). *The wasps.* Ann Arbor: University of Michigan Press.

Evans, M. A., & Evans, H. E. (1970). *William Morton Wheeler, biologist.* Cambridge, MA: Harvard University Press.

Fabre, J. H. (1915). *The hunting wasps.* New York, NY: Dodd, Mead and Company.

Fabre, J. H. (1919). *The mason wasps.* New York, NY: Dodd, Mead and company.

Feldman, J. A., & Ballard, D. H. (1982). Connectionist models and their properties. *Cognitive Science, 6*, 205–254.

Feynman, R. P. (1985). *"Surely you're joking, Mr. Feynman!": Adventures of a curious character.* New York, NY: W. W. Norton.

Filliat, D., & Meyer, J. A. (2003). Map-based navigation in mobile robots: I. A review of localization strategies. *Cognitive Systems Research, 4*, 243–282.

Finney, B. R. (1976). *Pacific navigation and voyaging.* Wellington, New Zealand: Polynesian Society.

Flavell, J. H. (1963). *The developmental psychology of Jean Piaget.* Princeton, NJ: Van Nostrand.

Fodor, J. A. (1968). *Psychological explanation: An introduction to the philosophy of psychology.* New York, NY: Random House.

Fodor, J. A. (1975). *The language of thought.* Cambridge, MA: Harvard University Press.

Fodor, J. A. (1980). Methodological solipsism considered as a research strategy in cognitive psychology. *Behavioral and Brain Sciences, 3*(1), 63–73.

Fodor, J. A. (1983). *The modularity of mind.* Cambridge, MA: MIT Press.

Fodor, J. A., & Pylyshyn, Z. W. (1988). Connectionism and cognitive architecture. *Cognition, 28*, 3–71.

Fogel, D. B. (1988). An evolutionary approach to the traveling salesman problem. *Biological Cybernetics, 60*(2), 139–144.

Fong, T., Nourbakhsh, I., & Dautenhahn, K. (2003). A survey of socially interactive robots. *Robotics and Autonomous Systems, 42*(3–4), 143–166.

Forbes, P. (2006). *The gecko's foot* (1st American ed.). New York, NY: W. W. Norton & Co.

Franks, N. R., & Sendova-Franks, A. B. (1992). Brood sorting by ants: Distributing the workload over the work surface. *Behavioral Ecology and Sociobiology, 30*(2), 109–123.

Franks, N. R., Sendova-Franks, A. B., & Anderson, C. (2001). Division of labour within teams of New World and Old World army ants. *Animal Behaviour, 62*, 635–642.

Franks, N. R., Wilby, A., Silverman, B. W., & Tofts, C. (1992). Self-organizing nest construction in ants: Sophisticated building by blind bulldozing. *Animal Behaviour, 44*(2), 357–375.

Freud, S. (1919/1976). The uncanny. *New Literary History, 7*(3), 619–645.

Frisch, K. v. (1966). *The dancing bees: An account of the life and senses of the honey bee* (2nd ed.). London, UK: Methuen.

Frisch, K. v. (1967). *The dance language and orientation of bees.* Cambridge, MA: Belknap Press of Harvard University Press.

Frisch, K. v. (1974). Decoding the language of the bee. *Science, 185,* 663–668.

Full, R. J., Earls, K., Wong, M., & Caldwell, R. (1993). Locomotion like a wheel. *Nature, 365*(6446), 495–495.

Full, R. J., & Tu, M. S. (1991). Mechanics of a rapid running insect: Two-, four-, and six-legged locomotion. *Journal of Experimental Biology, 156,* 215–231.

Gaines, J. R. (2005). *Evening in the palace of reason: Bach meets Frederick the Great in the Age of Enlightenment.* London, UK; New York, NY: Fourth Estate.

Gallistel, C. R. (1990). *The organization of learning.* Cambridge, MA: MIT Press.

Gallup, G. G. (1970). Chimpanzees: Self-recognition. *Science, 167,* 86–87.

Gardner, H. (1984). *The mind's new science.* New York, NY: Basic Books.

Gardner, M. (1982). *Logic machines and diagrams* (2nd ed.) Chicago, IL: The University of Chicago Press.

Gasperi, M., Hurbain, P., & Hurbain, I. (2007). *Extreme NXT: Extending the LEGO® MINDSTORMS® NXT to the next level.* Berkeley, CA; New York, NY: Apress.

Genesereth, M. R., & Nilsson, N. J. (1987). *Logical foundations of artificial intelligence.* Los Altos, CA: Morgan Kaufmann.

Gerkey, B. P., & Mataric, M. J. (2002). Sold!: Auction methods for multirobot coordination. *IEEE Transactions on Robotics and Automation, 18*(5), 758–768.

Gerkey, B. P., & Mataric, M. J. (2004). A formal analysis and taxonomy of task allocation in multi-robot systems. *International Journal of Robotics Research, 23*(9), 939–954.

Gibbs, R. W. (2006). *Embodiment and cognitive science.* Cambridge, UK: Cambridge University Press.

Gibson, J. J. (1966). *The senses considered as perceptual systems.* Boston, MA: Houghton Mifflin.

Gibson, J. J. (1979). *The ecological approach to visual perception.* Boston, MA: Houghton Mifflin.

Gladwin, T. (1970). *East is a big bird: Navigation and logic on Puluwat Atoll.* Cambridge, MA: Harvard University Press.

Glass, P. (1987). *Music by Philip Glass* (1st ed.). New York, NY: Harper & Row.

Goldman, A. I. (1993). *Readings in philosophy and cognitive science.* Cambridge, MA: MIT Press.

Goldstone, R. L., & Janssen, M. A. (2005). Computational models of collective behavior. *Trends in Cognitive Sciences, 9*(9), 424–430.

Goodale, M. A. (1988). Modularity in visuomotor control: From input to output. In Z. W. Pylyshyn (Ed.), *Computational processes in human vision: An interdisciplinary perspective* (pp. 262–285). Norwood, NJ: Ablex.

Goodale, M. A. (1990). *Vision and action: The control of grasping.* Norwood, NJ: Ablex.

Goodale, M. A. (1995). The cortical organization of visual perception and visuomotor control. In S. M. Kosslyn & D. N. Osherson (Eds.), *An invitation to cognitive science: Visual cognition* (Vol. 2, pp. 167–213). Cambridge, MA: MIT Press.

Goodale, M. A., & Humphrey, G. K. (1998). The objects of action and perception. *Cognition, 67,* 181–207.

Goodale, M. A., Milner, A. D., Jakobson, L. S., & Carey, D. P. (1991). A neurological dissociation between perceiving objects and grasping them. *Nature, 349*(6305), 154–156.

Goss, S., Aron, S., Deneubourg, J. L., & Pasteels, J. M. (1989). Self-organized shortcuts in the Argentine ant. *Naturwissenschaften, 76*(12), 579–581.

Grant, M. J. (2001). *Serial music, serial aesthetics: Compositional theory in post-war Europe.* New York, NY: Cambridge University Press.

Grasse, P. P. (1959). La reconstruction du nid et les coordinations interindividuelles chez Bellicositermes natalensis et Cubitermes sp. la théorie de la stigmergie: Essai d'interprétation du comportement des termites constructeurs. *Insectes Sociaux, 6*(1), 41–80.

Green, E. A. H., & Malko, N. A. (1975). *The conductor and his score.* Englewood Cliffs, NJ: Prentice-Hall.

Greeno, J. G., & Moore, J. L. (1993). Situativity and symbols: Response to Vera and Simon. *Cognitive Science, 17,* 49–59.

Gregory, R. L. (1970). *The intelligent eye.* London, UK: Weidenfeld & Nicolson.

Grenville, B. (2001). *The uncanny: Experiments in cyborg culture.* Vancouver, BC: Vancouver Art Gallery; Arsenal Pulp Press.

Grey Walter, W. (1950a). An electro-mechanical animal. *Dialectica, 4*(3), 206–213.

Grey Walter, W. (1950b). An imitation of life. *Scientific American, 182*(5), 42–45.

Grey Walter, W. (1951). A machine that learns. *Scientific American, 184*(8), 60–63.

Grey Walter, W. (1963). *The living brain.* New York, NY: W. W. Norton & Co.

Griffiths, P. (1994). *Modern music: A concise history* (Rev. ed.). New York, NY: Thames and Hudson.

Griffiths, P. (1995). *Modern music and after.* Oxford, UK; New York, NY: Oxford University Press.

Gutin, G., & Punnen, A. P. (2002). *The traveling salesman problem and its variations.* Dordrecht, The Netherlands; Boston: Kluwer Academic Publishers.

Haberlandt, K. (1994). *Cognitive psychology.* Boston, MA: Allyn and Bacon.

Harkleroad, L. (2006). *The math behind the music.* Cambridge, UK; New York, NY: Cambridge University Press.

Hartman, G., & Tornlov, S. (2006). Influence of watercourse depth and width on dam-building behaviour by Eurasian beaver (*Castor fiber*). *Journal of Zoology, 268*(2), 127–131.

Hastie, R. (2001). Problems for judgment and decision making. *Annual Review of Psychology, 52,* 653–683.

Haugeland, J. (1985). *Artificial intelligence: The very idea.* Cambridge, MA: MIT Press.

Hayes-Roth, B. (1985). A blackboard architecture for control. *Artificial Intelligence, 26*(3), 251–321.

Hayles, N. K. (1999). *How we became posthuman: Virtual bodies in cybernetics, literature, and informatics.* Chicago, IL: University of Chicago Press.

Hayward, R. (2001). The tortoise and the love-machine: Grey Walter and the politics of electroencephalography. *Science in Context, 14*(4), 615–641.

Healy, S. (1998). *Spatial Representation In Animals.* Oxford, UK: Oxford University Press.

Heidegger, M. (1962). *Being and time.* New York, NY: Harper.

Helmholtz, H. v., & Ellis, A. J. (1954). *On the sensations of tone as a physiological basis for the theory of music* (2d English ed.). New York, NY: Dover Publications.

Henle, M. (1977). The influence of Gestalt psychology in America. *Annals of the New York Academy of Sciences, 291*(1), 3–12.

Hermer, L., & Spelke, E. S. (1994). A geometric process for spatial reorientation in young children. *Nature, 370*(6484), 57–59.

Hess, R. H., Baker, C. L., & Zihl, J. (1989). The "motion-blind" patient: Low-level spatial and temporal filters. *The Journal of Neuroscience, 9,* 1628–1640.

Hildesheimer, W. (1983). *Mozart* (1st Vintage Books ed.). New York, NY: Vintage Books.

Hinchey, M. G., Sterritt, R., & Rouff, C. (2007). Swarms and swarm intelligence. *Computer, 40*(4), 111–113.

Hingston, R. W. (1929). *Instinct and intelligence*. New York, NY: The Macmillan Company.

Hingston, R. W. (1933). Instinct and intelligence in insects. *Journal of Personality, 1,* 129–136.

Hirose, M., & Ogawa, K. (2007). Honda humanoid robots development. *Philosophical Transactions of the Royal Society a-Mathematical Physical and Engineering Sciences, 365*(1850), 11–19.

Hirose, S. (1993). *Biologically inspired robots: Snake-like locomotors and manipulators.* Oxford, UK; New York, NY: Oxford University Press.

Hirose, S., Fukushima, E. F., & Tsukagoshi, S. (1995). Basic steering control methods for the articulated body mobile robot. *IEEE Control Systems Magazine, 15*(1), 5–14.

Hirose, S., & Mori, M. (2004). *Biologically inspired snake-like robots.* Paper presented at the IEEE International Conference On Robotics And Biomimetics, Shenyang, China.

Hirose, S., & Morishima, A. (1990). Design and control of a mobile robot with an articulated body. *International Journal of Robotics Research, 9*(2), 99–114.

Hjelmfelt, A., Weinberger, E. D., & Ross, J. (1991). Chemical implementation of neural networks and Turing machines. *Proceedings of the National Academy of Sciences of the United States of America, 88*(24), 10983–10987.

Hodges, A. (1983). *Alan Turing: The enigma of intelligence.* London, UK: Unwin Paperbacks.

Hofstadter, D. (1995). *Fluid concepts and creative analogies.* New York, NY: Basic Books.

Hofstadter, D. R. (1979). *Godel, Escher, Bach: An eternal golden braid.* New York, NY: Basic Books.

Holland, J. H. (1992). *Adaptation in natural and artificial systems.* Cambridge, MA: MIT Press.

Holland, J. H. (1998). *Emergence.* Reading, MA: Perseus Books.

Holland, O. (2001). From the imitation of life to machine consciousness. In T. Gomi (Ed.), *Evolutionary robotics* (pp. 1–38). New York, NY: Springer-Verlag.

Holland, O. (2003a). Exploration and high adventure: The legacy of Grey Walter. *Philosophical Transactions of the Royal Society of London Series a-Mathematical Physical and Engineering Sciences, 361*(1811), 2085–2121.

Holland, O. (2003b). The first biologically inspired robots. *Robotica, 21,* 351–363.

Holland, O., & Melhuish, C. (1999). Stigmergy, self-organization, and sorting in collective robotics. *Artificial Life, 5,* 173–202.

Hopfield, J. J. (1982). Neural networks and physical systems with emergent collective computational abilities. *Proceedings of the National Academy of Sciences, 79,* 2554–2558.

Hopfield, J. J. (1984). Neurons with graded response have collective computational properties like those of two state neurons. *Proceedings of the National Academy of Sciences USA, 81,* 3008–3092.

Hopfield, J. J., & Tank, D. W. (1985). "Neural" computation of decisions in optimization problems. *Biological Cybernetics, 52*(3), 141–152.

Horchler, A. D., Reeve, R. E., Webb, B., & Quinn, R. D. (2004). Robot phonotaxis in the wild: A biologically inspired approach to outdoor sound localization. *Advanced Robotics, 18*(8), 801–816.

Hornik, M., Stinchcombe, M., & White, H. (1989). Multilayer feedforward networks are universal approximators. *Neural Networks, 2,* 359–366.

Hurley, S. (2001). Perception and action: Alternative views. *Synthese, 129*(1), 3–40.

Hutchins, E. (1995). *Cognition in the wild.* Cambridge, MA: MIT Press.

Ichbiah, D. (2005). *Robots: From science fiction to technological revolution*. New York, NY: Harry N. Abrams.

Ingle, D. (1973). Two visual systems in the frog. *Science, 181*(4104), 1053–1055.

Inhelder, B., & Piaget, J. (1958). *The growth of logical thinking from childhood to adolescence*. New York, NY: Basic Books.

Inhelder, B., & Piaget, J. (1964). *The early growth of logic in the child*. New York, NY: Harper & Row.

Isacoff, S. (2001). *Temperament: The idea that solved music's greatest riddle* (1st ed.). New York, NY: Alfred A. Knopf.

Ito, K., & Murai, R. (2008). Snake-like robot for rescue operations: Proposal of a simple adaptive mechanism designed for ease of use. *Advanced Robotics, 22*(6–7), 771–785.

Jakobson, L. S., Archibald, Y. M., Carey, D. P., & Goodale, M. A. (1991). A kinematic analysis of reaching and grasping movements in a patient recovering from optic ataxia. *Neuropsychologia, 29*(8), 803–&.

Jansen, T. (2007). *The great pretender*. Rotterdam, The Netherlands: 010 Publishers.

Jeanne, R. L. (1996). Regulation of nest construction behaviour in *Polybia occidentalis*. *Animal Behaviour, 52*, 473–488.

Jefferies, M. E., & Yeap, W. K. (2008). *Robotics and cognitive approaches to spatial mapping*. Berlin, Germany; New York, NY: Springer-Verlag.

Johnson-Laird, P. N. (1983). *Mental models*. Cambridge, MA: Harvard University Press.

Johnson, S. (2001). *Emergence*. New York, NY: Scribner.

Jones, J. E., & Kamil, A. C. (2001). The use of relative and absolute bearings by Clark's nutcrackers, *Nucifraga columbiana*. *Animal Learning and Behavior, 29*(2), 120–132.

Jordà, S., Geiger, G., Alonso, M., & Kaltenbrunner, M. (2007). *The reacTable: Exploring the synergy between live music performance and tabletop tangible interfaces*. Paper presented at the Proceedings of the first international conference on "Tangible and Embedded Interaction" (TEI07), Baton Rouge, Louisiana.

Josephson, M. (1961). *Edison*. New York, NY: McGraw Hill.

Kalish, D., & Montague, R. (1964). *Logic: Techniques of formal reasoning*. New York, NY: Harcourt, Brace, & World.

Kaltenbrunner, M., Jordà, S., Geiger, G., & Alonso, M. (2007). *The reacTable*: A collaborative musical instrument*. Paper presented at the Proceedings of the Workshop on "Tangible Interaction in Collaborative Environments" (TICE), at the 15th International IEEE Workshops on Enabling Technologies (WETICE 2006), Manchester, UK.

Kamil, A. C., & Cheng, K. (2001). Way-finding and landmarks: The multiple-bearings hypothesis. *Journal of Experimental Biology, 204*(1), 103–113.

Kamil, A. C., & Jones, J. E. (1997). The seed-storing corvid Clark's nutcracker learns geometric relationships among landmarks. *Nature, 390*(6657), 276–279.

Kamil, A. C., & Jones, J. E. (2000). Geometric rule learning by Clark's nutcrackers (*Nucifraga columbiana*). *Journal of Experimental Psychology-Animal Behavior Processes, 26*(4), 439–453.

Karsai, I. (1999). Decentralized control of construction behavior in paper wasps: An overview of the stigmergy approach. *Artificial Life, 5*, 117–136.

Karsai, I., & Penzes, Z. (1998). Nest shapes in paper wasps: Can the variability of forms be deduced from the same construction algorithm? *Proceedings of the Royal Society of London Series B-Biological Sciences, 265*(1402), 1261–1268.

Karsai, I., & Wenzel, J. W. (2000). Organization and regulation of nest construction behavior in Metapolybia wasps. *Journal of Insect Behavior, 13*(1), 111–140.

Kelly, D. M., Spetch, M. L., & Heth, C. D. (1998). Pigeons' (*Columba livia*) encoding of geometric and featural properties of a spatial environment. *Journal of Comparative Psychology, 112*(3), 259–269.

Khinchin, A. I. A. (1957). *Mathematical foundations of information theory* (New Dover ed.). New York, NY: Dover Publications.

Kirkpatrick, S., Gelatt, C. D., & Vecchi, M. P. (1983). Optimization by simulated annealing. *Science, 220*(4598), 671–680.

Kivy, P. (1991). *Sound and semblance: Reflections on musical representation.* Ithaca, NY: Cornell University Press.

Koditschek, D. E., Full, R. J., & Buehler, M. (2004). Mechanical aspects of legged locomotion control. *Arthropod Structure & Development, 33*(3), 251–272.

Koenig, G. M. (1999). PROJECT 1 Revisited. On the Analysis and Interpretation of PR1 Tables. In J. Tabor (Ed.), *Otto Laske: Navigating new horizons* (pp. 53–72). Westport, CT: Greenwood Press.

Koffka, K. (1935). *Principles of Gestalt psychology.* New York, NY: Harcourt, Brace & World.

Köhler, W. (1947). *Gestalt psychology: An introduction to new concepts in modern psychology.* New York, NY: Liveright Publishing Corporation

Krumhansl, C. L. (1990). *Cognitive foundations of musical pitch.* New York, NY: Oxford University Press.

Kube, C. R., & Bonabeau, E. (2000). Cooperative transport by ants and robots. *Robotics and Autonomous Systems, 30*, 85–101.

Kube, C. R., & Zhang, H. (1994). Collective robotics: From social insects to robots. *Adaptive Behavior, 2*, 189–218.

Laporte, G., & Osman, I. H. (1995). Routing problems: A bibliography. *Annals of Operations Research, 61*, 227–262.

Lawler, E. L. (1985). *The traveling salesman problem: A guided tour of combinatorial optimization.* Chichester, West Sussex, UK; New York, NY: Wiley.

Lee, E. M. (1916). *The story of the symphony.* London, UK: Walter Scott Publishing Co., Ltd.

Lee, J. Y., Shin, S. Y., Park, T. H., & Zhang, B. T. (2004). Solving traveling salesman problems with DNA molecules encoding numerical values. *Biosystems, 78*(1–3), 39–47.

Leibovic, K. N. (1969). *Information processing in the nervous system.* New York, NY: Springer-Verlag.

Lepore, E., & Pylyshyn, Z. W. (1999). *What is cognitive science?* Malden, MA: Blackwell.

Levesque, H. J., & Lakemeyer, G. (2000). *The logic of knowledge bases.* Cambridge, MA: MIT Press.

Lévi-Strauss, C. (1966). *The savage mind.* Chicago, IL: University of Chicago Press.

Levin, I. (2002). *The Stepford wives* (1st Perennial ed.). New York, NY: Perennial.

Lindsay, P. H., & Norman, D. A. (1972). *Human information processing.* New York, NY: Academic Press.

Lippmann, R. P. (1987). An introduction to computing with neural nets. *IEEE ASSP magazine, April*, 4–22.

Livingstone, M., & Hubel, D. (1988). Segregation of form, color, movement and depth: Anatomy, physiology, and perception. *Science, 240*, 740–750.

Lobontiu, N., Goldfarb, M., & Garcia, E. (2001). A piezoelectric-driven inchworm locomotion device. *Mechanism and Machine Theory, 36*(4), 425–443.

Luce, R. D. (1999). Where is mathematical modeling in psychology headed? *Theory & Psychology, 9*, 723–737.

Lund, H., & Miglino, O. (1998). Evolving and breeding robots. In P. Husbands & J. A. Meyer (Eds.), *Evolutionary robotics. First European workshop, EvoRobot'98. Paris, France, April 16–17, 1998; proceedings (pp. 192–210)*. Heidelberg; Berlin, Germany: Springer-Verlag.

MacDorman, K. F., & Ishiguro, H. (2006). The uncanny advantage of using androids in cognitive and social science research. *Interaction Studies, 7*(3), 297–337.

MacIver, M. A. (2008). Neuroethology: From morphological computation to planning. In P. Robbins & M. Aydede (Eds.), *The Cambridge handbook of situated cognition* (pp. 480–504). New York, NY: Cambridge University Press.

MacKay, D. (1969). *Information, mechanism and meaning*. Cambridge, MA: MIT Press.

Marr, D. (1976). Early processing of visual information. *Philosophical Transactions of the Royal Society of London, 275*, 483–524.

Marr, D. (1982). *Vision*. San Francisco, CA: W. H. Freeman.

Marr, D., & Hildreth, E. (1980). Theory of edge detection. *Proceedings of the Royal Society of London, B207*, 187–217.

Marr, D., & Ullman, S. (1981). Directional selectivity and its use in early visual processing. *Proceedings of the Royal Society of London, B211*, 151–180.

Martin, F. G., Mikhak, B., Resnick, M., Silverman, S., & Berg, R. (2000). To mindstorms and beyond: Evolution of a construction kit for magical machines. In A. Druin & J. Hendler (Eds.), *Robots for kids: Exploring new technologies for learning* (pp. 9–33). San Francisco, CA: Morgan Kaufmann Publishers Inc.

Mataric, M. J. (1998). Using communication to reduce locality in distributed multiagent learning. *Journal of Experimental and Theoretical Artificial Intelligence, 10*(3), 357–369.

Matsuno, F. (2002). A mobile robot for collecting disaster information and a snake robot for searching. *Advanced Robotics, 16*(6), 517–520.

Maunsell, J. H. R., & Newsome, W. T. (1987). Visual processing in monkey extrastriate cortex. *Annual Review of Neuroscience, 10*, 363–401.

Maunsell, J. H. R., & van Essen, D. C. (1983). The connections of the middle temporal visual area (MT) and their relationship to a cortical hierarchy in the macaque monkey. Journal of *Neuroscience, 3*, 2563–2586.

McClelland, J. L., & Rumelhart, D. E. (1986). *Parallel distributed processing, Vol. 2*. Cambridge, MA: MIT Press.

McClelland, J. L., Rumelhart, D. E., & Hinton, G. E. (1986). The appeal of parallel distributed processing. In D. Rumelhart & J. McClelland (Eds.), *Parallel distributed processing Vol. 1* (pp. 3–44). Cambridge, MA: MIT Press.

McCulloch, W. S., & Pitts, W. (1943). A logical calculus of the ideas immanent in nervous activity. *Bulletin of Mathematical Biophysics, 5*, 115–133.

McGeer, T. (1990a). Passive dynamic walking. *International Journal of Robotics Research, 9*(2), 62–82.

McGeer, T. (1990b). *Passive walking with knees*. Paper presented at the Proceedings of the IEEE International Conference On Robotics And Automation, Cincinnati OH.

McLuhan, M. (1994). *Understanding media: The extensions of man* (1st MIT Press ed.). Cambridge, MA: MIT Press.

Medler, D. A. (1998). A brief history of connectionism. *Neural Computing Surveys, 1*, 18–72.

Melhuish, C., Sendova-Franks, A. B., Scholes, S., Horsfield, I., & Welsby, F. (2006). Ant-inspired sorting by robots: The importance of initial clustering. *Journal of the Royal Society Interface, 3*(7), 235–242.

Mellers, B. A., Schwartz, A., & Cooke, A. D. J. (1998). Judgment and decision making. *Annual Review of Psychology, 49*, 447–477.

Menzel, P., D'Aluisio, F., & Mann, C. C. (2000). *Robo sapiens: Evolution of a new species.* Cambridge, MA: MIT Press.

Meyer, D. E., Glass, J. M., Mueller, S. T., Seymour, T. L., & Kieras, D. E. (2001). Executive-process interactive control: A unified computational theory for answering 20 questions (and more) about cognitive ageing. *European Journal of Cognitive Psychology, 13*(1–2), 123–164.

Meyer, D. E., & Kieras, D. E. (1997a). A computational theory of executive cognitive processes and multiple-task performance .1. Basic mechanisms. *Psychological Review, 104*(1), 3–65.

Meyer, D. E., & Kieras, D. E. (1997b). A computational theory of executive cognitive processes and multiple-task performance .2. Accounts of psychological refractory-period phenomena. *Psychological Review, 104*(4), 749–791.

Meyer, D. E., & Kieras, D. E. (1999). Precis to a practical unified theory of cognition and action: Some lessons from EPIC computational models of human multiple-task performance. *Attention and Performance XVII, 17*, 17–88.

Meyer, L. B. (1956). *Emotion and meaning in music.* Chicago, Il : University of Chicago Press.

Milford, M. J. (2008). *Robot navigation from nature: Simultaneous localisation, mapping, and path planning based on hippocampal models.* Berlin, Germany: Springer-Verlag.

Miller, G. A. (2003). The cognitive revolution: A historical perspective. *Trends in Cognitive Sciences, 7*(3), 141–144.

Miller, N. Y., & Shettleworth, S. J. (2007). Learning about environmental geometry: An associative model. *Journal of Experimental Psychology-Animal Behavior Processes, 33*, 191–212.

Miller, N. Y., & Shettleworth, S. J. (2008). An associative model of geometry learning: A modified choice rule. *Journal of Experimental Psychology-Animal Behavior Processes, 34*(3), 419–422.

Minsky, M. (1972). *Computation: Finite and infinite machines.* London, UK: Prentice-Hall International.

Minsky, M. (1985). *The society of mind.* New York, NY: Simon & Schuster.

Minsky, M. (2006). *The emotion machine: Commensense thinking, artificial intelligence, and the future of the human mind.* New York, NY: Simon & Schuster.

Mitchell, M. (1996). *An introduction to genetic algorithms.* Cambridge, MA: MIT Press.

Mlinar, E. J., & Goodale, M. A. (1984). Cortical and tectal control of visual orientation in the gerbil: Evidence for parallel channels. *Experimental Brain Research, 55*(1), 33–48.

Mondada, F., & Floreano, D. (1995). Evolution of neural control structures: Some experiments on mobile robots. *Robotics and Autonomous Systems, 16*(2–4), 183–195.

Mondada, F., Franzi, E., & Ienne, P. (1994). Mobile robot miniaturisation: A tool for investigation in control algorithms In M. Thoma (Ed.), *Third International Symposium on Experimental Robotics* (pp. 501–513). Berlin, Germany: Springer-Verlag.

Moravec, H. (1999). *Robot.* New York, NY: Oxford University Press.

Morgan, L. H. (1868/1986). *The american beaver: A classic of natural history and ecology.* New York, NY: Dover Publications.

Mori, M. (1970). Bukimi no tani [the uncanny valley]. *Energy, 7*, 33–35.

Mouritsen, H. (2001). Navigation in birds and other animals. *Image and Vision Computing, 19*(11), 713–731.

Müller-Schwarze, D., & Sun, L. (2003). *The beaver: Natural history of a wetlands engineer.* Ithaca, NY: Comstock Publishing Associates.

Muybridge, E. (1887/1957). *Animals in motion.* New York, NY: Dover Publications.

Negroponte, N. (1995). *Being digital.* New York, NY: Vintage Books.

Newell, A. (1973). Production systems: Models of control structures. In W. G. Chase (Ed.), *Visual information processing* (pp. 463–526). New York, NY: Academic Press.

Newell, A. (1980). Physical symbol systems. *Cognitive Science, 4,* 135–183.

Newell, A. (1990). *Unified theories of cognition.* Cambridge, MA: Harvard University Press.

Newell, A., & Simon, H. A. (1961). Computer simulation of human thinking. *Science, 134*(349), 2011–2017.

Newell, A., & Simon, H. A. (1972). *Human problem solving.* Englewood Cliffs, NJ: Prentice-Hall.

Nilsson, N. J. (1984). *Shakey the robot.* Menlo Park, CA: Stanford Research Institute.

Nishikawa, K., Biewener, A. A., Aerts, P., Ahn, A. N., Chiel, H. J., Daley, M. A., et al. (2007). Neuromechanics: An integrative approach for understanding motor control. *Integrative and Comparative Biology, 47*(1), 16–54.

Nishiwaki, K., Kuffner, J., Kagami, S., Inaba, M., & Inoue, H. (2007). The experimental humanoid robot H7: A research platform for autonomous behaviour. *Philosophical Transactions of the Royal Society a-Mathematical Physical and Engineering Sciences, 365*(1850), 79–107.

Nolfi, S. (2002). Power and limits of reactive agents. *Neurocomputing, 42,* 119–145.

Nolfi, S., & Floreano, D. (2000). *Evolutionary robotics.* Cambridge, MA: MIT Press.

Norman, D. A. (2002). *The design of everyday things* (1st Basic paperback ed.). New York, NY: Basic Books.

Nyman, M. (1999). *Experimental music: Cage and beyond* (2nd ed.). Cambridge, UK; New York, NY: Cambridge University Press.

Oaksford, M., & Chater, N. (1998). *Rationality in an uncertain world: Essays on the cognitive science of human reasoning.* Hove, East Sussex, UK: Psychology Press.

Ohta, H., Yamakita, M., & Furuta, K. (2001). From passive to active dynamic walking. *International Journal of Robust and Nonlinear Control, 11*(3), 287–303.

O'Keefe, J., & Nadel, L. (1978). *The hippocampus as a cognitive map.* Oxford, UK: Clarendon Press.

Papert, S. (1980). *Mindstorms: Children, computers and powerful ideas.* New York, NY: Basic Books.

Papert, S. (1993). *The children's machine: Rethinking school in the age of the computer.* New York, NY: BasicBooks.

Parker, C. A. C., Zhang, H., & Kube, C. R. (2003). *Blind bulldozing: Multiple robot nest construction.* Paper presented at the Conference on Intelligent Robots and Systems, Las Vegas, Nevada.

Parker, L. E. (1998). ALLIANCE: An architecture for fault tolerant multirobot cooperation. *IEEE Transactions on Robotics and Automation, 14*(2), 220–240.

Parker, L. E. (2001). Evaluating success in autonomous multi-robot teams: Experiences from ALLIANCE architecture implementations. *Journal of Experimental & Theoretical Artificial Intelligence, 13*(2), 95–98.

Pelisson, D., Prablanc, C., Goodale, M. A., & Jeannerod, M. (1986). Visual control of reaching movements without vision of the limb. II. Evidence of fast unconscious processes correcting the trajectory of the hand to the final position of a double-step stimulus. *Experimental Brain Research, 62*(2), 303–311.

Pessin, A., Goldberg, S., & Putnam, H. (1996). *The twin Earth chronicles: Twenty years of reflection on Hilary Putnam's "The meaning of 'meaning'"*. Armonk, NY: M. E. Sharpe.

Petre, M., & Price, B. (2004). Using robotics to motivate 'back door' learning. *Education and Information Technologies, 9*, 147–158.

Pfeifer, R., & Scheier, C. (1999). *Understanding intelligence*. Cambridge, MA: MIT Press.

Phee, L., Accoto, D., Menciassi, A., Stefanini, C., Carrozza, M. C., & Dario, P. (2002). Analysis and development of locomotion devices for the gastrointestinal tract. *IEEE Transactions on Biomedical Engineering, 49*(6), 613–616.

Piaget, J. (1970a). *The child's conception of movement and speed*. London, UK: Routledge & Kegan Paul.

Piaget, J. (1970b). *Psychology and epistemology*. Harmondsworth, UK: Penguin Books.

Piaget, J. (1972). *The child and reality*. Harmondsworth, UK: Penguin Books.

Piaget, J., & Inhelder, B. (1969). *The psychology of the child*. London, UK: Routledge & Kegan Paul.

Piattelli-Palmarini, M. (1994). *Inevitable illusions: How mistakes of reason rule our minds*. New York, NY: Wiley.

Picard, R. W., Papert, S., Bender, W., Blumberg, B., Breazeal, C., Cavallo, D., et al. (2004). Affective learning: A manifesto. *Bt Technology Journal, 22*(4), 253–269.

Pleasants, H. (1955). *The agony of modern music*. New York, NY: Simon and Schuster.

Popper, K. (1978). Natural selection and the emergence of mind. *Dialectica, 32*, 339–355.

Port, R. F., & Van Gelder, T. (1995). *Mind as motion: Explorations in the dynamics of cognition*. Cambridge, MA: MIT Press.

Potter, K. (2000). *Four musical minimalists: La Monte Young, Terry Riley, Steve Reich, Philip Glass*. Cambridge, UK; New York, NY: Cambridge University Press.

Prochnow, D. (2007). *LEGO Mindstorms NXT hacker's guide*. New York, NY: McGraw-Hill.

Punnen, A. P. (2002). The traveling salesman problem: Applications, formulations, and variations. In G. Gutin & A. P. Punnen (Eds.), *The traveling salesman problem and its variations* (pp. 1–28). Dordrecht, The Netherlands; Boston: Kluwer Academic Publishers.

Putnam, H. (1967). Psychological predicates. In W. H. Capitan & D. D. Merrill (Eds.), *Art, mind, and religion* (pp. 37–48). Pittsburgh, PA: University of Pittsburgh Press.

Pylyshyn, Z. W. (1979). Metaphorical imprecision and the "top-down" research strategy. In A. Ortony (Ed.), *Metaphor and thought* (pp. 420–436). Cambridge, UK: Cambridge University Press.

Pylyshyn, Z. W. (1980). Computation and cognition: Issues in the foundations of cognitive science. *Behavioral and Brain Sciences, 3*, 111–169.

Pylyshyn, Z. W. (1981). The imagery debate: Analogue media versus tacit knowledge. *Psychological Review, 88*(1), 16–45.

Pylyshyn, Z. W. (1984). *Computation and cognition*. Cambridge, MA: MIT Press.

Pylyshyn, Z. W. (1987). *The robot's dilemma: The frame problem in artificial intelligence*. Norwood, NJ: Ablex.

Queller, D. C., & Strassmann, J. E. (2002). The many selves of social insects. *Science, 296*(5566), 311–313.

Quinlan, P. (1991). *Connectionism and psychology*. Chicago, IL: University of Chicago Press.

Reddy, M. J. (1979). The conduit metaphor: A case of frame conflict in our language about language. In A. Ortony (Ed.), *Metaphor and thought* (pp. 284–324). Cambridge, UK: Cambridge University Press.

Redish, A. D. (1999). *Beyond the cognitive map*. Cambridge, MA: MIT Press.

Reeve, R., Webb, B., Horchler, A., Indiveri, G., & Quinn, R. (2005). New technologies for testing a model of cricket phonotaxis on an outdoor robot. *Robotics and Autonomous Systems, 51*(1), 41–54.

Reeve, R. E., & Webb, B. H. (2003). New neural circuits for robot phonotaxis. Philosophical Transactions of the Royal Society of London Series a-Mathematical Physical and *Engineering Sciences, 361*(1811), 2245–2266.

Reich, S. (1974). *Writings about music*. Halifax, NS: Press of the Nova Scotia College of Art and Design.

Reich, S. (2002). *Writings on music, 1965–2000*. Oxford, UK; New York, NY: Oxford University Press.

Rescorla, R. A., & Wagner, A. R. (1972). A theory of Pavlovian conditioning: Variations in the effectiveness of reinforcement and nonreinforcement. In A. H. Black & W. F. Prokasy (Eds.), *Classical conditioning II: Current research and theory* (pp. 64–99). New York, NY: Appleton-Century-Crofts.

Resnick, M., Martin, F. g., Sargent, R., & Silverman, B. (1996). Programmable bricks: Toys to think with. *IBM Systems Journal, 35*(3–4), 443–452.

Robbins, P., & Aydede, M. (2009). *The Cambridge handbook of situated cognition*. Cambridge, UK; New York, NY: Cambridge University Press.

Robinson-Riegler, B., & Robinson-Riegler, G. (2003). *Readings in cognitive psychology: Applications, connections, and individual differences*. Boston, MA: Pearson Allyn & Bacon.

Robinson, J. (1994). The expression and arousal of emotion in music. In P. Alperson (Ed.), *Musical worlds: New directions in the philosophy of music* (pp. 13–22). University Park, PA: Pennsylvania State University Press.

Robinson, J. (1997). *Music and meaning*. Ithaca, NY: Cornell University Press.

Rock, I. (1983). *The logic of perception*. Cambridge, MA: MIT Press.

Rodman, H. R., & Albright, T. D. (1987). Coding of visual stimulus velocity in area MT of the macaque. *Vision Research, 27*, 2035–2048.

Rosen, C. (1988). *Sonata forms* (Rev. ed.). New York, NY: Norton.

Rosen, C. (2002). *Piano notes: The world of the pianist*. New York, NY: Free Press.

Ross, A. (2007). *The rest is noise: Listening to the twentieth century* (1st ed.). New York, NY: Farrar, Straus & Giroux.

Rumelhart, D. E., & McClelland, J. L. (1986). *Parallel distributed processing, Vol. 1*. Cambridge, MA: MIT Press.

Safa, A. T., Saadat, M. G., & Naraghi, M. (2007). Passive dynamic of the simplest walking model: Replacing ramps with stairs. *Mechanism and Machine Theory, 42*(10), 1314–1325.

Saito, M., Fukaya, M., & Iwasaki, T. (2002). Serpentine locomotion with robotic snakes. *IEEE Control Systems Magazine, 22*(1), 64–81.

Sawyer, R. K. (2002). Emergence in psychology: Lessons from the history of non-reductionist science. *Human Development, 45*, 2–28.

Schneider, W. (1987). Connectionism: Is it a paradigm shift for psychology? *Behavior Research Methods, Instruments, and Computers, 19*, 73–83.

Scholes, S., Wilson, M., Sendova-Franks, A. B., & Melhuish, C. (2004). Comparisons in evolution and engineering: The collective intelligence of sorting. *Adaptive Behavior, 12*(3–4), 147–159.

Schultz, A. C., & Parker, L. E. (2002). *Multi-robot systems: From swarms to intelligent automata*. Dordrecht, The Netherlands; Boston, MA: Kluwer Academic Publishers.

Schwarz, K. R. (1996). *Minimalists*. London: Phaidon.

Seeley, T. D. (1989). The honey bee colony as a superorganism. *American Scientist, 77*(6), 546–553.

Selfridge, O. G. (1956). Pattern recognition and learning. In C. Cherry (Ed.), *Information theory* (pp. 345–353). London, UK: Butterworths Scientific Publications.

Sendova-Franks, A. B., Scholes, S. R., Franks, N. R., & Melhuish, C. (2004). Brood sorting by ants: Two phases and differential diffusion. *Animal Behaviour, 68*, 1095–1106.

Shan, Y. S., & Koren, Y. (1993). Design and motion planning of a mechanical snake. *IEEE Transactions on Systems Man and Cybernetics, 23*(4), 1091–1100.

Shannon, C. E. (1948). A mathematical theory of communication. *The Bell System Technical Journal, 27*, 379–423, 623–656.

Sharkey, N. E. (1997). The new wave in robot learning. *Robotics and Autonomous Systems, 22*(3–4), 179–185.

Sharkey, A. J. C. (2006). Robots, insects and swarm intelligence. *Artificial Intelligence Review, 26*(4), 255–268.

Sharkey, N., & Sharkey, A. (2009). Electro-mechanical robots before the computer. *Proceedings of the Institution of Mechanical Engineers Part C-Journal of Mechanical Engineering Science, 223*(1), 235–241.

Sharp, P. E., Blair, H. T., & Cho, J. W. (2001). The anatomical and computational basis of the rat head-direction cell signal. *Trends in Neurosciences, 24*(5), 289–294.

Shipton, H. W. (1977). Obituary: W. Grey Walter. *Electroencephalography and Clinical Neurophysiology, 43*, iii–iv.

Siegelmann, H. T. (1999). *Neural networks and analog computation: Beyond the Turing limit*. Boston, MA: Birkhauser.

Simon, H. A. (1969). *The sciences of the artificial* (1st ed.). Cambridge, MA: MIT Press.

Simon, H. A. (1979). Information processing models of cognition. *Annual Review of Psychology, 30*, 363–396.

Simon, H. A. (1980). Cognitive science: The newest science of the artificial. *Cognitive science, 4*, 33–46.

Simon, H. A., & Newell, A. (1958). Heuristic problem solving: The next advance in operations research. *Operation Research, 6*, 1–10.

Singh, J. (1966). *Great ideas in information theory, language, and cybernetics*. New York, NY: Dover Publications.

Siqueira, P. H., Steiner, M. T. A., & Scheer, S. (2007). A new approach to solve the traveling salesman problem. *Neurocomputing, 70*(4–6), 1013–1021.

Skonieczny, K., & D'Eleuterio, G. M. T. (2008). Modeling friction for a snake-like robot. *Advanced Robotics, 22*(5), 573–585.

Smith, A. P. (1978). Investigation of mechanisms underlying nest construction in mud wasp Paralastor sp. (Hymenoptera Eumenidae). *Animal Behaviour, 26*(Feb.), 232–240.

Snow, C. P. (1969). *The two cultures and a second look*. London, UK: Cambridge University Press.

Solnit, R. (2003). *Motion studies: Time, space and Eadweard Muybridge*. London: Bloomsbury.

Solso, R. L. (1995). *Cognitive psychology* (4th ed.). Boston, MA: Allyn and Bacon.

Sovrano, V. A., Bisazza, A., & Vallortigara, G. (2003). Modularity as a fish (*Xenotoca eiseni*) views it: Conjoining geometric and nongeometric information for spatial reorientation. *Journal of Experimental Psychology-Animal Behavior Processes, 29*(3), 199–210.

Sparshoot, F. (1994). Music and feeling. In P. Alperson (Ed.), *Musical worlds: New directions in the philosophy of music* (pp. 23–36). University Park, PA: Pennsylvania State University Press.

Spenko, M. J., Haynes, G. C., Saunders, J. A., Cutkosky, M. R., Rizzi, A. A., Full, R. J., et al. (2008). Biologically inspired climbing with a hexapedal robot. *Journal of Field Robotics, 25*(4–5), 223–242.

Spetch, M. L., Rust, T. B., Kamil, A. C., & Jones, J. E. (2003). Searching by rules: Pigeons' (*Columba livia*) landmark-based search according to constant bearing or constant distance. *Journal of Comparative Psychology, 117*(2), 123–132.

Standage, T. (2002). *The Turk: The life and times of the famous eighteenth-century chess-playing machine.* New York, NY: Walker & Co.

Sternberg, R. J. (1996). *Cognitive psychology.* Fort Worth, TX: Harcourt Brace College Publishers.

Stewart, I. (1994). A subway named Turing. *Scientific American, 271*, 104–107.

Stillings, N., Feinstein, M. H., Garfield, J. L., Rissland, E. L., Rosenbaum, D. A., Weisler, S. E., et al. (1987). *Cognitive science: An introduction.* Cambridge, MA: MIT Press.

Stoerig, P., & Cowey, A. (1997). Blindsight in man and monkey. *Brain, 120*, 535–559.

Stone, G. O. (1986). An analysis of the delta rule and the learning of statistical associations. In D. E. Rumelhart & J. McClelland (Eds.), *Parallel distributed processing, Vol. 1* (pp. 444–459). Cambridge, MA: MIT Press.

Stuerzl, W., Cheung, A., Cheng, K., & Zeil, J. (2008). The information content of panoramic images I: The rotational errors and the similarity of views in rectangular experimental arenas. *Journal of Experimental Psychology-Animal Behavior Processes, 34*(1), 1–14.

Sugawara, K., & Sano, M. (1997). Cooperative acceleration of task performance: Foraging behavior of interacting multi-robots system. *Physica D, 100*(3–4), 343–354.

Sulis, W. (1997). Fundamental concepts of collective intelligence. *Nonlinear Dynamics, Psychology, and LIfe Sciences, 1*, 35–53.

Susi, T., & Ziemke, T. (2001). Social cognition, artefacts, and stigmergy: A comparative analysis of theoretical frameworks for the understanding of artefact-mediated collaborative activity. *Journal of Cognitive Systems Research, 2*, 273–290.

Swade, D. D. (1993). Redeeming Charles Babbage's mechanical computer. *Scientific American, 268*, 86–91.

Tarasewich, P., & McMullen, P. R. (2002). Swarm intelligence: Power in numbers. *Communications of the ACM, 45*(8), 62–67.

Taube, J. S., & Muller, R. U. (1998). Comparisons of head direction cell activity in the post-subiculum and anterior thalamus of freely moving rats. *Hippocampus, 8*(2), 87–108.

Thagard, P. (1996). *Mind: Introduction to cognitive science.* Cambridge, MA: MIT Press.

Theraulaz, G., & Bonabeau, E. (1995). Coordination in distributed building. *Science, 269*(5224), 686–688.

Theraulaz, G., & Bonabeau, E. (1999). A brief history of stigmergy. *Artificial Life, 5*, 97–116.

Theraulaz, G., Bonabeau, E., & Deneubourg, J. L. (1998). The origin of nest complexity in social insects. *Complexity, 3*(6), 15–25.

Theraulaz, G., Bonabeau, E., Nicolis, S. C., Sole, R. V., Fourcassie, V., Blanco, S., et al. (2002). Spatial patterns in ant colonies. *Proceedings of the National Academy of Sciences of the United States of America, 99*(15), 9645–9649.

Thorpe, W. H. (1963). *Learning and instinct in animals* (New ed.). London, UK: Methuen.

Tolman, E. C. (1948). Cognitive maps in rats and men. *Psychological Review, 55*, 189–208.

Touretzky, D. S., & Pomerleau, D. A. (1994). Reconstructing physical symbol systems. *Cognitive Science, 18*, 345–353.

Transeth, A. A., Leine, R. I., Glocker, C., & Pettersen, K. Y. (2008). 3-D snake robot motion: Nonsmooth modeling, simulations, and experiments. *IEEE Transactions on Robotics, 24*(2), 361–376.

Transeth, A. A., Leine, R. I., Glocker, C., Pettersen, K. Y., & Liljeback, P. (2008). Snake robot obstacle-aided locomotion: Modeling, simulations, and experiments. *IEEE Transactions on Robotics, 24*(1), 88–104.

Trullier, O., Wiener, S. I., Berthoz, A., & Meyer, J. A. (1997). Biologically based artificial navigation systems: Review and prospects. *Progress in Neurobiology, 51*(5), 483–544.

Turing, A. M. (1936). On computable numbers, with an application to the Entscheidungsproblem. *Proceedings of the London Mathematical Society, Series 2h, 42*, 230–265.

Turing, A. M. (1950). Computing machinery and intelligence. *Mind, 59*, 433–460.

Turkle, S. (1995). *Life on the screen: Indentity in the age of the Internet*. New York, NY: Simon & Schuster.

Tversky, A., & Kahneman, D. (1974). Judgment under uncertainty: Heuristics and biases. *Science, 185*(4157), 1124–1131.

Uexküll, J. v. (2001). An introduction to umwelt. *Semiotica, 134*(1–4), 107–110.

Ungerleider, L. G., & Mishkin, M. (1982). Two cortical visual systems. In D. Ingle, M. A. Goodale & R. J. W. Mansfield (Eds.), *Analysis of visual behavior* (pp. 549–586). Cambridge, MA: MIT Press.

VanLehn, K. (1991). *Architectures for intelligence*. Hillsdale, NJ: Lawrence Erlbaum Associates.

Varela, F. J., Thompson, E., & Rosch, E. (1991). *The embodied mind: Cognitive science and human experience*. Cambridge, MA: MIT Press.

Vera, A. H., & Simon, H. A. (1993). Situated action: A symbolic interpretation. *Cognitive Science, 17*, 7–48.

Vico, G. (1984). *The new science of Giambattista Vico* (Unabridged translation of the 3rd ed.). Ithaca, NY: Cornell University Press.

Vico, G. (1988). *On the most ancient wisdom of the Italians*. Ithaca, NY: Cornell University Press.

Vico, G. (1990). *On the study methods of our time*. Ithaca, NY: Cornell University Press.

Von Eckardt, B. (1993). *What is cognitive science?* Cambridge, MA: MIT Press.

von Frisch, K. (1974). *Animal architecture* (1st ed.). New York, NY: Harcourt Brace Jovanovich.

Vygotsky, L. S. (1986). *Thought and language* (Translation newly rev. and edited). Cambridge, MA: MIT Press.

Walton, K. (1994). Listening with imagination: Is music representational? In P. Alperson (Ed.), *Musical worlds: New directions in the philosophy of music* (pp. 47–62). University Park, PA: Pennsylvania State University Press.

Webb, B. (1996). A cricket robot. *Scientific American, 275*, 94–99.

Webb, B., & Consi, T. R. (2001). *Biorobotics: Methods and applications*. Menlo Park, CA: AAAI Press/MIT Press.

Webb, B., & Scutt, T. (2000). A simple latency-dependent spiking-neuron model of cricket phonotaxis. *Biological Cybernetics, 82*(3), 247–269.

Weiskrantz, L. (1986). *Blindsight: A case study and implications*. Oxford, UK: Oxford University Press; New York, NY: Clarendon Press.

Weiskrantz, L. (1997). *Consciousness lost and found*. Oxford, UK: Oxford University Press.

Weiskrantz, L., Warrington, E. K., Sanders, M. D., & Marshall, J. (1974). Visual capacity in hemianopic field following a restricted occipital ablation. *Brain, 97*(Dec), 709–728.

Wenzel, J. W. (1991). Evolution of nest architecture. In K. G. Ross & R. W. Matthews (Eds.), *The Social Biology of Wasps* (pp. 480–519). Ithaca, NY: Comstock Publishing Associates.

Wheeler, W. M. (1911). The ant colon as an organism. *Journal of Morphology, 22*(2), 307–325.

Wheeler, W. M. (1923). Social life among the insects. Lecture I. General remarks on insect societies. The social beetles. *The Scientific Monthly, 14*(6), 497–524.

Wheeler, W. M. (1926). Emergent evolution and the social. *Science, 64*(1662), 433–440.

Wiener, N. (1948). *Cybernetics: Or control and communciation in the animal and the machine.* Cambridge, MA: MIT Press.

Wiener, N. (1964). *God and Golem, Inc.: A comment on certain points where cybernetics impinges on religion.* Cambridge, MA: MIT Press.

Wilson, D. S., & Sober, E. (1989). Reviving the superorganism. *Journal of Theoretical Biology, 136*(3), 337–356.

Wilson, E. O., & Lumsden, C. J. (1991). Holism and reduction in sociobiology: Lessons from the ants and human culture. *Biology and Philosophy, 6*(4), 401–412.

Wilson, M., Melhuish, C., Sendova-Franks, A. B., & Scholes, S. (2004). Algorithms for building annular structures with minimalist robots inspired by brood sorting in ant colonies. *Autonomous Robots, 17*(2–3), 115–136.

Wilson, R. A. (2004). *Boundaries of the mind: The individual in the fragile sciences; Cognition.* Cambridge, UK; New York, NY: Cambridge University Press.

Wilson, R. A. (2005). *Genes and the agents of life: The individual in the fragile sciences, biology.* New York, NY: Cambridge University Press.

Wiltschko, R., & Wiltschko, W. (2003). Avian navigation: From historical to modern concepts. *Animal Behaviour, 65*, 257–272.

Winograd, T., & Flores, F. (1987). *Understanding computers and cognition.* New York, NY: Addison-Wesley.

Wisse, M., Hobbelen, D. G. E., & Schwab, A. L. (2007). Adding an upper body to passive dynamic walking robots by means of a bisecting hip mechanism. *IEEE Transactions on Robotics, 23*(1), 112–123.

Wisse, M., & Linde, R. Q. v. d. (2007). *Delft pneumatic bipeds.* Berlin, Germany; New York, NY: Springer-Verlag.

Wisse, M., Schwab, A. L., & van der Helm, F. C. T. (2004). Passive dynamic walking model with upper body. *Robotica, 22*, 681–688.

Wood, G. (2002). *Living dolls: A magical history of the quest for artificial life.* London, UK: Faber & Faber.

Wright, R. D., & Dawson, M. R. W. (1994). To what extent do beliefs affect apparent motion? *Philosophical Psychology, 7*, 471–491.

Wu, Q., & Sabet, N. (2004). An experimental study of passive dynamic walking. *Robotica, 22*, 251–262.

Wystrach, A., & Beugnon, G. (2009). Ants learn geometry and features. *Current Biology, 19*(1), 61–66.

Zeki, S. M. (1974). Functional organization of a visual area in the posterior bank of the superior temporal sulcus of the rhesus monkey. *Journal of Physiology, 236*, 549–573.

Zihl, J., von Cramon, D., & Mai, N. (1983). Selective disturbance of movement vision after bilateral brain damage. *Brain, 106*, 313–340.

Index

A

action, 2, 3, 10, 12, 14, 16, 23, 25, 28, 39, 57, 61, 64, 66, 68, 70, 72, 75–6, 87, 106, 147, 178, 201–5, 208, 231, 265, 314

adaptability, 174

affordance, 6, 130

analog-to-digital conversion, 107–8

analysis, 13, 82–5, 86, 88, 90, 94, 114, 119, 120, 124, 130–2, 143–4, 148, 161–2, 174, 224, 264, 308, 314, 316, 319, 327

annealing, 225, 320

antiSLAM, 94, 200, 263, 271, 282–6, 288, 289, 290–5, 297–308

ants, 8, 12, 19, 20, 75, 94, 226–8, 232–5, 261, 301, 313, 314, 316–7, 320, 326, 328–9

architecture, 22–3, 57, 59–60, 63, 65–6, 68, 76, 163, 197–200, 204, 206–12, 215, 217, 245, 247, 250, 252, 262, 282–4, 290–1, 295, 297, 300, 315, 317, 324

artificial intelligence, 2, 19, 27, 86, 311, 322

artificial neural networks, 37, 39, 80, 159, 177, 206, 305

associative learning, 278

automata, 173, 264–5, 267

autonomous robots, 138, 163, 208

B

beaver, 1–5, 19–20, 310, 314, 317

bees, 8, 269

behaviour-based robotics, 77

biomimetics, 119–21

blind bulldozing, 228, 234, 316

blindsight, 70, 314

Boolean Dream, 26–7

box pushing, 232–3, 236

brainstate-in-a-box, 257

Braitenberg Vehicle 2, 83, 93–100, 103, 105–10, 112–6, 137, 147, 182, 188, 198, 223, 236, 239, 263, 283–6, 288–91, 294, 298

bricolage, 26, 57, 58, 77–80, 88, 90–1, 93, 97, 115–6, 307

BricxCC, 2, 110–1, 113

C

classical sandwich, 201–8

cockroach, 121–2

cognition, 1–3, 5, 10, 20–2, 23, 25–7, 30, 32, 34, 39, 55, 57, 59, 61, 63, 67–8, 72–6, 80, 82, 87, 176, 199–202, 205, 235–6, 261, 263, 268, 282, 306–9, 313, 322, 324, 326–7, 329

cognitive development, 57, 63, 73, 78

cognitive revolution, 200, 322

cognitive science

classical, 1–3, 22–3, 25–30, 32–9, 41, 51–5, 57, 59, 60, 68, 71, 74, 76, 82, 85–8, 90, 93, 117, 119, 130, 200, 205, 209, 221, 269 embodied, 22–3, 26, 30, 38–40, 46, 59, 66, 71, 75, 87, 90–1, 93, 117, 119, 220–, 235, 261, 263, 268, 307, 308, 313

collective intelligence, 224, 259

collective memory, 224

communication, 16–8, 51, 83, 103, 134, 177, 230–3, 259, 321, 326

computation, 2–3, 71–3, 75, 266, 268, 279, 319, 321

conduit metaphor, 51, 325

consciousness, 2, 21, 174, 206, 220, 318

contralateral, 105–6, 112, 114, 284, 289, 290, 294

control structure, 17, 22–3, 36, 208, 227, 236

cooperative transport, 232–4

cues
    feature, 272–5, 278, 280, 282
    geometric, 272–6, 279, 284

cybernetic principles, 168, 170

cybernetics, 21, 39–40, 51, 164, 166–7, 176–7, 309–11, 313–5, 317, 319, 326, 329

D

decentralized control, 206

differential friction, 149

dodecaphony, 42–3

dorsal stream, 69–70

duplex theory, 69–71, 78

dynamical systems theory, 13–4

dynamics, 43, 123, 146, 178, 259–61, 314, 324

E

embodied agent, 39, 96, 106, 146

embodiment, 21–2, 26, 30, 34, 36, 39–41, 71, 104, 114–5, 146, 217, 221, 223–4, 236, 238, 242, 261–2, 267, 294, 308
    strong, 236–7

emergence, 12–3, 26, 40–1, 55, 59, 75, 118, 311, 324

engineering
    forward, 83–4, 86
    reverse, 82–3

evolutionary robotics, 279, 321

extended mind, 235–7, 261–2, 308, 313

F

feedback, 117, 132–4, 147, 149, 151, 155, 159, 166–7, 176–8, 220, 226–7

formal operations, 64, 67

formal rules, 26–7

frame of reference problem, 55, 83, 90–91, 161

functional analysis, 82–3

functionalism, 37

G

genetic algorithms, 135, 139, 162, 225, 227, 279–80, 311

Genghis, 209–10

geometric module, 277

goal seeking, 174

Grey Walter, William, 40, 83, 133–4, 159, 163–76, 178–80, 184, 189, 192, 197–8, 208, 212–3, 215, 218–20, 263–8, 311, 313, 317–8, 326

H

hippocampus, 269–71, 311, 322

holism, 13–4

holy numbers, 139, 162, 280

Homeostat, 132–4, 147, 149, 159, 162, 177–8

horizontal décalage, 78

hornet, 2, 7

I

ideal releaser, 11, 14

improvisation, 35–6, 50, 52–5

information processing, 22–3, 37, 39, 59–61, 68, 72–4, 78, 88, 131, 200, 235–6, 276

intelligence, 2–3, 5, 7, 9–12, 14, 19–21, 28, 49, 58, 72, 74–5, 87–8, 93, 96, 209, 227–30, 234, 237, 256, 259, 260–1, 310–1, 318, 326–8
    collective, 7
    swarm, 14, 227, 229

intentional cooperation, 231–2

interactions, 12–3, 48, 54, 57, 64, 68, 72, 80–2, 85, 87, 89, 131–2, 179, 209, 227, 230, 260–1, 268, 305, 306

ipsilateral, 106, 112, 114, 284, 289

Michael Dawson is a professor of psychology at the University of Alberta. He is the author of numerous scientific papers as well as the books *Understanding Cognitive Science* (1998), *Minds and Machines* (2004), and *Connectionism: A Hands-on Approach* (2005).

Brian Dupuis is a research assistant in psychology at the University of Alberta.

Michael Wilson is a biology undergraduate at the University of Alberta.